7

Topics in Medicinal Chemistry

Third World Diseases

Volume Editor: Richard L. Elliott

With contributions by

K. Arora · C.E. Barry III · H.I.M. Boshoff · J.N. Burrows · M. Carroll · I. Choi · R. Don · R.L. Edwards · K.D. England · T. von Geldern · F. Gomez de las Heras · T.D. Gruber · M.O. Harhay · T.H. Keller · J. Lee · D. Marquess · G.A. Marriner · T. Mukherjee · A. Nayyar · S. Nilar · I. Scandale · P.-Y. Shi · E. Uh · L.E. Via · D. Waterson · S.Y. Wong · Z. Yin

Editor
Dr. Richard L. Elliott
Box P.O. 98102
Seattle Washington
USA
richard.elliott@gatesfoundation.org

ISSN 1862-2461 e-ISSN 1862-247X
ISBN 978-3-642-23486-6 e-ISBN 978-3-642-23487-3
DOI 10.1007/978-3-642-23487-3
Springer Heidelberg Dordrecht London New York

Library of Congress Control Number: 2011936161

Typesetting and Production: SPi Publisher Services

Springer is part of Springer Science+Business Media (www.springer.com)

Volume Editor

Dr. Richard L. Elliott

Box P.O. 98102
Seattle Washington
USA
richard.elliott@gatesfoundation.org

Editorial Board

Topics in Medicinal Chemistry Also Available Electronically

Topics in Medicinal Chemistry is included in Springer's eBook package *Chemistry and Materials Science.* If a library does not opt for the whole package the book series may be bought on a subscription basis. Also, all back volumes are available electronically.

For all customers who have a standing order to the print version of *Topics in Medicinal Chemistry*, we offer the electronic version via SpringerLink free of charge.

If you do not have access, you can still view the table of contents of each volume and the abstract of each article by going to the SpringerLink homepage, clicking on "Browse by Online Libraries", then "Chemical Sciences," and finally by selecting *Topics in Medicinal Chemistry.*

You will find information about the

- Editorial Board
- Aims and Scope
- Instructions for Authors
- Sample Contribution

at springer.com using the search function by typing in *Topics in Medicinal Chemistry.*

Color figures are published in full color in the electronic version on SpringerLink.

Aims and Scope

Drug research requires interdisciplinary team-work at the interface between chemistry, biology and medicine. Therefore, the new topic-related series *Topics in Medicinal Chemistry* will cover all relevant aspects of drug research, e.g. pathobiochemistry of diseases, identification and validation of (emerging) drug targets, structural biology, drugability of targets, drug design approaches, chemogenomics, synthetic chemistry including combinatorial methods, bioorganic chemistry, natural compounds, high-throughput screening, pharmacological in vitro and in vivo investigations, drug-receptor interactions on the molecular level, structure-activity relationships, drug absorption, distribution, metabolism, elimination, toxicology and pharmacogenomics.

In general, special volumes are edited by well known guest editors.

In references *Topics in Medicinal Chemistry* is abbreviated *Top Med Chem* and is cited as a journal.

Preface to the Series

Medicinal chemistry is both science and art. The science of medicinal chemistry offers mankind one of its best hopes for improving the quality of life. The art of medicinal chemistry continues to challenge its practitioners with the need for both intuition and experience to discover new drugs. Hence sharing the experience of drug discovery is uniquely beneficial to the field of medicinal chemistry.

The series Topics in Medicinal Chemistry is designed to help both novice and experienced medicinal chemists share insights from the drug discovery process. For the novice, the introductory chapter to each volume provides background and valuable perspective on a field of medicinal chemistry not available elsewhere. Succeeding chapters then provide examples of successful drug discovery efforts that describe the most up-to-date work from this field.

The editors have chosen topics from both important therapeutic areas and from work that advances the discipline of medicinal chemistry. For example, cancer, metabolic syndrome and Alzheimer's disease are fields in which academia and industry are heavily invested to discover new drugs because of their considerable unmet medical need. The editors have therefore prioritized covering new developments in medicinal chemistry in these fields. In addition, important advances in the discipline, such as fragment-based drug design and other aspects of new lead-seeking approaches, are also planned for early volumes in this series. Each volume thus offers a unique opportunity to capture the most up-to-date perspective in an area of medicinal chemistry.

Dr. Peter R. Bernstein
Prof. Dr. Armin Buschauer
Prof. Dr. Gunda J. Georg
Prof. Dr. Yoshiaki Kiso
Dr. John Lowe
Dr. Hans Ulrich Stilz

Preface

The specter of diseases most prevalent in developing countries blatantly confronts us with one of the great ironies of the modern world, juxtaposing a tragic situation wherein rapid technological and scientific advances in biology and drug discovery have driven tremendous advances in the treatment of first-world diseases with a situation in the developing world wherein millions of people continue to die each year due to lack of safe and efficacious medicine, and where many of the drugs, when they are available, are ancient by modern standards and would not pass current requirements for safety and efficacy. Diseases of the developing world, often known by the moniker "neglected diseases," comprise a melee of infectious and parasitic diseases that, in terms of their biology and physiopathology, defy common description; their greatest commonality being the suffering and death they afflict on the poorest of the poor.

No single volume can reasonably expect to encompass the enormity of this subject. Thus, in this volume, we have focused on a selected representative set of diseases that make major contributions to the morbidity and mortality in developing countries, with the hope of raising general awareness of these diseases and highlighting some of the recent drug-discovery advances. Following a general overview chapter, five chapters focusing on tuberculosis, malaria, kinetoplastids, dengue fever, and diarrhea provide an in-depth look at the current status of drug discovery for these diseases. Unsurprisingly, each disease is encumbered by its own unique challenges and knowledge gaps, but share commonality in that, finally, state-of-the-art drug discovery tools and approaches are being rigorously applied to discover new therapies. The picture is no longer bleak, with a number of new drug candidates under active investigation and entering clinical trials. Still, much remains to be done; with many of the more "neglected" diseases (e.g., filarial diseases) being so under-investigated and under-resourced that, sadly, no chapter review was warranted. However, the recent increase in research activity and engagement of dedicated researchers throughout the world, collaborating together in academia, industry, and the nonprofit sector, is a new and uplifting trend which portends well for the future.

It is with the greatest respect that I acknowledge the important contributions each of the authors have made in this field, and thank them for contributing chapters to this volume. It is my hope that this volume will serve to inform and guide those working on or interested in diseases that afflict the world's poorest populations, and, more importantly, inspire a new generation of drug discoverers to devote their knowledge, skills, insight, creativity, and passion to develop new therapies for these diseases, such that we speak of these diseases as "neglected" no longer.

Richard L. Elliott

Contents

Top Med Chem 7: 1–46
DOI: 10.1007/7355_2011_12

Published online: 22 April 2011

Overview of Neglected Tropical Diseases

Federico Gomez de las Heras

Abstract Neglected tropical diseases (NTDs), caused by a host of pathogens including helminthes, protozoa, bacteria, and viruses, remain a difficult and challenging area for drug discovery, being limited by meager resources and funding, challenged by gaps in knowledge of pathogen biology and drug targets, and faced with limitations specific to the developing world, including stringent cost of good constraints, the lack of robust infrastructure to effectively deliver and administer drug therapy, and socioeconomic considerations in disease endemic areas. This chapter will endeavor to provide a general overview of scientific approaches for NTDs, and review some of the current challenges and advancements in diseases of the developing world (DDW) drug discovery, as well as discuss the interplay between international networks of organizations, foundations, and public and private partners involved in neglected disease drug discovery. This chapter will also review general aspects and considerations common to many or all of these diseases, including burden of disease, coinfections, pathogenicity, drug resistance, drug combination therapy, and mass drug administration programs. This chapter will briefly address disease treatment on a general level, including the treatment of coinfections and care of special groups of patients, such as children and pregnant women.

Keywords Diseases of the developing world, Neglected tropical diseases, New drugs discovery and development, Product development partnership (PDP), Public–private partnership (PPP)

Contents

F. Gomez de las Heras
Paseo de los Cerezos 5, 28016 Madrid, Spain
e-mail: fgomezdelasheras@yahoo.es

1 Introduction

Tropical diseases, neglected diseases, diseases of the developing world (DDW), diseases of poverty: these are all different names to identify a heterogeneous group of infectious diseases produced by viruses, bacteria, protozoa, and worms, which disproportionately affect people in low and medium income countries. The term "tropical" refers to diseases that occur solely, or principally, in tropical areas, such as malaria, leishmaniasis, Chagas disease, African trypanosomiasis, schistosomiasis, onchocerciasis, lymphatic filariasis, and dengue [1]. "Neglected" means that, in spite of the high mortality and/or morbidity produced by the disease, little scientific attention and resources have been dedicated to understanding and curing them [2]. The meaning of the term "neglected" is exemplified in the study of Trouillier et al. [3] who showed that of the 1993 new chemical entities marketed between 1975 and 1999, only 16 were for tropical diseases and tuberculosis. Neglected diseases include those caused by Helminths (e.g., schistosomiasis, onchocerciasis, lymphatic filariasis, hookworm, ascariasis, and trichuriasis); protozoa (malaria, kinetoplastids, plus amoebiasis, giardiasis); bacteria (tuberculosis, leprosy, Buruli ulcer, trachoma, and bacteria causing diarrhea), and viruses (yellow fever, Japanese encephalitis, chikungunya, and viruses causing diarrhea). The name "diseases of the developing world" includes all the above and refers to those that disproportionately affect low- and medium-income countries. The three most devastating diseases, HIV/AIDS, malaria, and tuberculosis (TB), collectively called "the big three", are now high on the international agenda and receive more attention and funding. Frequently, these diseases are classified as "the big three", on one side, and as neglected tropical diseases (NTDs), on the other. Nevertheless, in spite of the increased awareness and financial efforts by the Global Fund and other funding organizations, drugs for the treatment of malaria are still not available in all the endemic areas, and the drugs for HIV/AIDS and TB are still largely inaccessible in Africa and south/southeast Asia [4].

On their web pages, both WHO (http://www.who.int/neglected_diseases/diseases/en/) and the Special Programme for Research and Training in Tropical Diseases (TDR) (http://apps.who.int/tdr/svc/diseases); established in 1975 by the UNICEF, the United Nations Development Programme, the World Bank, and the WHO, to focus on neglected infectious diseases) include similar but not identical lists of neglected and tropical DDW. For the sake of clarity, we will use the terms "tropical" and "neglected" with the meanings given above, and when we refer to all of them we will use the term "DDW". HIV/AIDS will only be discussed in this chapter when it affects the course of other DDW or its treatment.

2 The Diseases

This chapter will focus on providing a general overview of DDW, to provide background information to those new to the field, and to serve as an introduction to the following chapters. The following chapters of this volume describe, in a disease-specific manner, the current status of DDW drug discovery of select diseases, specifically tuberculosis, malaria diseases produced by kinetoplastids, dengue, and diarrhea (Chaps. 2–6, respectively). In addition, interested readers may find additional information regarding disease burden, pathogenicity, unmet medical need, treatment, drug pipeline, etc. from the web pages and reports from the WHO, including a Global Health Observatory Map Gallery (http://gamapserver.who.int/mapLibrary/app/searchResults.aspx) and a WHO report on NTDs (http://whqlibdoc.who.int/publications/2010/9789241564090_eng.pdf). The public–private partnerships (PPPs) Web sites have additional reports and information dedicated to different aspects of these diseases (Table 4).

2.1 Disease Burden

Available information about the epidemiology of DDW is generally incomplete, oftentimes the reliability is uncertain, and the data difficult to compare because methodologies, criteria, and end points vary from study to study. Thus, increased disease surveillance and well-designed epidemiological data collection, along with improved diagnostic tools, are needed for a clearer and more accurate assessment of DDW disease burden. The most common parameter to define burden of disease (BOD) is the disability-adjusted life year (DALY), which can be thought of as one lost year of "healthy" life (http://www.who.int/healthinfo/global_burden_disease/metrics_daly/en/). The DALY is a complex parameter, calculated from the sum of years of life lost (YLL) due to mortality and years lived with disability (YLD) due to morbidity. Although imperfect, the DALY parameter allows one to quantify and compare the impact of disease on the health of populations across endemic areas

[5, 6]. This parameter is mainly driven by mortality rates [7], but also takes into consideration the impact of morbidity. Different studies using different methodological criteria report different BOD data [8]. BOD data included in Tables 1–3 are based on the WHO report "The Global Burden of Disease. 2004 Update", published in 2008.[1] Infectious and parasitic diseases are the second cause of death worldwide, second only to cardiovascular diseases and ahead of cancers (Table 1). The main health impact, in terms of DALYs, is produced by respiratory tract infections, diarrheal diseases, and the "big three" (Table 2). However, the combined impact of the most neglected diseases, including leishmaniasis, Chagas disease, human African trypanosomiasis (HAT), and the helminthes infection is comparable to that caused by each one of the big three [9] (Table 3).

Diarrhea is a common symptom of gastrointestinal infections caused by a wide range of pathogens, including viruses (e.g., rotavirus), bacteria (e.g., *Escherichia coli*, *Shigella*, *Campylobacter*, and *Salmonella*, along with *Vibrio cholerae* during epidemics), and protozoa (e.g., *Cryptosporidium*). The burden of diarrheal disease is the highest among DDW[2,3] (Table 2) [10], killing 1.8 million children with

Table 1 Leading causes of death, worldwide 2004

Disease	% of total deaths
Cardio vascular diseases	29.1
Infections and parasitic diseases	16.1
Cancers	12.6
Respiratory infections	7.2
Respiratory diseases	6.9

WHO – The global burden of disease. 2004 update. WHO Geneva 2008

Table 2 Global health impact of major diseases of the developing world[a]

Disease	Global prevalence Millions	Clinical incidence/year Millions	Global deaths (in developing countries) Millions	Global DALY (in developing countries) Millions
Lower resp. inf.	–	429.2	4.2 (2.9)	94.5 (76.9)
Diarrheal disease	–	4620.4	2.2 (1.8)	72.8 (59.2)
HIV/AIDS	31.4	2.8	2.0 (1.5)	58.5 (42.9)
Malaria	–	241.3	0.9 (0.9)	34.0 (32.8)
Tuberculosis	13.9	7.8	1.5 (0.9)	34.2 (22.4)
Dengue	50	9.0	0.019	0.7

[a]The global burden of disease, 2004 update WHO, Geneva 2008

[1]Global burden of disease 2004. WHO 2008 (http://www.who.int/healthinfo/global_burden_disease/2004_report_update/en/index.html).

[2]Diarrheal disease web page. WHO (http://www.who.int/mediacentre/factsheets/fs330/en/).

[3]Diarrhea: Why children are still dying and what can be done. The United Nations Children's Fund (UNICEF)/World Health Organization (WHO), New York, Geneva (2009).

Table 3 Impact of neglected tropical diseases[a,b]

Disease	Approximate global prevalence (millions)	Global no. of deaths (thousands)	Global DALYs (millions)
Intestinal nematode infections	151	6	4.0
Hook worm disease	576	0	1.1
Ascariasis	807	2	1.8
Trichuriasis	604	2	1.0
Lymphatic filariasis	65	0	5.9
Onchocerciasis	0.9	0	0.4
Schistosomiasis	2.61	41	1.7
Trachoma	6.3	0	1.3
Leishmaniasis	12	47	1.9
Human African Trypanosomiasis	<0.1	52	1.7
Chagas disease	65	11	0.4
Leprosy	0.4	5	0.2
Dengue	50	18	0.7

[a]The global burden of disease, 2004 update WHO, Geneva, 2008
[b]Ref. [1]

approximately 4.6 billion episodes a year. According to WHO 2004 estimates, rotavirus accounts annually for an estimated two million hospitalized cases of severe diarrheal disease in children, and kills an estimated 527,000 children per year. Rotavirus infection is ubiquitous, occurs worldwide, and affects nearly all children (95%) by the age of 5 years. Although the incidence of rotavirus disease is similar between industrialized and developing countries, adverse outcomes are more likely in children from the latter. The estimated number of deaths from diarrhea has fallen dramatically over the past two decades as a result of improved treatment with oral rehydration therapy, sanitation, and water interventions.

In spite of the tremendous toll of death and disability produced by these diseases (Tables 1–3), there are some promising signs of progress. As described in WHO's "World Malaria Report 2009",[4] more than one-third of the 108 malarious countries reported reductions in malaria cases of >50% in 2008 compared to 2000. Ten countries are implementing nationwide elimination programs, of which six entered the elimination phase in 2009. Eight countries are in the pre-elimination stage and a further nine countries have interrupted transmission and are in the phase of preventing reintroduction of malaria. Nevertheless, in 2008 there were an estimated 243 million cases of malaria, which produced estimated 863,000 deaths. The biggest problem is caused by *Plasmodium falciparum* in sub-Saharan Africa. However, *Plasmodium vivax* is geographically the most widely distributed cause of malaria, with up to 2.5 billion people at risk and many million clinical cases every year [11].

[4]World malaria Report 2009. WHO Geneva 2009. http://www.who.int/malaria/world_malaria_report_2009/en/index.html.

Tuberculosis also makes a large contribution to mortality and morbidity in developing nations. Thus, in 2008 there were an estimated 8.9–9.9 million new cases of TB, 1.1–1.7 million deaths among HIV-negative people, and an additional 0.45–0.62 million among HIV-positive people. Almost 30,000 cases of multidrug-resistant TB (MDR-TB) were identified in 2008 (11% of the estimated total number of cases). In the last few years, new and very virulent strains, including the extremely resistant XDR TB and the totally drug-resistant TB for which there is no treatment [12, 13], have been reported. Much of this mortality occurs in African children [14].[5] However, there have been some improvements in TB. Based on WHO's 2009 data,[6] in the last 10 years, the tuberculosis case-detection rate increased from an estimated 11% in 1995 to 63% in 2007, and data on treatment success rates under the current tuberculosis treatment paradigm, directly observed treatment short-course (DOTS) indicates consistent improvement, with successful treatment rising from 79% in 1990 to 85% in 2006.[7] Nonetheless, TB continues to be a major health burden in the developing world.

Dengue infection [15–17] is transmitted by mosquitoes and often produces a self-limited febrile illness called dengue fever (DF), which can lead to the severe forms of dengue: hemorrhagic fever (DHF) and dengue shock syndrome (DSS). Dengue is caused by four different dengue virus serotypes: DEN1, DEN2, DEN3, and DEN4. Severe forms of the disease occur in 5–10% of DF patients. Patients with DHF require hospitalization, and 30–40% of children with DHF progress to DSS. The case-fatality rate for DHF/DSS can be as high as 5%. Importantly, those who recover from DF possess life-long immunity to the dengue virus serotype that infected them. An estimated 2.5 billion people are at risk for dengue infection and DHF, primarily in urban tropical and subtropical regions, but also increasingly in rural areas. In the latter part of the twentieth century, endemic areas suffered unprecedented global dengue epidemics with large numbers of fatalities. WHO estimates 500,000 cases of DHF annually.

The diseases produced by helminthes are responsible for approximately 200,000 deaths annually, although some experts estimate much higher numbers [18]. Virtually all of the 1.4 billion people living on less than US $1.25 per day (the World Bank's poverty line) are infected with helminthes, which ruin their health, hinder education, and impair agricultural productivity [19–21]. The health toll of helminth infections is immense, with the soil-transmitted helminthes (e.g., ascariasis, roundworm), trichuriasis (whipworm, and hookworm) alone accounting for 4 million DALYs, with lymphatic filariasis and onchocerciasis contributing another 5.9 million and 0.4 million DALYs, respectively. Fortunately, large-scale treatment programs have been established in response to these diseases. For example, in 2007, 546 million

[5]Guidance for National Tuberculosis Programmes on the Management of Tuberculosis in Children. World Health Organization, Geneva, WHO/HTM/TB/2006.371, WHO/FCH/CAH/2006.7.

[6]Global tuberculosis control: a short update to the 2009 report. WHO Geneva 2009 (http://www.who.int/tb/publications/global_report/2009/update/en/index.html).

[7]World Health Statistics 2009. WHO Geneva 2009 (http://www.who.int/whosis/whostat/2009/en/index.html).

people were treated to prevent transmission of lymphatic filariasis. These mass drug administration (MDA) programs will be discussed in more detail below.

The disease burden figures for leishmaniasis [22–24], endemic in 88 countries with two million new cases per year (1.5 million for cutaneous leishmaniasis and 500,000 for visceral leishmaniasis); HAT, endemic in 36 countries with 60 million people at risk and 70,000 existing cases[8] [25–27]; and Chagas disease [28, 29] with 200,000 new cases per year, are dramatic. Significant reduction in infection rates and mortality for Chagas disease has been achieved in recent years through vector control measures. For example, in the Southern Cone, between 1990 and 2000 the number of new cases per year decreased from 700,000 to 200,000 and the number of deaths from more than 45,000 to 21,000. Likewise, the burden of HAT in Africa has decreased significantly the last 10 years, with reported number of new cases each year in the 10,000 range for 2008. Unfortunately, disease control has been more complicated for leishmaniasis due to a number of factors, including the nature and habitat of the vector and the nature of the disease.

2.2 *Pathogenicity*

Pathogenicity is caused by a variety of mechanisms. Some pathogens directly damage cells and tissues, and in other cases pathogenicity is produced by the host's immune response to pathogen invasion. Three diseases, malaria, Chagas, and HAT, are discussed to exemplify various pathogenic mechanisms.

An important part of the strategy of parasites to evade the host immune system is their complex cellular life cycle. The *Plasmodium* life cycle is reviewed in Chap. 3 of this book dedicated to Antimalarial Drug Discovery. The preerythrocytic stages of the life cycle are completely asymptomatic. Unfortunately, this preerythrocytic stage is currently experimentally very difficult to study and therefore has remained poorly understood. To complicate matters, there are significant differences between the cell cycles of the four *Plasmodia* species producing malaria in humans, including adhesion molecules, duration of their life cycle, and their behavior within the human liver. In the case of *P. vivax* and *Plasmodium ovale* infections, some of the parasites during the initial liver stage of the infection form dormant hypnozoites, which can be later reactivated and produce disease relapses without new mosquito infecting bites.

Malaria pathogenicity is multifactorial and is poorly understood [30–32]. Some of the factors contributing to pathogenicity include synchronized lysis of both infected and noninfected red blood cells (RBCs), which produces fever and anemia. Infected RBCs adhere to the endothelium of peripheral blood vessels, which results in considerable occlusion of blood flow and consequently impaired oxygen delivery.

---

[8]WHO African trypanosomiasis (sleeping sickness) Fact sheet No 259. Revised August 2006. (www.who.int/mediacentre/factsheets/fs259/en).

Massive sequestration of the parasite in the brain is believed to be the underlying cause of coma in cerebral malaria. The reason for parasite sequestration is unknown, but it has been speculated that binding to the endothelium is a way to circumvent spleen-dependent destruction. Two receptors, CD36 and chondroitin sulfate A (CSA), seem to provide stable stationary adherence. A single parasite protein, *P. falciparum* erythrocyte membrane protein 1 (PfEMP1), which is expressed at the infected erythrocyte surface, mediates parasite binding to these receptors. In pregnant women, RBCs sequester in the placenta, causing premature delivery, low birth weight, and increased mortality in the new born and anemia in the mother. Other pathogenic effects of malaria include anemia, hypoglycemia, and metabolic acidosis. The two latter are strongly associated, suggesting that parasite and/or host metabolism may be contributing to both. Hypoglycemia, which produces deleterious effects on the central nervous system (CNS), may be caused by the high consumption of glucose by the parasite and/or inadequate hepatic gluconeogenesis. The metabolic acidosis, predominantly lactic acidosis, leads to a syndrome of respiratory distress and is recognized as the main determinant for death or survival in malaria patients. Lactic acidosis in children with severe malaria may be due to the increased production of lactic acid by parasites, to decreased lactic acid clearance by the liver, or most likely by the combined effects of several factors that reduce oxygen delivery to tissues.

Chagas disease infection [28, 29, 33] mainly occurs by inoculation of *Trypanosoma cruzi* by triatomine insects, and also through blood transfusion, congenital transmission, and organ transplantation. The disease manifests in two phases, starting with the acute phase in which parasitization of cardiac muscle and brain may cause acute myocarditis and meningoencephalitis, respectively. The second phase of chronic disease generally occurs 10–20 years after acute illness and mainly targets the heart and the digestive tract, resulting in significant morbidity and mortality. Chronic cardiomyopathy is the most common and serious manifestation of chronic disease, and digestive tract pathology manifests as megadisease (megaesophagus and megacolon).

There are two hypotheses to account for Chagas disease pathogenicity [28, 29]. In the first, pathogenicity in the chronic phase is attributed to the presence of *T. cruzi* parasites in blood and tissue. The second suggests that tissue damage is due to the host immune response. Experimental data seem to support both hypotheses. The autoimmune hypotheses [28, 29] proposes that autoantibodies and autoreactive cellular response to host cell components are elicited either by similar host and parasite epitopes or to cell contents released by host cell lysis. According to this theory, it is the cellular response that is responsible for cardiac tissue damage. However, the discovery of parasite persistence in chronic disease argues against a purely autoimmune hypothesis. The high prevalence of parasites in the blood and tissue of chronic Chagas disease patients, the association of *T. cruzi* antigens with cardiac inflammation, and the correlation of parasite presence with disease severity point to a primary role of the parasite in the pathogenesis of chronic Chagasic cardiomyopathy. If this is the case, it suggests that etiologic treatment should be effective for chronic disease.

Recent studies demonstrating that such treatment improved clinical cardiac outcomes in chronic phase Chagas patients support this idea [34].

Understanding the genesis of tissue damage is important for the design of therapeutic strategies for Chagas disease, e.g., etiologic treatment or modulation of the host immune response [28, 29]. Despite the controversy over the primary cause of pathogenesis, acute and chronic inflammation mediated by the host immune system is undoubtedly important in development of tissue damage. A strong innate immune response and polyclonal activation of B and T cells follow the initial contact between host and parasite, with the induction of cytokines, particularly IFN-γ and TNF-α, and expression of adhesion molecules that promote CD8 dominant leukocyte recruitment. This inflammatory response leads to an acute myocarditis, and the sustained production of IFN-inducible cytokines establishes a facilitative environment for continued inflammation.

In the case of HAT [27], approximately 5–15 days after infection by the tsetse fly vector, the parasites spread in the host bloodstream and invade the lymph nodes and systemic organs, including the liver, spleen, heart, endocrine system, and eyes in what is termed the early, stage 1, or *hemolymphatic* stage. If untreated, the parasites inevitably cross the blood–brain barrier (BBB) and enter the CNS, which marks the late, stage 2, or *encephalitic* stage of the disease. The stage of the disease has important implications for the effectiveness of treatment, as drugs for the treatment of stage 2 disease need to cross the BBB to have effect.

2.3 Epidemiologic Overlap of DDW: Coinfection

In many regions of developing countries, a variety of DDW occurs in the same area and frequently presents themselves as coinfections with other DDWs. A significant proportion of the population is poly-parasitized, introducing additional complications for disease treatment and cure [20, 35].

The coinfections TB–AIDS [36] and severe malaria–AIDS [37] are frequent and very severe conditions [38]. HIV infection may alter the natural history and progression of other DDWs, impede rapid diagnosis, and reduce the efficacy of antiparasitic treatment. Tropical infections may facilitate the transmission of HIV and accelerate progression from asymptomatic HIV infection to AIDS.

Helminthes are the most common co-parasites found in HIV-, tuberculosis-, and malaria-infected patients [39, 40]. Ninety percent of all malaria burden and 33% of all LF burden occur in Africa, where they are co-endemic in many areas and are transmitted by a number of common vectors [35]. The major Anopheles species that transmit malaria are also vectors for filariasis. Malaria–filariasis coinfection occurs, but with low prevalence (between 0.4% and 4.3%). Helminth coinfection increases the clinical severity of malaria, the risk of women of having malaria, and also increases the susceptibility to HIV/AIDS [18]. Mixed infections of *P. falciparum* and *P. vivax* are rarely reported, but, as demonstrated by PCR, can be as high as 30% of all *Plasmodium* infections [31].

In sub-Saharan Africa, over 75% of cases of TB are HIV-associated. TB is the leading cause of AIDS-related deaths in low-income countries, and it has been shown that HIV infection adversely affects the outcome of pulmonary tuberculosis and facilitates the progression from latent to active tuberculosis [41, 42].

HIV/*Leishmania* coinfection occurs in areas of Southern Europe, Brazil, and regions of Africa and Asia where Leishmaniasis is endemic. Visceral leishmaniasis (VL) promotes the clinical progression of AIDS, and is an important opportunistic infection in HIV immune-compromised individuals which decreases life expectancy. HIV reduces the likeliness to therapeutic response to VL and enhances the probability of relapse. Both diseases decrease the cellular immune response since both agents damage similar immune resources. Infection with schistosomes makes people more susceptible to HIV infection by interfering with immune responses and by increasing the risk of transmission [43].

A major consequence of polyparasitism is anemia [41]. Malaria, hookworm, schistosomiasis, VL, and trypanosomiasis are major causes of anemia. The anemia produced by hookworm and malaria can synergize to produce profound reductions in hemoglobin and adverse health consequences, particularly to African pregnant women, children, and individuals with HIV.

In summary, polyparasitism may weaken the immune response, thus facilitating the establishment of a second or third infection and worsening the outcome of both diseases. Polyparasitism may introduce additional complications for the simultaneous treatment of the several coinfections, including adverse drug–drug interactions and complex administration regimens, leading to reductions in drug compliance.

2.4 Unmet Needs

The drug armamentarium for DDW is limited. For some diseases, like Dengue, there is no specific treatment or vaccine available and control is dependent largely on public health measures directed against the vectors [37]. No new drug has been developed for TB in 40 years [4]. Antimalarial treatment is mostly dependent on the safety and efficacy of artemisinin-based combinations. Artemisinins are effective against blood stages of *P. vivax*, although relapses occur due to chronic liver infections by *P. vivax* hypnozoites that can currently only be treated with primaquine, a drug that must be used for at least 2 weeks and produces hemolysis in patients with glucose-6-phosphate dehydrogenase (G6PD) deficiency. New medicines that safely eliminate hypnozoites from the liver are greatly needed. Drugs to treat HAT are antiquated, scarce, highly toxic, and have developed parasites resistant to them [44]. For onchocerciasis, the only available drug is ivermectin. Although currently there are no established reports of ivermectin resistance in onchocerciasis patients, emergence of resistant strains would be catastrophic. For lymphatic filariasis and onchocerciasis, an effective macrofilaricide is not available. The main therapeutic target for LF is the microfilariae (mf), which are sensitive to combinations of albendazole, ivermectin, and diethylcarbamazine [45].

Compliance with drug treatment regimens for DDW is often difficult. For example, the recommended TB treatment is the DOTS which involves administration of four drugs; isoniazid (I), rifampicin (R), pyrazinamide (Z), and ethambutol (E), for 2 months, followed by the administration of I and R for four additional months. A shortening of this regimen or the use of fixed dose drug combination pills would be a significant advantage [46]. There is sometimes compliance problems even with the 3-day schedule used to treat uncomplicated malaria, as when the fever drops treatment is sometimes stopped prematurely. This leads to relapses and risks the generation of resistance. A one-dose cure would be ideal.

Monotherapy and widespread use of drugs for decades have encouraged the selection of drug resistance. Multidrug resistant (MDR) strains in TB, defined as resistance to both R and I, was estimated by the WHO in 2004 as 4.3% of all cases. Extensively drug-resistant TB (XDR-TB), defined as resistance to R, I, fluoroquinolones and either aminoglycosides or capreomycin or both, is a serious threat to TB control [47–49].[9] In malaria, except for the artemisinins, for which only some decreased sensitivity has been recently reported in southeast Asia, the use of all the other kinds of antimalarials is compromised by resistance.

Side effects and toxicity are serious limitations for drugs treating DDW, as the armamentarium of drugs is in general several decades old. For example, melarsoprol, an arsenic-containing drug used for the treatment of the CNS stage of HAT, kills 5% of treated patients. Amphotericin B, used for the treatment of Leishmaniasis, requires in-patient care in a relatively well-equipped hospital for 30 days, because of the need for i.v. administration and the risk of potentially serious side effects (especially renal toxicity). Some of these diseases, like malaria, mainly affect children under 5 years and pregnant women, and thus the safety requirements for these two groups of patients are particularly demanding.

For prophylaxis, the definitive tool to control DDW would be an efficacious and safe prophylactic vaccine which generates lifelong protective immunity. Unfortunately, such vaccines are often lacking, and with the exception of the vaccines for TB and rotavirus prevention relies primarily on vector control and chemoprophylaxis. In the case of malaria, chemoprophylaxis is mostly done with malarone and mefloquine. Another prophylactic strategy for malaria is intermittent preventive treatment (IPT) for pregnant women (IPTp), infants (IPTi), and children (IPTc). IPT consists of the administration of a therapeutic course of an antimalarial drug, most frequently sulfadoxine–pyrimethamine (SP), at predetermined intervals, regardless of infection status, taking advantage of regular visits to their health services [50] IPT will be discussed in further depth in Sect. 4.2.3.

Drug–drug interactions can pose serious problems for the treatment of DDW, and the simultaneous treatment of two or more diseases is not possible in some cases. For example, rifampicin, a drug used in the DOTS regimen for the treatment of TB, induces a number of cytochrome P450 isozymes (e.g., CYP3A4, CYP1A2, CYP2C9, and CYP2C19), which accelerate the metabolism of many other drugs.

[9]Anti-tuberculosis drug Resistance. Fourth Global report WHO Geneva 2008.

Thus, simultaneous treatment of TB and HIV is frequently difficult because of the interaction of R with some anti-HIV drugs, particularly protease inhibitors [51].

Another unmet need for DDW is with pediatric formulations. Oftentimes, pediatric formulations are not available, and thus adult medicines have to be converted into formulations of lower strength, which not infrequently leads to a misdosing and the use of unproven formulations. Although pediatric formulations for some DDW diseases are becoming available (e.g., pediatric Coartem for malaria), more flexible pediatric formulations for many DDW drugs are still needed. These needs are discussed in more detail in Sect. 4.2.3.

Many DDW occur in the same areas, and thus coinfection with multiple pathogens is frequent. Many DDW pathogens modify the host immune response to favor their own parasitization, and simultaneous infection by two or more pathogens usually complicates the disease outcome. A particularly deadly synergism is produced by the coinfection HIV–AIDS with TB, leishmania, and Chagas disease, particularly when CD4 counts fall below 200 cells/ml [29].

Relapse, latency, and reactivation are issues with a number of DDW. Some parasites, such as *Mycobacterium tuberculosis*, can assume dormant forms which can persist in the body within granulomas, macrophages, adipocytes, and other cells and tissues, long after disappearance from body fluids. *Mycobacterium tuberculosis*, which can maintain latency for years, can be reactivated by a number of factors, such as an immune deficiency. *Plasmodium vivax* can persist as the hypnozoite form within the liver and produce relapse of the disease, with no new infecting mosquito bite, after months or years of the initial infection. Thus, elimination of the hypnozoite form is critical to achieving radical elimination of *P. vivax* malaria [30]. Many people infected with VL are asymptomatic and represent an infectious pool for continued disease transmission. Thus, these factors complicate therapeutic treatment, disease control, and elimination efforts, and are an important consideration for drug discovery.

3 Current Status

3.1 General Socioeconomic Factors

The successful introduction of new DDW drugs and vaccines in developing countries should take into account that new interventions have to be: (a) available to the global market place, development agencies, health services, and patients in developing countries; (b) affordable for the end-user and for those institutions or governments who finance procurement and distribution; (c) acceptable to government regulatory and policy makers, development agencies, health services, and end-users, which presupposes technological, economic, and social acceptability; (d) socially and culturally acceptable to the population to be treated; and (e) readily adoptable by government policy makers, development agencies, health services, and end-users [52].

Disease in the developing world is both a cause and a consequence of poverty. Areas in which DDWs are endemic, including sub-Saharan Africa, are among the poorest in the planet. There is a clear association between disease and poverty. At the individual and family levels, disease is the cause of suffering and lost salaries. In countries where much of the population lives on less than US $1.25 a day, buying much needed drugs is out of reach for many households. DDW also have longer term consequences, killing millions of children in endemic countries, impairing childhood growth, and having a profound effect on demography and in the education and schooling of children with a negative impact on cognitive development and learning ability, as well as adverse effects on future earnings and productivity. Many NTDs are not only poverty-promoting conditions but are also disfiguring and stigmatizing.

The dramatic economic effects of DDW on the population living in endemic areas has been well described [5, 7, 30, 53–55]. In countries with intense malaria transmission, the economic impact of that disease slows economic growth by 1.3% of the GDP per year, which translates into a gross domestic product cost in sub-Saharan Africa of $12 billion per year. In Ghana, the cost of care per patient with Buruli ulcer ranges from 94% to 242% of their annual earnings. In Thailand and South Vietnam, an average family pays more than an average monthly salary to treat a child with dengue hemorrhagic fever. Thus, in many areas DDW treatment is either unaffordable for the individual or family, or comes at great cost and sacrifice.

The greatest burden of TB falls on productive adults who, once infected, are weakened and often unable to work. The burden of taking care of sick individuals usually falls to other family members and, in addition to putting them at greater risk of infection, can lower their productivity. Besides loss of productivity, the cost of treating TB can be as high 8–20% of annual household income, varying by region [56].

Molyneux, Hotez et al. have estimated the cost of treatment of AIDS, malaria, TB and other neglected diseases [19]. They estimate that the cost for an integrated control of seven neglected diseases in Africa (ascariasis, trichuriasis, hookworm infection, schistosomiasis, lymphatic filariasis, trachoma, and onchocerciasis) using a combination of four drugs donated by Pharmaceutical companies (albendazole, ivermectin, azithromycin, and praziquantel) could be achieved at very high cost-effectiveness.

3.2 *New Commitments to Combat Neglected Diseases*

With regards to the big three diseases (TB, malaria, HIV/AIDS), donors, international agencies, nongovernmental development agencies, and governments have responded by creating a number of disease-specialized initiatives, such as UNAIDS (http://www.unaids.org.), and within the WHO, Stop TB (http://www.stoptb.org) and Roll Back Malaria (http://www.rollbackmalaria.org). These developments stimulated the establishment of the Global Fund to Fight AIDS, TB, and malaria (http://www.theglobalfund.org), with a financing mechanism that provides up to 5 years of funding to projects relating to these three diseases. In 2009, the Global Fund launched the Affordable Medicines Facility-malaria (AMFm), which will subsidize the purchase of drugs and manage the price increases along the supply chain, allowing the price gap to be bridged in both the public and private sectors.

Since the beginning of the twenty-first century, there has been an explosion in the number of PPPs, also called product development partnerships (PDPs), created to address specific health problems of DDW. The Initiative on Public–Private Partnerships for Health listed 92 such partnerships in its "partnerships database" (http://www.ippph.org/index.cfm?page=/ippph/partnerships). These PPPs are dedicated to finding and/or implementing solutions for DDW, including developing new drugs and vaccines, diagnostics, use of impregnated bed nets, and access to affordable medications. The main PPPs dedicated to the development of new drugs for DDW are listed in Table 4.

The availability of funding streams that were not available in the past, including the new wave of scientific and corporate responsibility by pharmaceutical companies, academic institutions, governments, and philanthropies, has stimulated activity in neglected diseases R&D. Recent funding programs include The US Leadership against HIV/AIDS, Tuberculosis, and Malaria Act of 2003 [http://www.olpa.od.nih.gov/actions/public/108session1/pl108-25.asp]), and the President's Emergency Plan for AIDS Relief (http://www.avert.org/pepfar.htm). The main philanthropic donations include those from the Bill and Melinda Gates Foundation (http://www.gatesfoundation.org), the Welcome Trust (http://www.wellcome.ac.uk/), The Rockefeller Foundation (http://www.rockefellerfoundation.org/), and The World Bank (http://www.worldbank.org/). A number of pharmaceutical companies, such as GlaxoSmithKline (GSK), Novartis, Astra-Zeneca, Sanofi-Aventis, Pfizer, Ely-Lilly, Bayer, Ranbaxy, Shin Poong, Sigma-Tau, BMS, and others, are contributing drug development know-how, in-kind contributions, and drug donations, and are committed to discover new DDW drugs. The development of vaccines and diagnostics have similar network of PPPs.

The Millenium Goals of Development (MGD) [57] (http://www.un.org/millenniumgoals/; http://www.un.org/millenniumgoals/pdf/MDG_Report_2009_ENG.pdf) has defined specific targets for reduction the burden of these diseases.

3.3 *Current Model for the R&D of New Drugs for DDW: PPPs*

Until recently, drugs have been discovered and developed almost exclusively by pharmaceutical companies following market-driven principles. However, in the last quarter of the twentieth century, the market has failed to deliver the required drugs to control DDWs. Except for the big three, the pipelines for the development of new drugs for DDW are meager and insufficient to cope with attrition to replace drugs which become useless because of resistance. Low returns on investments discouraged drug companies from investing in research and development for DDWs, and academia had greater focus on basic science and scientific publications than translational research for new therapies. Governments of high income countries mainly focused their R&D funding in diseases affecting their populations, and

Table 4 DDW diseases, parasites, vectors, genomes, PPPS

Disease	Parasite (genome sequence ref.)	Vector	Main PPPs (PDP) (web pages of interest)
Malaria	*P. falciparum* (150)	Mosquitoes of *Anofeles aedes*, etc. species	Medicines for Malaria Venture (MMV) (http://www.mmv.org)
	P. vivax (151)		
	P. ovale, *P. malariae*		Roll back malaria (http://www.rollbackmalaria.org)
	P. knowlesi (152)		http://www.who.int/malaria/en
Tuberculosis	*Mycobacterium tuberculosis* (153)		Global Alliance for tuberculosis drug Development (TB Alliance)
	Mycobacterium bovis (154)		(http://www.tballiance.org)
			Stop TB (http://www.stoptb.org)
Leishmaniasis cutaneous visceral	*L. major* (155)	Phlebotomine sand fly	Drug for Neglected Diseases Initiative (DNDI) (http://www.dndi.org)
	L. tropica, *L. mexicana*, *L. braziliensis* (156)		
	L. panamensis, *L. guayanensis* *L. donovani* *L. infantum* (156) *L. chagasi*		Institute of One World Health (IOWH) (http://www.oneworldhealth.org)
Human African trypanosomiasis (HAT)	*Trypanosoma brucei rhodesiense* (157)	Tse Tse fly	DNDi (http://www.dndi.org)
	Trypanosoma brucei gambiense[19]		
Chagas disease	*Trypanosoma cruzi* (158)	Triatomines: *T. infestans*, *T. dimidiata*, *T. brasiliensis*, *T. maculata* *Panstrogylus* and *Rhodnius* spp.	DNDi (http://www.dndi.org)
Intestinal nematode infections		Fecal-contaminated environment	http://www.who.int/wormcontrol
Hookworm dis.	*Ancylostoma duodenale* (159)		
	Necator americanus (159)		
Trichuriasis	*Trichuris trichuria*[20]		
Ascariasis	*Ascaris lumbricoides*[20]		
Lymphatic filariasis	*Wuchereria bancrofti* (in process at Broad Institute)	Mosquitoes	Global program to eliminate lymphatic filariasis
	Brugia malayi (160)		http://www.filariasis.org
	Brugia timori		

(*continued*)

Table 4 (continued)

Disease	Parasite (genome sequence ref.)	Vector	Main PPPs (PDP) (web pages of interest)
Onchocerciasis	*Onchocerca volvulus* (in process at Broad Institute and at Sanger Institute)	Simulium black fly	African program for onchocerciasis control (http://www.apoc.bf/en)
Schistosomiasis	*S. mansoni* (162) *S. japonicum* (161) *S. mekongi*, *S. hematobium*	Fresh water and amphibious snails	Schistosomiasis Control Initiative (http://www.schisto.org)
Dengue	Dengue viruses DEN1, DEN2, DEN3, DEN4	*Aedes aegipti* *Aedes albopictus*	Scientific working group on dengue

Web pages of interest: http://www.apps.who.int/tdr/index.html and http://www.who.int/neglecteddiseases/diseases/edn; http://www.sanger.ac.uk/projects/helminths; http://www.ebi.ac.uk/parasites/parasites-genome.html
[19]http://www.sanger.ac.uk/Projects/T_b_gambiense/.
[20]http://www.sanger.ac.uk/search?t=Parasitic+helminth+genomes&db=allsanger&search=Search.

governments of low income countries in endemic areas have neither resources nor infrastructure to address the discovery and development of new drugs and vaccines. A new model to discover and develop those drugs was needed. That model, based in a network of partnerships, started to take shape in the last decade of the twentieth century, and currently is in the implementation and optimization phase. From the scientific point of view, the leading partners are both academia and the pharmaceutical industry. Funding comes primarily from governments and a few philanthropic foundations, particularly the Bill and Melinda Gates Foundation (BMGF), who contributes to helping establish strategic directions and priorities and providing resources in an unprecedented way. The PPPs have fundamentally modified the way in which new drugs for neglected diseases are developed, and are working to establish a robust pipeline for many of the DDW diseases, particularly malaria, TB, diarrhea, and the kinetoplastid diseases.

Generally, PPPs operate as virtual pharmaceutical companies [58, 59] on a not-for-profit basis, spreading the considerable risk (and cost) of drug development among all the partners. PPPs are frequently focused on specific diseases (e.g., malaria, TB, apicomplexans) and approaches (drugs, vaccines, diagnostic, etc.). The PPPs rely on experienced staff and expert advisors to implement and oversee their R&D programs, in collaboration with their partners. Project teams work against defined target product profiles (TPPs) within well-defined time-frames and budgets, and project progress is reviewed regularly by internal staff and the appropriate scientific advisory committee (SAC), consisting of scientists with expertise in the DDW diseases and drug discovery and development. Those projects not fulfilling predetermined development milestones are discontinued, and the product pipeline is managed in a portfolio approach. Thus, the scientific operations and business plan of a PPP are run much in the way of any for-profit pharmaceutical concern. The portfolio has projects at various stages of preclinical and clinical development and aims to be balanced from the points of view of innovation and risk, segment of the development process, delivery time, cost, and to deliver a sustainable flow of new products. Upstream activities, such as target identification and validation, tend to be performed in the academic setting, whereas partnerships with industry are often formed for preclinical and clinical development. Drugs developed are typically out-licensed by the PPP to industrial manufacturers to be subsequently distributed, at reduced prices, in the developing world. Some of the key PPPs, focused on the development of new drugs for DDW, are listed in Table 4. In the 1970s, 1980s, and 1990s, when little effort was focused on these diseases, the Special Programme for Research and Training in Tropical Diseases (TDR) sponsored by UNICEF, UNDP, World Bank, and WHO (http://apps.who.int/tdr/) (http://www.who.int/tdr/publications/tdrnews/news77/drug.htm) maintained the R&D effort for new DDW drugs and vaccines. The impact that PPPs have had on DDW R&D is considerable. They are key intermediaries between research funders and executers, contributing to coordinating projects, avoiding duplications, leveraging opportunities, catalyzing activity, and defining the TPPS and the therapeutic and scientific strategy of their specific disease area, and, within any IP limitations, contribute to the dissemination of knowledge and transfer of technology.

In summary, the PPPs rely on a collaborative partnership with multiple funders, including governments, supranational bodies (e.g., WHO, World Bank), and philanthropic foundations (e.g., BMGF, Wellcome Trust). Some pharmaceutical companies can be included in this group because of their important in-kind donations. The PPPs use this funding to coordinate the development of new drugs (e.g., MMV, GATB, DNDi, iOWH) and vaccines (e.g., GAVI, MVI, AERAS, IAVI). The PPPs select, fund, and oversee the academic and industrial groups which carry out research projects. Although the PPPs have made significant progress, additional efforts are needed to be made in developing and capitalizing on the potential of disease-endemic countries, in ensuring financial sustainability, and capitalizing on the full potential of the academic and industrial scientific community.

In the public sector, a number of academic centers have been created in recent years to discover and/or develop new drugs for DDW. These centers are the core around which other partners join to complete the set of skills needed to discover new drug candidates. This new model tries to simplify the current one, which was developed in the last half of the twentieth century by the interplay of pharma industry and regulatory agencies. A comprehensive list of academic and public institutes carrying out drug screening for tropical diseases has been reported [60]. Examples include The Sandler Center for Basic Research in Parasitic Diseases at UCSF (http://www.edu/mckerrow/slide.html), and the Drug Discovery Unit at the University of Dundee (http://www.drugdiscovery.dundee.ac.uk/tropical/overview/).

The R&D of new drugs for DDW has undergone continuous innovation to explore new ways to discover and develop new drugs, to challenge old paradigms including IP, confidentiality of chemical structures and data (e.g., pre-competitive information), to increase efficiency of the long and complex drug-discovery process, and to look for new approaches to share risks and lower costs. Ironically, drug discovery for DDW diseases has challenged some of the basic medicinal chemistry principles and tenants, including concepts of acceptable functionality (e.g., endoperoxides, nitro azoles, boron) and acceptable ranges of physiochemical properties (Lipinsky rules) for drug candidates.

In the private sector, GlaxoSmithKline has recently transformed its Tres Cantos (Spain) Centre dedicated to the Discovery of new drugs for DDW into an open DDW Medicines Development Campus [61] to translate ideas and approaches coming from GSK, academia, or other companies into medicines for DDW. Likewise, a number of other pharmaceutical companies, including the Novartis Institute for Tropical Diseases (NITD; Singapore), focused on dengue, malaria, and tuberculosis, and AstraZeneca Bangalore Research Institute in India, focused on tuberculosis, have established dedicated research centers to work on DDW drug discovery. In a recent trend, a number of companies have published HTS data and structures from their DDW research, including GSK's HTS whole-cell hits (13,000 compounds from the corporate two million compound collection) for malaria [62], and TB and malaria compounds and data from Novartis. In addition, GSK has spearheaded the concept of an open patent pool, wherein IP can be used freely for the development of new drugs for DDW. Thus, there is a growing trend among big Pharma to more fully engage and partner in DDW drug discovery, which bring much welcomed resources and know-how to the field.

3.4 Incentives

In spite of the tremendous effort carried out in the last 10 years, the R&D pipeline of new drugs for DDW is inadequate. The drug discovery and development process is long (10–12 years) and expensive, and attrition rates are high [63] (about 50–80% for the early discovery process from target to candidate and about 90% from candidate to launching). Thus, a large number of targets, leads, candidates, and compounds in clinical development are required to ensure a sufficient number of new drugs to effectively treat (and ideally eliminate or eradicate) DDW.

The genomes of many of the key pathogens are known (Table 4), and the biological understanding of parasites and vectors has progressed rapidly. However, the translation of this knowledge into new drugs for DDW has been slow. Increasing the speed and productivity of the process will require a change to the academic culture that rewards publications in high-impact journals over translational research. New approaches and mechanisms are needed to attract more companies, both big pharma and biotechs, to DDW R&D, and to foster academia and industry collaboration to close the gap between scientific knowledge and third-world disease drug R&D.

With these goals in mind, a number of push and pull mechanisms are being implemented. Push mechanisms increase direct investment in research and development and subsidize costs incurred in the discovery of drugs for unprofitable markets. The most successful push mechanism is provided by the public–private partnerships (PPPs, PDPs). Pull mechanisms, such as patent extensions, advance market commitments, and price and patent buyouts are designed to provide incentives at the end of the innovation pathway. Additional powerful pull mechanisms include the orphan drug legislation (ODL) and the Priority Review Voucher.

In the last 20 years, ODL [64] promoted R&D investments to develop new pharmaceutical products for the treatment of rare diseases in the European Union, Japan, the USA, and other countries. Incentives provided by this legislation facilitated the discovery and development of new drugs, which because of their small market do not offer the possibility to recover the investment. From the economic point of view, the development of medicines for tropical diseases has the same financial barriers as the rare diseases and may benefit from similar incentives.

The priority review voucher is a prize awarded by the USA FDA to any company that obtains approval for a treatment (drug or biologic) for an NTD selected from a specific list [65], The voucher, which is transferable and can be sold, entitles the bearer to a priority review for another product. The incentive is a reduced review time, from the target 10 months to a reduced period of 6 months. Actual FDA review timelines, however, can be longer than the target review period, particularly for new products that have not previously been approved. To be able to get to the market four or more months earlier can be a significant advantage, the exact value of which depends on the product sales and other aspects such as the value created for a company if the faster review provides them the possibility of being introduced in the market ahead of a similar, competing product.

4 Current Treatments

Chemotherapy, in addition to vaccines and vector control, plays a major role in treating and helping control or eliminate diseases in the developing world. In many cases, chemotherapy remains the only or most effective option for both clinical management and overall control for these diseases. However, many of the current drugs were discovered more than 30–40 years ago, and many of them have problems of toxicity, poor efficacy, or have lost efficacy due to the emergence and spread of drug resistance. The current treatments for DDW are discussed in the corresponding chapters of this book. For further information, the reader is referred to the following references: malaria [30, 32, 66, 67]; tuberculosis [51, 68, 69]; leishmaniasis [22–24, 70]; HAT [26, 27, 71]; Chagas disease [28, 29, 33]; lymphatic filariasis [20, 72–74]; and onchocerciasis [20, 75, 76].

An important issue is the quality of drugs for DDW. Counterfeits of branded medicines, which often contain no active ingredients or contain less than the required active ingredient (owing to inappropriate standards of manufacture, formulation, packaging, and stability), result in further morbidity and an increased risk of resistance development [77]. It has been estimated that up to 15% of all drugs sold are fake, and in parts of Africa and Asia this figure exceeds 50% [78] A study published in 2008 showed that approximately half of packets of artesunate tablets, collected in several countries of southeast Asia, were fakes and contained no or small quantities of artesunate (up to 12 mg per tablet as opposed to 50 mg per genuine tablet). Chemical analysis demonstrated a wide diversity of other ingredients, including banned pharmaceuticals and toxic compounds [79, 80]. Not surprisingly, drug counterfeiting is greatest in regions where regulatory and enforcement systems for medicines are weakest.

4.1 Nondrug Approaches

In addition to drugs, other nondrug approaches such as vector control and vaccines are used to combat DDW. Current vector control measures include the use of insecticide impregnated bed-nets and indoor residual spraying of insecticides. Other interventions under investigation include the development of genetically modified mosquitoes and mosquitocidal or transmission-blocking vaccines. One final solution to control infectious and parasitic diseases is long lasting and efficacious protective vaccines. However, for most of these diseases, vaccines are either nonavailable, poorly effective, or too expensive.

4.1.1 Vaccines

Vaccines for DDW currently exist, with variable efficacies, for tuberculosis [81, 82], rotavirus-induced diarrhea, cholera, bacteria meningococcus, and a few other bacterial and viral diseases [83], Vaccine development programs are ongoing against HIV, malaria, leishmaniasis, tuberculosis, dengue, schistosomiasis, hookworm, and other infectious diseases. Current status of these programs have been reviewed [84–86].[10]

Based on a 2009 WHO report, more than 80 candidate vaccines are in late stages of clinical testing, and about 30 of them aim to protect against major diseases [87]. Significant progress has been made toward the development of vaccines for malaria [58], TB [57], and dengue [88] The RTS,S/AS01 malaria vaccine, currently in Phase 3 trials, could be licensed by 2012. Two candidate vaccines against dengue virus have been evaluated in children, and one candidate vaccine is currently being evaluated in a large-scale trial. Two new vaccines against rotavirus causing diarrhea, namely Rotarix by GlaxoSmithKline and RotaTeq by Merck, have been recently launched.[11]

Conventional vaccine methods have largely failed against parasitic infections mainly due to the complexity of the parasite life cycle, the ability of the parasite to evade the immune system, and difficulties in identifying and eliciting the desired protective immune responses.

Vaccines for some diseases, like those produced by the trypanosomatids, are very difficult to develop because of their extreme degree of antigenic variation.

[10]State of the art of new Vaccines Research and Development. WHO Geneva 2006.

[11]http://www.path.org/files/VAC_rotavirus_fs.pdf.

Strain variation must also be taken into account; for example, to be effective, a dengue vaccine must protect against all four dengue virus serotypes. The immunization strategies followed to design and develop vaccines for DDW, such as live attenuated strains, dead microorganism, recombinant proteins DNA vaccines, viral vector vaccines, and the use of new more potent adjuvants, are reviewed in some recent reviews [85–87].[12] A number of PPPs, such as the International AIDS Vaccine Initiative (http://www.iavi.org), the Malaria Vaccine Initiative (http://www.malariavaccine.org), and the Aeras Global Tuberculosis Vaccine Foundation (http://www.aeras.org), are specialized in the development of vaccines for DDW.

4.1.2 Vector Control and Transmission Blocking

Vector-borne diseases occur in more than 100 countries and affect about half of the world population, resulting in a high BOD, estimated at about 56 million DALYs [89, 90]. The need for vector control interventions is highest for malaria, dengue, lymphatic filariasis, and the kinetoplastid diseases. Vector control methods include traps, screens, insecticide impregnated materials, indoor residual spraying of insecticides, environmental management, and biological control of vectors, which may in the future include the deployment of genetically modified vectors and mosquitocidal vaccines [91]. The vectors for the tropical diseases are summarized in Table 4.

In diseases like dengue, in which there are no vaccines or treatment available, vector control remains the main intervention for disease prevention and control. Vector control programs have had some significant successes, for example, in the transmission blocking of Chagas disease [29, 33].[13] However, on a global basis, the interventions for control of insects have not been as efficacious as initially planned due to insufficient knowledge of vectors and their environments, poor implementation of existing interventions on large scale, different attractiveness of nets and traps for different species of flies and mosquitoes, resistance of insects to insecticides, etc. Some WHO publications review current situation of vector control programs.[14,15] The genomes of a number of vectors of human pathogens have been sequenced, which will help to further understand vector biology and should lead to new control strategies and a better understanding of the mechanisms of insecticide resistance [92]. Some databases, such as VectorBase (http://www.vectorbase.org), host vector genome information.

[12]See footnote 10.

[13]Innovative Vector control Interventions. Special Programme for Research & Training in Tropical Diseases Sponsored by UNICEF/UNDP/World Bank/WHO. Geneva 2009.

[14]http://www.who.int/malaria/publications/vector-control/en/.

[15]Technical consultation on specifications and quality control of netting materials and mosquito nets. Updated WHO specifications for netting materials and mosquito nets. WHO Geneva 2005.

4.1.3 Diagnostics

Simple, reproducible, sensitive, and affordable diagnostic tools for many of the DDW are needed to map the incidence and prevalence of the diseases, to assess the need for treatment, to evaluate the efficacy of clinical trials and control measures, and to identify the presence of parasite in the blood of blood banks and in tissues for transplantations. The cost of the diagnostic is an important consideration, particularly in impoverished endemic areas. Blood-borne NTDs, particularly Chagas, HIV, malaria, and leishmaniasis, pose a significant threat to blood bank's supply [93]. Some of these diseases have an acute phase or a blood stage in which the parasite is easily accessible and can be easily identified by optical microscopy. However, in other diseases, like the chronic phase of Chagas disease, the persistent/latent phases of TB, the liver phases of *P. vivax* malaria, or the encephalitic phase of HAT, the parasite is not easily accessible and detectable and an accurate diagnostic becomes problematic.

Diagnosis of acute Chagas disease is made by detecting parasites in blood, usually by direct microscopy. Standard diagnosis in chronic disease, where parasitemia is low or transient, relies on PCR methods or on identifying the immune response of the patient to the parasite-serological testing. PCR may become in the future the diagnostic of choice for some of these difficult cases; however, at present, PCR is mainly used for research and not widely used in clinical settings, particularly in impoverished areas of endemic countries [26]. A key issue in HAT treatment is the correct staging of the disease so that the early and late stages can be distinguished reliably. This is critical because the current treatment of late-stage disease, when the parasites have invaded the CNS, is very toxic. Because there are no reliable markers, all patients who are suspected of having CNS involvement must undergo a painful lumbar puncture to examine the CSF [27]. Diagnosis of TB continues to be based on clinical presentation and the Mantoux test, a semiquantitative tuberculin sensitivity test that has remained essentially unchanged since its development 100 years ago. Diagnosis of TB in children is particularly difficult because children rarely develop cavitatory lung disease or expectorate sputum, and a greater proportion of cases are extrapulmonary [14]. Serological assays to diagnose TB have so far proved disappointing, and nucleic acid amplification tests for *M. tuberculosis* still lack sensitivity [94]. Some promising results have been recently obtained with the new Xpert MTB/RIF assay, a nucleic acid amplification-based diagnostic system that detects *M. tuberculosis* and rifampin (RIF) resistance in under 2 h [95].

4.2 Drug Approaches

Most drug approaches to treat diseases caused by parasites are aimed at inhibiting the growth of the microorganism (static) or to kill it (cidal). However, in the case of lymphatic filariasis and onchocerciasis, caused by filarial nematodes such as

Wuchereria bancrofti and *Onchocerca volvulus*, there is an alternative option [96]. These nematodes have evolved a symbiosis with intracellular bacteria of the genus *Wolbachia*, which are required for nematode embryogenesis and survival. The essential role of these bacteria in the biology of the nematode and their demonstrated involvement in the pathogenesis of filariasis make Wolbachia a promising novel chemotherapeutic target for the control of filarial infection and disease. In fact, treatment of *O. volvulus* with tetracycline antibiotics eliminates the endosymbiont *Wolbachia* from the adult female worm. A combination of a single dose IVM with multiple doses of doxycycline (DXC) resulted in a prolonged elimination of mf from the skin of infected individuals.

4.2.1 MDA and Integrated Control Programs

Over the past 20 years, the mainstay of control and elimination efforts for some highly prevalent neglected diseases, such as soil transmitted diseases (STH), schistosomiasis, LF, onchocerciasis and Trachoma, has been MDA to communities of inexpensive, effective, and safe drugs with annual or twice yearly campaigns [97]. The MDA strategy assumes that once the prevalence of the infections is reduced to below a critical threshold, the level transmission will remain low and reemergence of the disease as a public health problem is unlikely. Elimination of these diseases requires additional measures, such as vector control, improved hygiene, environmental sanitation, and health education. MDA programs rely on hundreds of thousands of community drug distributors working together with international organizations, the local government, and the companies donating the drugs. For example, in 2006, more than 350,000 community-directed distributors and 31,000 health workers in 15 countries participated in Ivermectin distribution, within Merck's Mectizan (IVM) Donation Program. The dimension of these MDA programs is illustrated by the GlaxoSmithKline donation in 2009 to The Global Programme to Eliminate Lymphatic Filariasis (GPELF) of 425 million albendazole treatments to 28 countries, and over 1.4 billion albendazole tablets since the program was created in 1998. The program, lead by the WHO, aims to eliminate LF as a public health problem by 2020. Since the beginning of the program, GlaxoSmithKline (GSK) and Merck & Co. Inc. are committed to donate albendazole and Mectizan®, respectively, for as long as necessary to eliminate LF [98].

Lymphatic filariasis [72, 74, 99] is produced by adult worms (macrofilariae) living in the lymphatic system and by secondary infections. After mating, the female worm produces thousands of larvae of microfilariae (mf), which appear in the peripheral blood at times that coincide with the biting activity of mosquito vectors. Chronic complications include lymphedema or elephantiasis of the limbs, damage of the genital organs, and damage of the kidneys and the lymphatic system. There are no drugs against macrofilaria, but there are efficacious microfilaricides.

MDA for LF is conducted through the annual distribution of single doses of two microfilaricidal drugs for at least 5 years. Transmission interruption is achieved by killing of the microfilaria when they are produced once a year. In most instances, between two and six rounds of effective MDA are sufficient to clear microfilaremia and to eliminate this disease as a public health problem. The drugs used are combinations of diethylcarbamazine plus albendazole, or ivermectin plus albendazole in areas, like sub-Saharan Africa, in which onchocerciasis is co-endemic. MDA is administered to all eligible individuals of the entire at-risk population [20, 72–75]. The GPELF started in 2000. Since then, active transmission has been interrupted in seven countries and more than 1.9 billion people have been treated, thereby preventing disease in 6.6 million newborns and averting 1.4 million cases of hydrocele and 0.8 million cases of lymphedema. Such activities translate into 32 million DALYs averted [20, 72–75]. In 2007, a total of 546 million people at risk for lymphatic filariasis were treated. However, in spite of the success of ongoing MDA programs, mathematical modeling suggest that interruption of transmission may be hard to achieve within the proposed time-frame of 6 years and with current percentage of coverage of the target population. It is thought that at least two more years with two additional rounds of treatment would be desirable.

Onchocerciasis [20, 75, 100] is also called "river blindness" because eye lesions may lead to serious visual impairment and blindness, and the larvae of the black fly vector lives in rivers. The adult worms (macrofilaria), which in the human body live in fibrous subcutaneous nodules, produce millions of microfilaria (mf) that migrate under the skin and through the eyes, giving rise to a variety of dermal and ocular abnormalities. The MDA strategy for onchocerciasis is to treat the entire population of meso- and hyperendemic communities with an oral dose of Ivermectin (IVM) once a year to all eligible individuals. The Onchocerciasis Control Program (OCP) (1974–2002) achieved cessation of transmission in nearly all areas of 10 of the 11 West African countries in which it operated. As in LF, IVM acts primarily on mf by killing circulating mf and reducing their production, although it does not kill adult female worms. Given the very long lifespan of adult female worms of up to 15 years, it has been estimated that MDA should be continued for a minimum of 20 years but, depending on the coverage and other factors, control may take 25–35 years or more.

The Schistosomiasis [101, 102] Control Initiative (SCI) [19, 20] contributed by mid-2008 to the administration of approximately 40 million treatments in six countries, reaching over 20 million individuals, although approximately ten times this number of people are thought to need treatment. Treatment is a combination of praziquantel and albendazole. Fortunately, no signs of emerging resistance to PZQ have yet been observed.

The almost complete disappearance of LF from several countries after MDA demonstrated that the control of these diseases through MDA is possible. One concern, however, is the suboptimal coverage of MDA programs, with suggestions that to achieve program goals prolongation of treatment with additional rounds of MDA during additional years will be needed, or MDA should be intensified and administered more frequently. Another concern is the possible generation of resistance associated with the massive use of these drugs. Tolerance to IVM has been detected in Sudan and Ghana after more than ten rounds of IVM mass drug treatment. Genetic analysis of adult worms demonstrated that IVM treatment causes a genetic selection resulting in less genetic diversity, which may well reflect the selection toward IVM tolerant parasites.

Despite the success of MDA programs, there are currently a number of barriers to disease elimination using MDA. For example, the use of ivermectin for LF or onchocerciasis control is counter-indicated in regions in which the filarial nematode *Loa loa* is co-endemic, due to potentially life-threatening side effects due to rapid and massive killing of these microfilaria. Likewise, the use of DEC is counter-indicated for use in onchocerciasis-endemic areas due to severe side effects that can lead to blindness. Also, given the long lifetimes of the LF and onchocerciasis adult worms (macrofilaria), estimated to be approximately 5–7 years and at least 15 years, respectively; many years of MDA treatment to kill the microfilaria is required, and even brief treatment intermission can substantially set back elimination programs. These considerations point to the need for a safe and efficacious macrofilaricide, particularly one that can be safely used in *L. loa*-endemic areas.

Regarding many of the filarial diseases mentioned above, there is an extensive overlap of geographic areas in which these diseases are co-endemic. Advantageously, many antihelminthic drugs are active against several parasites so that multiple neglected diseases can be targeted simultaneously. Thus, a strategy has recently been put forward [1, 18, 19, 21, 76] to integrate programs for the control of seven NTDs – ascariasis, trichuriasis, hookworm, lymphatic filariasis, onchocerciasis, schistosomiasis, and trachoma – using existing drugs. For sub-Saharan Africa, integrated control means the combined use of three anthelmintics: albendazole, ivermectin and praziquantel, and the antibacterial azithromycin. In Asia and the Caribbean, where onchocerciasis is not present, diethylcarbamazine (DEC) can replace ivermectin. The triple combination of albendazole, ivermectin, and praziquantel has been studied and has been shown to be safe. Combination of these drugs may add significant benefit to the treatment of other worm infections, such as ascariasis and trichuriasis, and would help to reduce anemia as well as the progression of disease resulting from HIV/AIDS, tuberculosis, and malaria. At the same time, combinations of ivermectin and azithromycin prevent blindness by reducing the incidence of onchocerciasis and trachoma, respectively.

Since the drugs are donated by Pharma companies, Molyneux et al. estimate that over 500 million individuals could benefit from this MDA preventative chemotherapy, reducing tens of millions of DALYs annually, for just US $200 million per year for 5 years, translating to a cost of about $0.40 per person per year.

Integration of programs for the big three diseases with those focused on the neglected diseases could dramatically reduce the number of life years lost from premature death and disability in Africa. The different programs could take advantage of the community-based healthcare systems set up to deliver drugs, bed nets, vaccines, etc.

4.2.2 Resistance

One of the main challenges for anti-infective drugs is that typically microorganisms (viruses, bacteria, fungi, protozoans, etc.) will become resistant to them, sooner or later. Antimicrobial drug resistance develops when spontaneously occurring parasite mutants with reduced susceptibility to a drug are selected and transmitted. Thus, anti-infective drug therapy constitutes an ongoing race for new drugs with novel mechanisms of action against the continuous generation of drug-resistant parasites. If we discontinue our research effort, the microbes will not stop their fight for survival and we will gradually lose our armamentarium of anti-infective drugs. The strategy to delay resistance involves the generation of new effective antimicrobials and improving the use of current drugs, particularly in the use of drug combinations.

Many of the drugs which in the second half of the twentieth century have been the gold standard for the treatment of a DDW, like the antimalarial chloroquine, are no longer recommended due to the development of parasite resistance. Drug resistance can reach high levels; for example, the proportion of resistance to at least a TB single drug reaching 56% in Baku City, Azerbaijan, the proportion of MDR-TB as high as 19.4% in the Republic of Moldova [51], and over 60% of patients with visceral leishmaniasis in Bihar State, India, not responding to treatment with previous gold standard pentavalent antimonials [70].

Microorganisms use a number of strategies to generate resistance. Point mutations on the target protein can reduce drug's affinity and effectiveness. For example, *P. falciparum* has generated resistant mutants to antifolates and atovaquone through mutations in dihydrofolate reductase and dihydropteroate synthase, and in the mitochondrial cytochrome bc1, respectively.

Microorganisms may also become resistant by over-expressing the target protein and by modifying the transport properties of the microorganism membrane to block the entry of drug or to facilitate their export outside of the parasite cell. For example, in the case of the quinoline antimalarials, chloroquine and mefloquine, the parasite has become resistant through mutations in transporters involved in exporting drug out of the parasite [66]. The *P. falciparum* chloroquine resistance transporter (*Pf*CRT) and *P. falciparum* multidrug resistance channel (*Pf*mdr-1) reduce the parasite drug concentration below effective levels, resulting in the reduction or lack of drug efficacy. An ABCG-like transporter, LiABCG6, localized mainly at the plasma membrane in *Leishmania* parasites, seems to be responsible for drug efflux in this protozoan. When over-expressed, this transporter confers significant resistance to the leishmanicidal agents miltefosine and sitamaquine [103]. Pentamidine, which accumulates in the *Leishmania* mitochondrion, is rendered less effective due to the development of a resistance phenotype resulting in blocking mitochondrial accumulation of the drug within the parasites [104]. Similarly, drug resistance to current trypanocides for the treatment of HAT is due to the loss of this selective drug uptake capacity [105].

A fourth mechanism of resistance involves modulating the metabolic enzymes which activate or deactivate drugs. The gold standard drugs for the treatment of Chagas disease, Benznidazole and Nifurtimox, are prodrugs containing a nitro group which needs to be activated within the parasite by a NADH-dependent, mitochondrially localized nitroreductase (NTR) [106]. The downregulation of this enzyme may generate resistance. Loss of a single copy of this gene in *T. cruzi*, either through in vitro drug selection or by targeted gene deletion, is sufficient to cause significant cross-resistance to benznidazole, nifurtimox, and other nitro-heterocyclic drugs. In *Trypanosoma brucei*, a causal agent for HAT, loss of a single NTR allele confers similar cross-resistance without affecting growth rate or the ability to establish an infection. In some cases, the mechanism of resistance is not clear or may include more than one of the above strategies. The mechanisms of resistance, and of action, of anti-leishmanial pentavalent antimonials are related to drug metabolism, thiol metabolism, and drug efflux [66]. Other mechanisms of drug resistance found in bacteria, such as the transference of resistance genes, appear to be less important in DDW infections.

A number of comprehensive reviews about resistances of individual diseases, including malaria [66, 107–109], tuberculosis [51, 110, 111],[16] leishmaniasis [70, 103, 104], Chagas disease [106], sleeping sickness [105], and other diseases have been published. A collaborative effort between the Global Malaria Program of the WHO and the World Wide Antimalarial Resistance Network (WWARN) has recently been launched. The aim is to create a global network that will provide quality information on antimalarial drug resistance [112]. As exemplified by recent reports of artemisinin resistance on the Thai-Cambodian border, drug resistance is a constant ever-present threat for DDW which must be carefully considered when developing and evaluating new drugs. Careful field monitoring and sensitive diagnostic tests play an important role in evaluating the development and spread of drug resistance, and provide critical information to guide DDW drug discovery.

4.2.3 The Special Cases of Pregnant Women and Children: IPT

Children and pregnant women are important target population for many of the DDW [113]. Available evidence suggests that, in African populations, systemic blood-borne parasitoses of mothers are associated with enhanced susceptibility to infection of their children. Thus, children born to mothers with filariasis or schistosomiasis are infected earlier, and offspring of mothers with placental *P. falciparum* at delivery are themselves at higher risk of developing parasitaemia during infancy [114].

With each malaria episode, the human host builds partial immunity. After repeated malaria infections, people living in endemic areas reach a state of semi-immunity in which, when infected, they experience attenuated or asymptomatic forms of malaria. However, people with little or no exposure to *Plasmodium* infection, such as children under 5 years or people who have lost their semi-immunity after years absent from endemic areas, are susceptible to the fully virulent form of disease. Children in their first years of life have no immunity to malaria, and thus most of the mortality is produced in this age group. Pregnant women are also more susceptible to malaria than nonpregnant women, and this susceptibility is greatest in the first and second pregnancies [115]. Erythrocytes infected with *P. falciparum* accumulate in the placenta

---

[16] See footnote 9.

through adhesion to molecules such as CSA. Common complications of malaria in pregnancy are miscarriage, stillbirth, congenital malaria, and low birth weight. Administration of drugs to the two main population target groups, pregnant women and children, is complicated by special safety requirements.

Children have particular requirements concerning formulation [89]. Pediatric formulations are not always available, particularly for the very young. Thus, tablets need to be fractionated in ½ and ¼ or smaller fractions, or prepared as extemporaneous formulations to adjust to patient weight, which often leads to sub or overdosing that may affect efficacy and safety of treatment. This practice also frequently leads to a lack of standardization and the use of unproven formulations. Furthermore, there have been no appropriate studies on the relationship of age to the pharmacokinetic (PK) properties and effects of DDW drugs. Although pediatric formulations for some DDW diseases are becoming available (e.g., Coartem, a fix dose combination of artemether and lumefantrine and current gold standard for malaria treatment), more flexible dosing systems for children are needed.

In endemic settings, prophylaxis has been shown to protect children from episodes of malaria, anemia, and death. Despite its beneficial impact, mass implementation of chemoprophylaxis raises two main concerns. The first about whether immunity in treated individuals develops as in untreated ones, and the second about the development of resistance [116].

In order to facilitate compliance and help infants to acquire natural immunity against malaria, IPT has been shown to be an effective intervention by delivering antimalarial prophylaxis at defined intervals, typically during regular visits to health services [50, 117]. IPT consists of the administration of a therapeutic course of an antimalarial drug at predetermined intervals, regardless of infection status. The purpose of IPT is to clear any current infection and prevent new ones. The intervals between doses are longer than the time to clear the drug from the bloodstream, which allows for infections between doses. The ultimate goal is, then, to simultaneously reduce frequency of infection and life-threatening illness, while allowing immunity to build up. Compared to continuous chemoprophylaxis, IPT reduces the number of times an individual has to be given the antimalarial, which is delivered at routine health visits: IPT during pregnancy (IPTp) is linked to ongoing routine antenatal care and IPT for infants (IPTi) is delivered through the Expanded Programme of Immunization (EPI). IPTp and IPTi are currently implemented routinely in most malaria endemic countries.

Various studies in Africa have shown that IPTi with sulfadoxine–pyrimethamine (SP) given at the time of routine vaccinations in the first year of life reduces the incidence of clinical malaria by between 20% and 59% [118, 119]. IPT of malaria in pregnancy (IPTp) with SP reduces the incidence of low birth weight, preterm delivery, intrauterine growth-retardation, and maternal anemia [120, 121]. However, the public health benefits of IPTp are declining due to SP resistance [122, 123]. The combination of azithromycin and chloroquine has been suggested as a potential alternative to SP for IPTp [124].

4.2.4 Combinations

To maximize effectiveness, minimize adverse effects, and delay development of resistance, the recommended treatment for most DDW is a combination of two or more drugs. Drugs selected for combination should have different mode of action, matched pharmacokinetics, and ideally a synergistic or at least an additive effect. Combination regimens can reduce toxicity by allowing for lower doses of the individual drugs and may also increase the spectrum of activity compared to that of the individual drugs [23].

Concerning delay of drug resistance, if the target of the first drug is an enzyme with a single point mutation rate of 10^{-7}, and the second target enzyme has a mutation rate of 10^{-8}, the probability of having a microorganism with the two point mutations and thus resistant to the two drugs is 10^{-15}. In the case of malaria, it has been estimated that symptomatic individuals harbor up to about 10^{12} parasites [125]. Thus, in this situation the chance of developing resistance to a single agent is high, but the likelihood of developing resistance to two compounds with different targets is very low.

Ideally the PKs, particularly the half lives ($t_{1/2}$) of the combined drugs, should be similar. This guarantees that none of the drugs will be left alone in low blood concentrations, which may favor generation of resistance. Another strategy (followed in the antimalarial ACT combinations) is to combine drugs in which one of the products has a very rapid parasite killing, even if it has a short half-life, and the other drug has a long half-life. The first drug would rapidly reduce the number of parasites to below the number in which resistant strains may have been selected, and the second drug would be present in sufficient concentration during a number of parasite cell cycles to eliminate the residual parasites. Drug combination treatments are the standard for the big three diseases; however, for some of the neglected diseases, because of limitation in the number of drugs, treatment is sometimes done with combinations (e.g., LF) and other times with monotherapy (e.g., praziquantel for schistosomiasis).

Since 2001, the WHO-recommended treatments for malaria are the artemisinin based combinations (ACTs). This is the therapy of choice for minimizing widespread drug resistance in *P. falciparum*, and, at the same time, preventing recrudescence due to artemisinin monotherapy. There are currently five fixed dose ACTs available or in late-stage development [30] Issues of ACTs are affordability, as ACTs are nearly ten times as expensive as previous first-line choices, and that all of them are susceptible to the same mechanism of drug resistance. Although small increases in the inhibitory concentration (IC50) of artemisinin-based drugs have already been detected in the Thai–Cambodian border, no artemisinin-resistant isolate has yet been characterized. In malaria, drug combinations have been used for many years. Sulfadoxin-pyrimethamine (SP) is an antimalarial combination of two antifolates; malarone, a drug used for malaria prophylaxis, is a synergistic combination of atovaquone and proguanil. Combination of azithromycin and chloroquine, the only combination in Phase III trials that does not contain an endoperoxide, is currently in Phase III clinical for IPT of malaria during pregnancy.

Azithromycin reverses the clinically observed chloroquine resistance phenotype, by a mechanism that is not yet known [30].

The recommended combination for TB is the DOTS, mentioned before. All the new anti-TB drugs in the development are being tested clinically within the DOTS paradigm or related regimens. Even in settings with moderate rates of MDR tuberculosis, DOTS can rapidly reduce the transmission and incidence of both drug-susceptible and drug-resistant tuberculosis. However, further interventions, such as drug-susceptibility testing and standardized or individualized treatment regimens, are needed to reduce mortality rates for MDR tuberculosis [126].

Because of the life cycle of helminthes producing LF and onchocerciasis, and the lack of drugs to eliminate macrofilaria, treatment aims to control these diseases by blocking transmission and eliminating microfilaria through MDA programs [20]. Treatment consists of an annual administration of a single dose of the combination ALB-IVM or Diethylcarbamazine (DEC)-ALB in Latin America and southeast Asia, where there is no onchocerciasis. A combination of four drugs, ALB, IVM, PZQ, and AZM, has been proposed for the integrated treatment of the helminth-produced diseases, plus trachoma [18, 19].

Drug combinations investigated for HAT produced by *T. brucei gambiense* include pentamidine and suramin, eflornithine and melarsoprol, and melarsoprol and nifurtimox. Combinations investigated for stage 2 *T. brucei rhodesiense* include suramin and eflornithine, and suramin and metronidazole [26].

Drug combinations can be administered as individual pills for each drug or, preferably, as fixed-dose pills in which the drugs are formulated in a single tablet. The fixed-dose or single-pill therapy is superior to a multipill regimen in terms of compliance, The combined drugs should not adversely affect the uptake, distribution, or safety of each other. The drugs must not interact chemically with each other, and the final co-formulation should be stable under conditions of high temperature and humidity. The fixed-dose combination should be available in different doses and formulations for small children and pregnant women.

Another approach to achieve poly-pharmacology consists of combining two independent pharmacophores in the same molecule, so that a single compound may be able to hit multiple targets. Some significant early stage successes have been achieved following this multi-target-directed ligands (MTDLs) strategy [44]. However, combining two pharmacophores in a single molecule often introduces additional problems concerning the synthesis and cost of a large molecule, as well as introducing further complications regarding the physiochemical properties of the molecule, such as solubility, absorption, and bio-distribution.

5 New DDW Drugs

In the last third of the twentieth century, the R&D of diseases endemic in developing countries was, for the most part, neglected. However, in the first decade of the twenty-first century, the situation has started to change, as significant advances have been achieved in the science and funding commitment to the area. Hopefully, this should enable a flow of desperately needed new drugs to reach the market in the coming years. General research strategies for DDW drug discovery will be briefly outlined below and are covered in greater depth in the following chapters.

5.1 Drug Discovery Strategies

5.1.1 Target-Based Approaches to Lead Identification

The strategies for screening new compounds against parasitic diseases can be either target-based or whole-organism (phenotypic screening)-based [58]. Other strategies, based on opportunistic, piggy-back approaches and computer-based, rational design, are also being followed. The target-based approach consists in identifying potential drug targets, validating them, setting up an assay that can be developed for high-throughput screening (HTS), as well as secondary assays, and screening collections of compounds to obtain leads. A good target should be essential, selective, testable, and "druggable". A fully validated drug target will be useless if no reasonable drug lead can be identified, and a target that must be inhibited nearly 100% for 24 h/day is not considered druggable. Target-based approaches for DDW have been hindered due to significant knowledge gaps and the lack of research tools to identify and validate proposed targets. The HTS of biochemical targets is often unsuccessful [127], as even when the target is essential and validated the target-based approach offers no guarantee that enzymatic inhibitors identified in in vitro biochemical screens will have whole cell activity, due to cellular uptake problems or whole cell metabolic instability, for example. In addition, the "essentiality" of a target can be dependent on the assay conditions under which target essentiality is explored, and may not reflect the conditions found within the host during infection.

With the increasing availability of the complete genome sequences of many parasites (Table 4), comparative genomics applications are now driving the development of new bioinformatics tools to map entire metabolic pathways *in silico* [128].[17] Methods for genetic, cellular, biochemical, and pharmacological target identification and validation are also being developed. Systems performing similar functions [129, 130], such as energy metabolism, signal transduction, and protein and amino acids recycling, through the different stages of the cell cycle, typically contain clusters of related classes of targets (e.g., kinases and proteases), which can be screened in a systematic way against collections of compounds particularly designed to interact with them.

Validation of the essentiality of a target is usually done by reverse genetics, which involves disruption of the target gene, by either targeted gene knockout or RNA interference, and by chemical genetic technologies, which involves the

[17] http://www.sanger.ac.uk.

demonstration that a small compound (e.g., a selective enzyme inhibitor) leads to parasite's death.

Regarding target selectivity, the cellular organization and biochemistry of parasites producing DDW are frequently different from those of mammals. Those different pathways and targets are, in principle, good candidates to suitable drug targets, as selectivity against host targets may be easier to achieve. For example, N1, N8-bisglutathionyl spermidine (trypanothione), rather than glutathione, is the major redox reactive metabolite in trypanosomatids. This has focused attention on this molecule and the enzymes involved in its metabolism (e.g., trypanothione reductase, TR), as potential targets [131]. Carbohydrate metabolism (i.e., glycolysis) of trypanosomatids has been considered an excellent target owing to the unusual compartmentation of the glycolytic enzymes in the glycosomes.

Comparative genomics allows the identification of gene products that are selective with respect to the human homolog. Many parasites lack pathways to synthesize molecules that are essential. For example, all parasites are purine auxotrophs, and many of them have lost the ability to salvage pyrimidines, relying on *de novo* pyrimidine biosynthesis. All apicomplexan and kinetoplastid parasites lack at least some components of the urea cycle, and several parasites have also developed alternative strategies for sterol and lipid synthesis. These differences represent a key basis by which safe and effective drugs can be developed against these parasites.

Apicomplexan parasites (*Plasmodia, Cryptosporidia, Toxoplasma*) have a relic plastid named apicoplast, originating from secondary endosymbiosis of a red algae, which carries out some unique biochemical and metabolic functions. Some of those pathways have a striking resemblance to prokaryotic metabolism, Approximately 15% of the *P. falciparum* genome encodes proteins destined for the apicoplast. The most extensively studied is the 1-deoxy-d-xylulose 5-phosphate pathway for isoprenoid biosynthesis, which is the target of fosmidomycin, now in Phase II clinical trials in combination with clindamycin or azithromycin. In addition, the apicoplast has a unique fatty-acid biosynthesis pathway and a crucial role in compartmentalizing heme biosynthesis. Housekeeping enzymes associated with this organelle are particularly attractive as drug targets, such as protein synthesis, which is prokaryotic in nature and provides targets for drugs such as doxycyclin, azithromycin, and other antibiotics. Apicomplexans also have other unique pathways, such as the shikimate pathway, whose enzymes are present in bacteria, fungi, and apicomplexan parasites but are absent in mammals. Particularly, the chorismate synthase (CS), which catalyzes the seventh step in this pathway, is an attractive target [132].

Mycolic acids, key components of the wall of mycobacteria, are high molecular weight α-alkyl, β-hydroxy fatty acids. Biochemical and genetic experimental data showed that the product of the *M. tuberculosis inhA* structural gene (InhA) is the primary target of isoniazid, the most prescribed antitubercular agent. InhA is an NADH-dependent enoyl-ACP(CoA) reductase specific for long-chain enoyl thioesters which elongates acyl fatty acid precursors of mycolic acids [133]. The enoylreductases from *M. tuberculosis (InhA)* and *P. falciparum (FabI)* are targets for the development of antitubercular and antimalarial agents.

5.1.2 Whole Cell Approaches to Lead Identification

The whole cell or phenotypic screening involves testing of compounds against in vitro cultures of the pathogen whole cells. Positives hits in this screening are compounds that can be transported to the compound's site of action within the pathogen, but provides no information regarding the mode of action (MoA) or selectivity. This approach has yielded most of the current antimicrobials and is particularly suitable for unicellular microorganisms with relatively simple culture requirements, such as *P. falciparum* and African trypanosomes [60]. For intracellular parasites, for example amastigotes of *Leishmania* spp. or *T. cruzi*, the feasibility of HTS is more limited. For the last decades, screening has been based on time-consuming assays involving radioactive substrates, dyes, and cell counting. Recently, efforts have been made to standardize quick, one-step assays in 96-, 384-, and 1536-well microtitre plates for *P. falciparum* and/or *T. brucei* using RNA-binding dyes and an inverted confocal imaging platform [134], fluorescent-active cell sorting in combination with the nuclear dye YOYO-1, luciferase-driven ATP-bioluminescence, the DNA-intercalating dye, 4,6-diamidino-2-phenylindole [135], and nuclear dyes such as SYBR Green I and nitro blue tetrazolium (NTB) [62].

The main attractiveness of the whole cell-screening approach is that all the targets expressed in the proteome from the particular cultured parasite stage are available, including targets of known and unknown essentiality, both single proteins and complete biochemical pathways. This approach also allows the identification of unselective compounds inhibiting more than one target, which are expected to generate resistances at a slower pace. Challenges of the cell-based screening approach include finding leads which are selective with respect to human cells and in discovering the mechanism of action (MOA) of hundreds or thousands of leads.

In addition to chemical genomics, a number of biological and genetic approaches have been developed to help determine MOA. These include the generation of resistant mutants to identify single-nucleotide polymorphisms (SNPs), to implicate the target, and the generation of strains under-expressing and over-expressing proteins to look for minimum inhibitory concentration (MIC) shifts with the hit compound. Technologies based on genomics or systems-based methods are also being used to help determine the MOA of hit compounds.

Whole cell hits often represent a good starting point for a hit to lead project, as by definition the compound has the physiochemical properties required to penetrate the parasite and access the target. Naturally, the selectivity of the hits must be determined to exclude compounds which may be acting through nonspecific mechanisms, such as general cell membrane disruption or promiscuous DNA binding, for example. Potency is usually an easier property to optimize than uptake for many of the DDW pathogens, as pathogen penetration is related to the physiochemical properties (solubility, lipophilicity, molecular size, etc.) of the compound such that modification of these properties frequently requires substantial modification of the lead and concomitant loss of the microbial inhibitory properties. Whole-organism screening is available at some academic and research institutions worldwide, some of which are part of the compound evaluation network coordinated by the WHO-based Tropical Disease Research (WHO/TDR) Programme [59].

Two Pharma companies (GlaxoSmithKline [62] and Novartis [136]) and an academic consortium [137] have recently carried out HTS of large compound collections against whole cells of *P. falciparum*. Novartis tested 1.7 million compounds and identified a diverse collection of approximately 6,000 small molecules comprised of >530 distinct scaffolds, all of which show potent antimalarial activity (<1.25 μM). GlaxoSmithKline screened 1.9 million compounds against asexual blood stage *P. falciparum* and identified more than 13,500 active compounds inhibiting more than 80% at a concentration lower than 2 μM. Most of these compounds were active against multidrug-resistant *P. falciparum* parasites, and not cytotoxic against a human liver-cancer cell line. Most of the new hits were proprietary to GSK, and the chemical structures and antimalarial activity have now been made accessible to the DDW community for the first time [62]. A large part of these compounds are predicted to target kinase enzymes of *P. falciparum*, which may allow the exploitation of the large collections of protein kinase inhibitors available for other therapeutic areas such as solid-tumor cancers, inflammation, arthritis, diabetes, and cardiovascular disease. The third large screening, a multidisciplinary effort done by a consortium of academic groups, tested 310,000 selected compounds led to more than 1,100 hits which inhibited parasite growth by 80% at a concentration of 7 μM. A selection of 172 compounds having novel chemical structures seems to be active against resistant strains and to act on parasite targets distinct from those affected by the current drugs. These leads may afford the next generation of drugs and also are valuable research tools to discover and validate new targets.

5.1.3 Other Approaches

Other lead identification strategies involve a piggy-back approach, which leverages the synthetic chemistry efforts already applied to orthologous targets under investigation in other disease settings. Good examples are the cysteine protease inhibitors to treat osteoporosis, which have demonstrated antimalarial (e.g., Falcipains [138]) and antitrypanosome activity, the kinases [139] and histone deacetylases [140]. The challenge in this approach is to identify molecules with sufficient selectivity with respect to homolog host targets to ensure safety. Another piggy-back approach involves the use of drugs already approved for use in different diseases to be tested for inhibition of parasite whole cell growth or phenotypic screens. This strategy may afford a quick win, since many of the studies required for first time in human, such as GLP toxicology and PK, as well as some data required for registration have already been collected. Some examples under exploration are the antitubercular activity of the antimalarial mefloquine [141, 142] or the use of the antimicrobial Nitazoxanide for the treatment of rotavirus-induced diarrhea [143].

The so-called rational approach for the identification and optimization of leads involves the use of molecular modeling approaches based on the availability of target and/or target-inhibitor structural information coming from X-ray crystals or suitable structural models. Pathways such as nucleoside biosynthesis have been highlighted by genome sequence analysis, leading to the identification of three potential drug targets. Screening against dihydroorotate dehydrogenase (DHODH) has yielded potent and selective compounds which are being optimized with the availability of high-resolution structural data of the enzyme–ligand complex [144, 145]. Rational design approaches have also been followed to design inhibitors of adenosine deaminase based on detailed electronic calculations of the reaction transition state [146]. Also, inhibitors designed to be active against the transition state of purine nucleoside phosphorylase have been shown to be active against *P. falciparum* [147, 148].

5.2 R&D Drug Pipeline for DDWs

The R&D pipelines of the different diseases vary widely from each other, reflecting the nature of the disease and the degree of attention and funding received by each disease. The pipelines of new drugs for malaria, TB, and diseases produced by the trypanosomatids are now arguably stronger than ever. However, a critical analysis reveals that, on a risk-adjusted basis, current pipelines are insufficient to address the unmet needs, replace drug rendered useless by resistance, and deliver an armamentarium of new DDW drugs.

Each PPP manages a variable but significant part of the corresponding global disease pipelines in collaboration with their partners. The updated pipelines of the different diseases can be seen in the web pages of the corresponding PPPs (Table 4). Concerning the TB pipeline, ten compounds have progressed into clinical development, including six new compounds specifically developed for tuberculosis [54]. Over 90% of current malaria drug discovery projects are targeting the asexual blood stages of *P. falciparum*. Five new fixed-dose ACTs have been introduced or will be available to patients over the next 1–2 years [30, 149]. The pipelines for the development of the kinetoplastids are also better populated than ever.[18]

5.3 Issues and Opportunities

Some of the issues to discover new DDW drugs derive from the complex biology of these parasites. Some parasites, particularly protozoans, such as plasmodia, trypanosomes and leishmania, have complex life cycles including sexual and asexual cycles within both the human host and the invertebrate vector, which adopt different forms depending on the particular stage of the life cycle. The proteomics of each

[18]http://www.dndi.org.

one of these forms may be different, and the number and type of potential drug targets present in each one of the cell cycle stages may also be different (see, for example, the *P. falciparum* life cycle discussed in the antimalarial drug discovery, Chap. 3). Furthermore, DDW pathogens invade multiple cells, organs, and tissues within the human host. Thus, parasites such as *Leishmania*, *M. tuberculosis*, and *Plasmodia* parasitize blood cells; *T. brucei* accesses the CNS; parasitic helminthes like *O. volvulus* may live in the eyes; hookworms, *Ascaris* and *Trichuris* live in the intestines; and *Wuchereria* or *Brugia* filarial species reside in the lymph nodes. Thus, the design of drugs to kill these highly specialized parasites requires not only targeting proteins or processes which are essential for the parasite, but also that the target is present during the time of drug exposure and that the active site of the target parasite stage can be reached by the drug.

Plasmodia species are very specific for their host. Human malaria is produced by five *Plasmodium* species (*P. falciparum*, *P. vivax*, *P. ovale*, *P. malariae*, and *P. knowlesi*), and mice malaria is produced by four different *Plasmodium* species (*P. berghei*, *P. chabaudi*, *P. vinkei*, and *P. yoelii*). Because of this specificity and the availability of in vitro cultures of *P. falciparum*, all current antimalarials were discovered in the past by performing the in vitro evaluation against the human parasite and the in vivo evaluation against the mice parasites. Thus, this model of malaria drug discovery is based on the assumption that the in vivo mouse activity against rodent *Plasmodium* strains will translate into activity in humans against the human relevant *Plasmodium* strains. This example highlights the need for more reliable, relevant, and user-friendly assays which are more predictive of outcome in human clinical trials. The discovery of drugs for each disease has its own unique challenges and issues. For example, *M. tuberculosis* grows very slowly, the development of macrofilaricidal drugs is hindered by the difficulty in obtaining adult worms for compound evaluation, drugs such as Praziquantel may have a strong immune component to its action, and the latent nature of many of the pathogens causing these diseases (e.g., TB, *P. vivax*, VL) complicates the development of new drugs.

Many of the parasites producing these diseases, including *M. tuberculosis*, *M. bovis*, *M. avium*, and *Dengue virus*, require working in a Biosafety level 3 laboratory. Thus, the staff working with these microorganisms must have access to the required training to work in these specially designed laboratories. Other parasites, such as *T. brucei*, *Leishmania donovani*, *Leishmania infantum*, and *P. falciparum*, require biosafety level 2 labs in the USA, and even higher biosafety measures in Europe. The requirement for specialized facilities and training mean a higher entry barrier to discover new drugs against these pathogens.

From the scientific point of view, the main challenges for DDW drug discovery are the limited number of leads and targets with demonstrated essentiality, selectivity, and druggability; the quality and availability of research tools for some microorganisms, such as those used for genetic modification, to validate and prioritize drug targets; and appropriate and predictive in vitro assays and animal models.

There are also many new opportunities for DDW drug discovery. Scientific breakthroughs, such as rapid and inexpensive genetic sequencing and molecular biology methods, are driving rapid advances in the basic science. New computational tools and informatics tools facilitate data analysis and hypothesis generation. System biology approaches can be used to obtain a more integrated view of pathogen biology and pathology. Modern drug discovery technologies which are of common use in the R&D of other human diseases, including HTS, robotized and miniaturized screening assays, molecular modeling, genetic and proteomic studies, and molecular epidemiology, are now beginning to have wide-scale application to DDW. New funding and an increasing wave of social responsibility are attracting more scientists, project managers, and a network of organizations focused on new drugs and vaccines R&D to the area. All of these factors create a tremendous opportunity for the discovery of new DDW drugs.

5.4 New R&D Tools

In the last decade, high-throughput genome sequencing and complementary techniques such as microarray and proteomics have generated, and will continue to generate, ever-increasing amounts of data. The genomes of many of the microorganisms producing DDW have been published (Table 4), including *P. falciparum* [150], *P. vivax* [151], *P. knowlesi* [152], *M. tuberculosis* [153], *M. bovis* [154], *Leishmania major* [155], and related species [156] such as (*L. tropica*, *L. mexicana*, *L. braziliensis*, and *L. infantum*), *T. brucei rhodesiense* [157], *T. b. gambiense*,[19] *T. cruzi* [158], *Ancylostoma duodenale and Necator americanus* [159], *Trichuris tricluria and Ascaris lumbricoides*,[20] *Wuchereria bancrofti* (in process at the Broad Institute), *Brugia malayi* [160] and *B. timori*, *O. volvulus*, *Schistosoma mansoni* [161, 162], *S. japonicum*, *S. mekongi*, *S. hematobium*, and those of their corresponding invertebrate vectors, However, this information overload has failed to generate, for a variety of reasons, the expected bonanza of druggable targets and new drugs for DDW [163]. Databases to accommodate and integrate the diverse data generated by different projects coming from both parasites and vectors are needed [92, 164, 165]. These databases should provide researchers with tools to formulate complex, biologically based queries.[21] In parallel to this sequencing effort, a number of tools have been developed to manage that huge amount of information, such as *Artemis* and *ACT* (http://www.sanger.ac.uk/resources/software/act/), which are Wellcome Trust Sanger Institute free tools. The former is a genome viewer and annotation tool, and *ACT* is a DNA sequence comparison viewer, *GeneDB* (http://www.genedb.org/) is a web accessible database. Additional databases include *Drug Target Portfolio* (http://www.TDRtargets.org), a globally accessible database sponsored by an international network of researchers, populated with a prioritized list of potential drug targets; and *PlasmoDB* (http:// www.PlasmoDB.org), which integrates sequence information, automated

[19] http://www.sanger.ac.uk/Projects/T_b_gambiense/.

[20] http://www.sanger.ac.uk/search?t=Parasitic+helminth+genomes&db=allsanger&search=Search.

[21] http://www.ebi.ac.uk/parasites/paratable.html.

analyses, and annotation data emerging from the *P. falciparum* genome work. Other databases containing relevant genomic information include: http://www.sanger.ac.uk; http://www.ebi.ac.uk/parasites/parasite-genome.html; and http://www.compbio.dfci.harvard.edu/tgi/tgipage.html.

In silico computational methods represent a powerful tool to propose models, new structures, to carry out "virtual" HTS, and to prioritize targets and leads. Some limitations of this technology include our current capacity to explore the stored biological information, the quality and level of genomic annotation, the level of database integration, and the performance and user-friendliness of existing analytic and mining tools. *In silico* systems have been used for the identification of putative drug targets and lead compounds in *Plasmodium* species based on the filtering of protein and chemical properties [166, 167].

The screening of compound libraries for new drug leads has been stymied by current assay limitations and knowledge gaps. For malaria, in vitro culture systems for blood-stage and liver-stage *P. vivax* are urgently required. Only recently have high-throughput assays to test intracellular parasites, such as Chagas disease or visceral Leishmaniasis, become feasible in a high-throughput format. Assay conditions under which a screen is run can be critical, as highlighted by a recent report from NITD [168] in which carbon-source-dependent hits for TB devoid of in vivo activity were identified. Thus, there is now greater consideration of assay conditions developed to more closely mimic the environment found in the human host. For example, TB, drug screening is now being performed under low oxygen tension and/or low nutrient concentration, thus mimicking conditions thought to occur within the granuloma. To facilitate these efforts, scientific and experimental tools and data should be made available for scientists in the field, including strain collections of epidemiologically representative wild-type and resistant strains, cDNAs, antibodies, and crystal structures. Assays should be standardized and be made available to other researchers [169].

As mentioned above, the predictive value of animal models for human disease needs improvement. While the modeling of simpler parameters such as dengue viremia in a mouse model is feasible, the modeling of a complex disease process such as dengue hemorrhagic shock is far more complicated. Similarly, parasitemia in a mouse model of malaria is a useful parameter in the drug discovery process but, because of differences between the human and mouse parasites and immunological responses, a mouse malaria model may not capture drug efficacy in real life situations in humans. Modern genetic engineering and biological approaches are now being applied to improve the animal models used in DDW drug discovery. For example, malaria models have been developed in SCID mice carrying human erythrocytes to evaluate efficacy of antimalarials against the human parasite [170–172]. This, coupled with the development of flow cytometry methods [173], should allow for a more accurate assessment of potential drug candidates. In addition, chimeric mice with human livers are being developed to study the liver phase of the malaria infection and the eradication of *P. vivax* [174] and *P. ovale* hypnozoites. In the TB field, easy to use and reproducible animal models mimicking the conditions of human TB persistence and latency are much needed. Some new methods are being developed [175, 176]. Many of the animal models still used in DDW drug discovery are suboptimal but, by necessity, will continue to be utilized until more predictive models are developed.

5.5 Potential Synergy Across Targets/Diseases

Disease-specific targets relevant to the drug discovery of DDW are discussed in the following chapters of this book. Some targets and leads, which may be broadly applicable across multiple diseases, are briefly discussed here.

Certain metabolic pathways and potential drug targets may be similar across a number of parasites, particularly those performing the same biochemical function such as proteases, kinases, and enzymes involved in the oxidative and reductive metabolism. The biological information and knowledge of one pathway may be used for the design of inhibitors of a similar pathway in another microorganism. The kinome of *P. falciparum* has revealed profound divergences, at several levels, between the kinases of the parasite and those of its host [177]. The *P. falciparum* kinases have been well studied, focusing on target identification and validation and on structure-based design. These studies explored the possibility of interfering with the *Plasmodium* kinases regulating transmission to the mosquito vector as well as the host kinases that may be required for parasite survival [178]. Some metabolic functions, particularly those associated with the synthesis of proteins, DNA, RNA, lipids, and glycosides, may share cofactors or have transporters and efflux pumps belonging to the same families, offering the possibility of synergies across diseases.

Thus, it is reasonable to expect that, for example, a set of kinase inhibitors may inhibit several enzymes from a range of pathogens depending on the degree of parasite kinase homology. In this manner, hits from one HTS may provide hits and leads for multiple programs. This may be particularly true within parasite families, such as within the group of trypanosomatids. Not surprisingly, hits common across DDW pathogens, even within a parasitic family, typically diverge during the hit-to-lead and lead optimization stage, such that it is unlikely that a single compound will represent an optimized drug for multiple indications.

Some compounds exploit biological similarities between different classes of microorganisms, such as the nitroimidazoles, which are thought to be bio-reduced to produce toxic reactive intermediates. Thus, the nitro-aromatics nifurtimox and benznidazole are being used for the treatment of Chagas disease, nifurtimox is being used in combination with melarsoprol for the treatment of stage 2 HAT, and nitroimidazoles PA-824 and OCP67683 are in development for the treatment of TB. Likewise, the oxaboroles, representing a new drug chemotype, are being studied by Anacor and partners for their potential use for the treatment of TB, malaria, HAT, and Chagas disease, in partnership with TB Alliance, MMV, and DNDi, respectively. The unique stereoelectronic properties of the boron atom provides the oxaboroles multiple modes for high-affinity and selective interactions with the target molecules, while still maintaining overall good drug properties such as good chemical stability, solubility, molecular size, and polar surface area.

5.6 *Criteria for Progression*

In terms of program management, drug discovery for DDW is no different from drug discovery for first-world diseases. Programs should identify clear objectives and well-defined criteria for progression toward the clinical candidate, which is often formally defined in the TPPs. Drug PDPs (e.g., MMV, TB Alliance, DNDi) have TPP definitions for their programs in their respective web pages (Table 4). Criteria for lead progression have recently been discussed [179]. A clinical candidate typically meets most, though not necessarily all, of the criteria as defined by the TPP, including in vivo efficacy in animal models, good potency and duration of action, acceptable route of administration (usually PO), a well-defined PK/PD relationship, and, based on human in vitro properties and animal models, good predicted human PK properties.

6 Future Outlook/Conclusion

After decades of neglect, the situation regarding DDW is improving significantly. A new awareness of the dimension of the problem, and new funding streams, has prompted the creation of PPPs which have stimulated all aspects of neglected diseases. More pharmaceutical companies, research institutions, and academic groups are becoming engaged in and committed to third-world diseases, forming consortia of academic and industrial groups that are working together to discover new drugs and vaccines for DDW.

In the long term, prophylaxis by vaccination may be the definitive solution. However, the flow of vaccines will be slow and will take, at least, the next 10–20 years. For some diseases, this time will be even longer. Thus, although vaccines will increase their importance in the control of DDW in the next decade, drugs will have a primary role for at least the next two decades.

Lessons learned over several decades in the discovery and development of new drugs and the technologies that have been developed need to be rigorously applied to neglected diseases drug research. Target selection and validation, lead finding, and predictive animal models need to receive greater attention. As the DDW drug discovery portfolio matures over the next 5–10 years, consistent and sustained funding will be needed to progress these leads into preclinical and clinical development. Clinical trial capacity will need to be greatly improved in the disease-endemic areas to allow for complete clinical testing. With respect to HIV/AIDS, malaria, and TB, the recently established European Developing Country Clinical Trial Partnership should hopefully provide a lead [180].[22]

An aspirational goal would be to eliminate or eradicate all DDW. The feasibility of this goal is dependent on our knowledge of these diseases, the state of our tools and methods, and our commitment to overcoming these diseases. For some of these diseases, eradication may be achievable. In November 2007, the Bill and Melinda Gates Foundation set with a goal for the final goal of eradicating malaria. In order to achieve this goal, antimalarial drug discovery must continue working on the development of new medicines to tackle emerging drug resistance, block disease transmission, and, in the case of *P. vivax* infection, target the dormant liver stage of the parasite [30].

Further improvements can be expected with increased collaboration and integration of health implementation programs at the country and regional level. For example, integration of treatments aimed at the control or elimination of helminth infections may additionally benefit patients by reducing anemia, worm burdens, and susceptibility to HIV/AIDS, tuberculosis, and malaria [4]. Implementation of MDA programs for the combined treatment of helminth-caused diseases, and the development of synergies with control programs for other co-endemic diseases should be pursued [19, 48].

Lymphatic filariasis, trachoma, and onchocerciasis all appear to be in global decline, with good prospects for eventual elimination. Expansion in funding for HIV/AIDS and malaria provides grounds for optimism for the control of these diseases. Meanwhile, the rise of drug-resistant TB poses continuous challenges for physicians and highlights the urgency for new shorter acting TB drugs.

The discovery engine for new DDW drugs is revving up. DDW expertise continues to grow, and key discoveries have been made in the last decade, including genome sequencing of parasites and large-scale whole-cell HTS screens. Efforts to improve assays, animal models, and critical research tools are underway. The social and political commitment to DDW is growing, new research and development partners are being engaged, and funding sources are enabling new clinical candidates to be discovered and enter the R&D pipeline. No doubt many challenges remain, but we have come quite a way in the last decade, and there are many indications that the future for DDW drug discovery is auspicious. Let us hope that someone writing a similar article in 10 years time will say the same.

Acknowledgments The support of Maribel Ardid and Nick Camack from GlaxoSmithKline is gratefully acknowledged.

[22] http://www.edctp.org.

References

1. Hotez PJ, Fenwick A, Savioli L, Molyneux DH (2009) Lancet 373:1570–1575
2. Vanderelst D, Speybroeck N (2010) PLoS Negl Trop Dis 4(1):e576
3. Trouiller P, Olliaro P, Torreele E, Orbinski J, Laing R, Ford N (2002) Lancet 359:2188–2194
4. Matter A, Keller KH (2008) Drug Discovery Today 13(7/8):347–352
5. Mathers CD, Ezzati M, Lopez AD (2007) PLoS Negl Trop Dis 1(2):e114
6. King CH, Bertino AM (2008) PLoS Negl Trop Dis 2(3):e209
7. Conteh L, Engels T, Molyneux DH (2010) Lancet 375:239–247
8. Lopez AD, Mathers CD, Ezzati M, Jamison DT, Murray CJL (2006) Global burden of disease and risk factors. Oxford University Press and The World Bank, New York
9. Molyneux DH (2008) Trans R Soc Trop Med Hyg 102(6):509–519
10. Boschi-Pinto C, Velebit L, Shibuya K (2008) Bull World Health Org 86(9):657–736
11. Mueller I, Galinski MR, Baird JK, Carlton JM, Kochar DK et al (2009) Lancet Infect Dis 9(9):555–566
12. Velayati AA, Farnia P, Masjedi MR, Ibrahim TA, Tabarsi P et al (2009) Eur Respir J 34(5):1202–1203
13. Velayati AA, Masjedi MR, Farnia P, Tabarsi P, Ghanavi J et al (2009) Chest 136(2):420–425
14. Brent AJ, Anderson ST, Kampmann B (2008) Trans R Soc Trop Med Hyg 102:217–218
15. Kyle JL, Harris E (2008) Annu Rev Microbiol 62:71–92
16. Halstead SB (2007) Lancet 370:1644–1652
17. Gould EA, Solomon T (2008) Lancet 371:500–509
18. Hotez PJ, Molyneux DH, Fenwick A, Ottesen E, Sachs SE et al (2006) PLoS Med 3(5):e102
19. Molyneux DH, Hotez PJ, Fenwick A (2005) PLoS Med 2(11):e336
20. Smits HL (2009) Expert Rev Anti Infect Ther 7(1):37–56
21. Hotez PJ, Molyneux DH, Fenwick A, Kumaresan J, Sachs SE et al (2007) N Engl J Med 357:1018–1027
22. Ameen M (2007) Expert Opin Pharmacother 8(16):2689–2699
23. Den Boer ML, Alvar J, Davidson RN, Ritmeijer K, Balasegaram M (2009) Expert Opin Emerg Drugs 14(3):395–410
24. Reithinger R, Dujardin JC, Louzir H, Pirmez C, Alexander B et al (2007) Lancet Infect Dis 7:581–596
25. Fevre EM, Wissmann BV, Welburn SC, Lutumba P (2008) PLoS Negl Trop Dis 2(12):e333
26. D'Silva C (2007) Drugs Future 32(2):149–160
27. Kennedy PGE (2008) Ann Neurol 64:116–127
28. Urbina JA, Docampo R (2003) Trends Parasitol 19(11):495–501
29. Von A, Zaragoza E, Jones D, Rodríguez-Morales AJ, Franco-Paredes C (2007) J Infect Dev Countries 1(2):99–111
30. Wells TNC, Alonso PL, Gutteridge WE (2009) Nat Rev Drug Discovery 8:879–891
31. Miller LH, Baruch DI, Marsh K, Doumbo OK (2002) Nature 415:673–679
32. Greenwood BM, Fidock DA, Kyle DE, Kappe SHI, Alonso PL et al (2008) J Clin Invest 118(4):1266–1276
33. Tarleton RL, Reithinger R, Urbina JA, Kitron U, Gürtler RE (2007) PLoS Med 4(12):e332
34. Tarleton RL (2001) Int J Parasitol 31(5–6):549–553
35. Muturi EJ, Jacob BG, Kim CH, Mbogo CM, Novak RJ (2008) Parasitol Res 102:175–181
36. Maher D, Harries A, Getahun H (2005) Trop Med Int Health 10:734–742
37. Van Eijk AM, Ayisi JG, ter Kuile FO, Misore AO, Otieno JA et al (2002) Am J Trop Med Hyg 67(1):44–53
38. Harms G, Feldmeier H (2005) Infect Dis Clin North Am 19(1):121–135
39. Druilhe P, Tall A, Sokhna C (2005) Trends Parasitol 21(8):359–362
40. Raso G, Luginbühl A, Adjoua CA, Tian-Bi NT, Silué KD et al (2004) Int J Epidemiol 33(5):1092–1102

41. Borkow G, Weisman Z, Leng Q, Stein M, Kalinkovich A, Wolday D, Bentwich Z (2001) Scand J Infect Dis 33(8):568–571
42. Elliott AM, Kyosiimire J, Quigley MA, Nakiyingi J, Watera C et al (2003) Trans R Soc Trop Med Hyg 97(4):477–480
43. Boraschi D, Alemayehu MA, Aseffa A, Chiodi F, Chisi J et al (2008) PLoS Negl Trop Dis 2(6):e255
44. Cavalli A, Bolognesi ML (2009) J Med Chem 52:7339–7359
45. Solomon AW, Nayagam S, Pasvol G (2009) Trans R Soc Trop Med Hyg 103:647–652
46. Duncan K (2003) Tuberculosis 83:201–207
47. Espinal MA (2003) Tuberculosis 83:44–51
48. Wright A, Zignol M, Van Deun A, Falzon D, Gerdes SR et al (2009) Lancet 373 (9678):1861–1873
49. Johnson R, Streicher EM, Louw GE, Warren RM (2009) Curr Issues Mol Biol 8:97–112
50. Aponte JJ, Schellenberg D, Egan A, Breckenridge A, Carneiro I et al (2009) Lancet 374 (9700):1533–1542
51. Ma Z, Lienhardt C, McIlleron H, Nunn AJ, Wang X (2010) Lancet 375:2100–2109. doi:10.1016/S0140-6736(10)60359-9
52. Mahoney RT, Krattiger A, Clemens JD, Curtiss R (2007) Vaccine 25:4003–4011
53. Trouiller P, Torreele E, Olliaro P, White N, Foster S et al (2001) Trop Med Int Health 6(11):945–951
54. Gallup JL, Sachs JD (2001) Am J Trop Med Hyg 64(1–2 Suppl):85–96
55. Sachs J, Malaney P (2002) Nature 415:680–685
56. Laxminarayan R, Klein E, Dye C, Floyd K, Darley S et al (2007) Economic benefit of tuberculosis control. The World Bank, Washington, DC
57. Sachs JD (2005) N Engl J Med 352(2):115–117
58. Caffrey CR, Steverding D (2008) Expert Opin Drug Discov 3(2):173–186
59. Nwaka S, Ridley RG (2003) Nat Rev Drug Discov 2(11):919–928
60. Nwaka S, Hudson A (2006) Nature Rev Drug Discov 5:941–955
61. Andrew Witty (2009) Harvard University, 13 February 2009
62. Gamo F-J, Sanz LM, Vidal J, de Cozar C, Alvarez E et al (2010) Nature 465:305–312
63. Kola I, Landis J (2004) Nature Rev Drug Discov 3:711–716
64. Villa S, Compagni A, Reich MR (2009) Int J Health Plann Manage 24(1):27–42
65. Ridley DB, Grabowski HG, Moe JL (2006) Health Affairs 25(2):313–324
66. Woodrow CJ, Krishna S (2006) Cell Mol Life Sci 63(14):1586–1596
67. Barnes KI, Watkins WM, White NJ (2008) Trends Parasitol 24(3):127–134
68. Kumari S, Ram VJ (2004) Drugs Today 40(6):487–500
69. Van den Boogaard J, Kibiki GS, Kisanga ER, Boeree MJ, Aarnoutse RE (2009) Antimicrob Agents Chemother 53(3):849–862
70. Croft SL, Sundar S, Fairlamb AH (2006) Clin Microbiol Rev 19(1):111–126
71. Barrett MP, Burchmore RJ, Stich A, Lazzari JO, Frasch AC et al (2003) Lancet 362 (9394):1469–1480
72. Perera M, Whitehead M, Molyneux D, Weerasooriya M, Gunatilleke G (2007) PLoS Negl Trop Dis 1(2):e128
73. Bockarie M, Taylor MJ, Gyapong JO (2009) Expert Rev Anti Infect Ther 7(5):595–605
74. Mathew N, Kalyanasundaram M (2007) Expert Opin Ther Patents 17(7):767–789
75. Molyneux DH, Bradley M, Hoerauf A, Kyelem D, Taylor MJ (2003) Trends Parasitol 19 (11):516–522
76. Van den Enden E (2009) Expert Opin Pharmacother 10(3):435–451
77. Newton PN, Green MD, Fernández FM, Day NPJ, White NJ (2006) Lancet 6(9):602–613
78. Cockburn R, Newton PN, Agyarko EK, Akunyili D, White NJ (2005) PLoS Med 2(4):e100
79. Dondorp AM, Newton PN, Mayxay M, Van Damme W, Smithuis FM et al (2004) Trop Med Int Health 9(12):1241–1246

80. Newton PN, Fernandez FM, Plançon A, Mildenhall DC, Green MD et al (2008) PLoS Med 5(2):e32
81. Skeiky YAW, Sadoff JC (2006) Nat Rev Microbiol 4:469–476
82. Sacarlal J, Aide P, Aponte JJ, Renom M, Leach A et al (2009) J Infect Dis 200(3):329–336
83. Velazquez FR (2009) Pediatr Infect Dis J 28(3):S54–S56
84. Da'dara AA, Harn DA (2005) Expert Rev Vaccines 4(4):575–589
85. Crampton A, Vanniasinkam T (2007) Infect Genet Evolution 7(5):664–673
86. Hotez PJ, Brown AS (2009) Biologicals 37(3):160–164
87. WHO, UNICEF, World Bank (2009) State of the world's vaccines and immunization, 3rd edn. WHO, Geneva
88. Webster DP, Farrar J, Rowland-Jones S (2009) Lancet Infect Dis 9(11):678–687
89. Ernest TB, Elder DP, Martini LG, Roberts M, Ford JL (2007) J Pharm Pharmacol 59(8):1043–1055
90. Tolle MA (2009) Curr Probl Pediatr Adolesc Health Care 39(4):97–140
91. Billingsley PF, Foy B, Rasgon JL (2008) Trends Parasitol 24(9):396–400
92. Megy K, Hammond M, Lawson D, Bruggner RV, Birney E et al (2009) Infect Genet Evol 9:308–313
93. Alter HJ, Stramer SL, Dodd RY (2007) Semin Hematol 44(1):32–41
94. Dinnes J, Deeks J, Kunst H, Gibson A, Cummins E et al (2007) Health Technol Assess 11(3): 1–196
95. Blakemore R, Story E, Helb D, Kop J, Banada P et al (2010) J Clin Microbiol 48(7): 2495–2501. doi:10.1128/JCM.00128-10
96. Johnston KL, Taylor MJ (2007) Curr Infect Dis Rep 9(1):55–59
97. Hotez PJ (2009) Clin Pharmacol Ther 85(6):659–664
98. Ottesen EA, Hooper PJ, Bradley M, Biswas G (2008) PLoS Negl Trop Dis 2(10):e317
99. Bockarie M, Taylor MJ, Gyapong JO (2007) Expert Rev Anti Infect Ther 7(5):595–605
100. Udall DN (2007) Clin Infect Dis 44(1):53–60
101. Brooker S (2006) Trans R Soc Trop Med Hyg 101:1–8
102. Utzinger J, Raso G, Brooker S, De Savigny D, Tanner M et al (2009) Parasitol 136(13): 1859–1874
103. Castanys-Muñoz E, Pérez-Victoria JM, Gamarro F, Castanys S (2008) Antimicrob Agents Chemother 52(10):3573–3579
104. Basselin M, Denise H, Coombs GH, Barrett MP (2002) Antimicrob Agents Chemother 46 (12):3731–3738
105. Delespaux V, de Koning HP (2007) Drug Resist Updates 10(1–2):30–50
106. Wilkinson SR, Taylor MC, Horn D, Kelly JM, Cheeseman I (2008) Proc Natl Acad Sci U S A 105(13):5022–5027
107. White NJ (2004) J Clin Invest 113(8):1084–1092
108. Hyde JE (2005) Trends Parasitol 21(11):494–498
109. Mackinnon MJ (2005) Acta Trop 94(3):207–217
110. Shah NS, Wright A, Bai G-H, Barrera L, Boulahbal F et al (2007) Emerg Infect Dis 13(3): 380–387
111. Aziz MA, Wright A, Laszlo A, De Muynck A, Portaels F et al (2006) Lancet 368:2142–2154
112. Sibley CH, Philippe J, Guerin PJ, Ringwald P (2010) Trends Parasitol 26(5):221–224
113. Greenwood B, Alonso P, Ter Kuile FO, Hill J, Steketee RW (2007) Lancet Infect Dis 27:169–174
114. Broen K, Brustoski K, Engelmann I, Luty AJF (2007) Mol Biochem Parasitol 151:1–8
115. Rogerson SJ, Hviid L, Duffy PE, Leke RFG, Taylor DW (2007) Lancet Infect Dis 7:105–117
116. Aguas R, Lourenco JML, Gomes MGM, White LJ (2009) PLoS ONE 4(8):e6627
117. Hutton G, Schellenberg D, Tediosi F, Macete E, Kahigwa E et al (2009) Bull World Health Organ 87(2):123–129. doi:10.1590/S0042-96862009000200014
118. Gysels M, Pell C, Mathanga DP, Adongo P, Odhiambo F et al (2009) Malaria J 8:191. doi:10.1186/1475-2875-8-191

119. Moorthy VS, Reed Z, Smith PG (2009) Malaria J 8:23. doi:10.1186/1475-2875-8-23
120. Menendez C, Bardají A, Sigauque B, Sanz S, Aponte JJ et al (2010) PLoS ONE 5(2):e9438. doi:10.1371
121. Olliaro PL, Delenne H, Cisse M, Badiane M, Olliaro A et al (2008) Malaria J 7:234. doi:10.1186/1475-2875-7-234
122. O'Meara WP, Breman JG, McKenzie FE (2005) Malaria J 4:33. doi:10.1186/1475-2875-4-33
123. Harrington WE, Mutabingwa TK, Muehlenbachs A, Sorensen B, Bolla MC et al (2009) PNAS 106(22):9027–9032
124. Chico RM, Pittrof R, Greenwood B, Chandramohan D (2008) Malaria J 7:255. doi:10.1186/1475-2875-7-255
125. White NJ, Pongtavornpinyo W (2003) Proc R Soc Lond B 270:545–554
126. DeRiemer K, García-García L, Bobadilla-del-Valle M, Palacios-Martínez M (2005) Lancet 365(9466):1239–1245
127. Payne DJ, Gwynn MN, Holmes DJ, Pompliano DL (2007) Nature Rev Drug Discov 6(1): 29–40
128. Chaudhary K, Roos D (2005) Nat Biotechnol 23(9):1089–1091
129. Winzeler EA (2006) Nat Rev Microbiol 4:145–151
130. Dharia NV, Chatterjee A, Winzeler EA (2010) Curr Opin Investig Drugs 11(2):131–138
131. Fairlamb AH, Cerami A (1992) Annu Rev Microbiol 46:695–729
132. Dias B, Marcio V, Ely F, Palma MS, de Azevedo WF et al (2007) Curr Drug Targets 8(3): 437–444
133. Oliveira JS, Vasconcelos IB, Moreira IS, Santos DS, Basso LA (2007) Curr Drug Targets 8(3):399–411
134. Cervantes S, Prudhomme J, Carter D, Gopi KG, Li Q (2009) BMC Cell Biol 10:45
135. Baniecki ML, Wirth DF, Clardy J (2007) Antinicrob Agents Chemother 51(2):716–723
136. Plouffe D, Brinker A, McNamara C, Henson K, Kato N et al (2008) PNAS 105(26): 9059–9064
137. Guiguemde WA, Shelat AA, Bouck D, Duffy S, Crowther GJ et al (2010) Nature 465(20): 311–315
138. Joachimiak MP, Chang C, Rosenthal PJ, Cohen FE (2001) Mol Med 7(10):698–710
139. Doerig C, Billker O, Haystead T, Sharma P, Tobin AB et al (2008) Trends Parasitol 24(12): 570–577
140. Patel V, Mazitschek R, Coleman B, Nguyen C, Urgaonkar S et al (2009) J Med Chem 52(8): 2185–2187
141. Bermudez LE, Kolonoski P, Wu M, Aralar PA, Inderlied CB et al (1999) Antimicrob Agents Chemother 43(8):1870–1874
142. Danelishvili L, Wu M, Young LS, Bermudez LE (2005) Antimicrob Agents Chemother 49(9):3707–3714
143. Rossignol JF, Abu-Zekry M, Hussein A, Santoro MG (2006) Lancet 368(9530):124–129
144. Baldwin J, Michnoff CH, Malmquist NA, White J, Roth MG (2005) J Biol Chem 280(23): 21847–21853
145. Gujjar R, Marwaha A, El Mazouni F, White J, White KL et al (2009) J Med Chem 52(7): 1864–1872
146. Larson ET, Deng W, Krumm BE, Napuli A, Mueller N et al (2008) J Mol Biol 381(4): 975–988
147. Taylor Ringia EA, Schramm VL (2005) Curr Top Med Chem 5(13):1237–1258
148. Madrid DC, Ting LM, Waller KL, Schramm VL, Kim K (2008) J Biol Chem 283(51): 35899–35907
149. Olliaro P, Wells TN (2009) Clin Pharmacol Ther 85(6):584–595
150. Gardner MJ, Hall N, Fung E, White O, Berriman M et al (2002) Nature 419(6906):498–511
151. Carlton JM, Adams JH, Silva JC, Bidwell SL, Lorenzi H et al (2008) Nature 455(7214): 757–763
152. Pain A, Bohme U, Berry AE, Mungall K, Finn RD et al (2008) Nature 455:799–803

153. Cole ST, Brosch R, Parkhill J, Garnier T, Churcher C et al (1998) Nature 393:537–544
154. Garnier T, Eiglmeier K, Camus JC, Medina N, Mansoor H et al (2003) PNAS 100(13): 7877–7882
155. Ivens AC, Peacock CS, Worthey EA, Murphy L, Aggarwal G et al (2005) Science 309 (5733):436–442
156. Peacock CS, Seeger K, Harris D, Murphy L, Ruiz JC et al (2007) Nat Genet 39(7):839–847
157. Berriman M, Ghedin E, Hertz-Fowler C, Blandin G, Renauld H et al (2005) Science 309 (5733):416–422
158. El-Sayed NM, Myler PJ, Bartholomeu DC, Nilsson D, Aggarwal G et al (2005) Science 309 (5733):409–415
159. Hu M, Chilton NB, Gasser RB (2002) Int J Parasitol 32(2):145–158
160. Ghedin E, Wang S, Spiro D, Caler E, Zhao Q et al (2007) Science 317(5845):1756–1760
161. Zhou Y, Zheng H, Chen Y, Zhang L, Wang K et al (2009) Nature 460(7253):345–351
162. Berriman M, Haas BJ, LoVerde PT, Wilson RA, Dillon GP et al (2009) Nature 460 (7253):352–358
163. Winzeler EA (2008) Nature 455:751–756
164. Hertz-Fowler C, Hall N (2004) Methods Mol Biol 270:45–74
165. Bahl A, Brunk B, Coppel RL, Crabtree J, Diskin SJ et al (2002) Nucleic Acids Res 30(1): 87–90
166. Saïdani N, Grando D, Valadić H, Bastien O, Maréchal E (2009) Infect Genet Evol 9(3): 359–367
167. Joubert F, Harrison CM, Koegelenberg RJ, Odendaal CJ, de Beer TAP (2009) Malaria J 8:178. doi:10.1186/1475-2875-8-178
168. Pethe K, Sequeira PC, Agarwalla S, Rhee K, Kuhen K et al (2010) Nat Commun 1(5):1–8
169. Fidock DA, Rosenthal PJ, Croft SL, Brun R, Nwaka S (2004) Nat Rev Drug Discov 3:509–520
170. Ito M, Kobayashi K, Nakahata T (2008) Curr Top Microbiol Immunol 324:53–76
171. Pearson T, Greiner DL, Shultz LD (2008) Curr Top Microbiol Immunol 324:25–51
172. Angulo-Barturen I, Jimenez-Dıaz MB, Mulet T, Rullas J, Herreros E et al (2008) PLoS ONE 3(5):e2252
173. Jiménez-Díaz MB, Mulet T, Gómez V, Viera S, Alvarez A et al (2009) Cytometry A 75(3): 225–235
174. Galinski MR, Barnwell JW (2008) Malaria J 7(Suppl 1):S9. doi:10.1186/1475-2875-7-S1-S9
175. Apt A, Kramnik I (2009) Tuberculosis (Edinb) 89(3):195–198
176. Pichugin AV, Yan BS, Sloutsky A, Kobzik L, Kramnik I (2009) Am J Pathol 174(6): 2190–2201
177. Doerig C, Meijer L (2007) Expert Opin Ther Targets 11(3):279–290
178. Geyer JA, Prigge ST, Waters NC (2005) Biochim Biophys Acta 1754(1–2):160–170
179. Nwaka S, Ramirez B, Brun R, Maes L, Douglas F et al (2009) PLoS Negl Trop Dis 3(8):e440
180. Olliaro P, Smith PG (2004) J HIV Ther 9(3):53–56

Top Med Chem 7: 47–124
DOI: 10.1007/7355_2011_13

Published online: 29 April 2011

The Medicinal Chemistry of Tuberculosis Chemotherapy

Gwendolyn A. Marriner, Amit Nayyar, Eugene Uh, Sharon Y. Wong, Tathagata Mukherjee, Laura E. Via, Matthew Carroll, Rachel L. Edwards, Todd D. Gruber, Inhee Choi, Jinwoo Lee, Kriti Arora, Kathleen D. England, Helena I.M. Boshoff, and Clifton E. Barry III

Abstract The development of effective chemotherapy for the treatment of tuberculosis (TB) began in the 1940s and has been reinvigorated recently due to concern regarding the emergence of highly drug-resistant TB strains. This chapter explores the medicinal chemistry efforts that gave rise to current frontline and second-line drugs in global use today and attempts to comprehensively summarize ongoing discovery and lead optimization programs being conducted in both the private and the public sector. TB has a large number of disease-specific considerations and constraints that introduce significant complexity in drug discovery efforts. Conceptually, the disease encompasses all the drug discovery challenges of both infectious diseases and oncology, and integrating these considerations into programs that often demand collaboration between industry and academia is both challenging and rewarding.

Contents

This work was funded in part by the Division for Intramural Research, NIAID.

G.A. Marriner, A. Nayyar, E. Uh, S.Y. Wong,
T. Mukherjee, L.E. Via, M. Carroll, R.L. Edwards, T.D. Gruber, I. Choi, J. Lee, K. Arora, K.D. England, H.I.M. Boshoff, and C.E. Barry III (✉)
Tuberculosis Research Section, National Institute of Allergy and Infectious Disease, NIH, Bethesda, MD, USA

1 Introduction

1.1 TB: A Global Epidemic

Tuberculosis (TB) is the second major cause of death due to an infectious disease in adults worldwide with nine million new cases and close to 1.8 million deaths annually [1]. TB is caused by *Mycobacterium tuberculosis* (MTb), an airborne pathogen transmitted among humans which infects macrophages in the lungs. Two possible outcomes follow macrophage infection: (1) the infected macrophage can be recognized by effectors of the innate immune system and eradicated; or (2) the bacilli may further multiply in the cell, ultimately leading to its destruction and the infection of new macrophages drawn to the site of infection. The second scenario may initiate T cell-mediated adaptive immunity enabling the host to eradicate the bacilli at the initial site of infection. Failure of adaptive immunity to eradicate the bacilli leads to uncontrolled growth of the organism and subsequent spread through the lymphatic system to secondary sites. These sites may be in the lung or in some cases in extra-pulmonary sites, which is manifested as clinical disease with various degrees of severity which, if not treated, kills more than 50% of patients. There is also an intermediate situation wherein adaptive immunity may be able to contain the growth of the organism by controlling its metabolism for years, even decades, until waning of immunity allows reactivation of this latent form of disease [2]. More recently, researchers have begun to suspect that latent disease may not be a single metastable state but rather a subtle guerilla war with waxing and waning local battles on a small scale resulting in a "spectrum" of subclinical active disease [2].

Based on the global incidence of a positive response in the tuberculin skin test, which is associated with adaptive immunity, it is estimated that about two billion people are latently infected with MTb [1]. Of these two billion people, about 10% will develop active tuberculosis in their lifetime, although HIV infection dramatically increases this risk to a 10% annual risk of conversion to active TB [3]. TB is the major cause of death in HIV-infected individuals. A person with contagious pulmonary TB infects 10–15 more people on average, which created tuberculosis epidemics in the developed world until the advent of chemotherapy. Isolation of patients resulted in limited success in TB control through the establishment of sanatoria in the mid-nineteenth century. Sanatoria patients would occasionally achieve spontaneous resolution of the disease, although subsequent relapse rates were high, highlighting the chronic and dynamic nature of this infection.

Albert Calmette and Camille Guérin produced a strain of *Mycobacterium bovis* (bacille Calmette-Guérin, BCG) by serial passaging of an isolate of the related bacillus that causes TB in cattle on potato–bile–glycerin media until it was no longer virulent in laboratory animals. Vaccination started in the 1920s, but the efficacy of the vaccine varied greatly depending on factors such as geographic location and strain of BCG used for vaccination [3, 4]. Large-scale clinical trials throughout the world have shown that the vaccine protects against severe forms of TB in children but does not protect against the development of adult pulmonary TB [3]. Thus, BCG vaccination has not reduced the global incidence of TB. Disturbingly, recent data suggest that this vaccine applied now for decades on a global scale may also have accelerated the development of even more virulent forms of TB [5].

In accord with the Millennium Development Goals established by the United Nations, the World Health Organization's "Stop TB" Strategy aims to halve the prevalence of, and mortality due to, TB compared to that seen in 1990 by the year 2015 and to have reduced the incidence of new TB cases to one per million by 2050. Very few think that these goals remain realistic, given our current progress with available tools. Our only chance of achieving such progress lies in the development of better diagnostic methods and new drugs to combat both drug-sensitive and drug-resistant disease [6].

1.2 The Medical History of Current TB Chemotherapy

Effective chemotherapy for tuberculosis began in 1940s with the discovery and use of streptomycin (STR, Fig. 1; **1a**) and *para*-aminosalicylic acid (PAS, Fig. 1; **2a**) [7–9]. The first randomized controlled study of STR treatment for TB by the British Medical Research Council (BMRC) showed that streptomycin was effective in the short term but that ultimately so many patients developed STR-resistant TB and hearing loss that at 5 years, no net clinical benefit was seen [10]. Contemporaneously, PAS was found to be bacteriostatic against MTb (including STR-resistant strains) in experimental models and able to prevent the development of STR

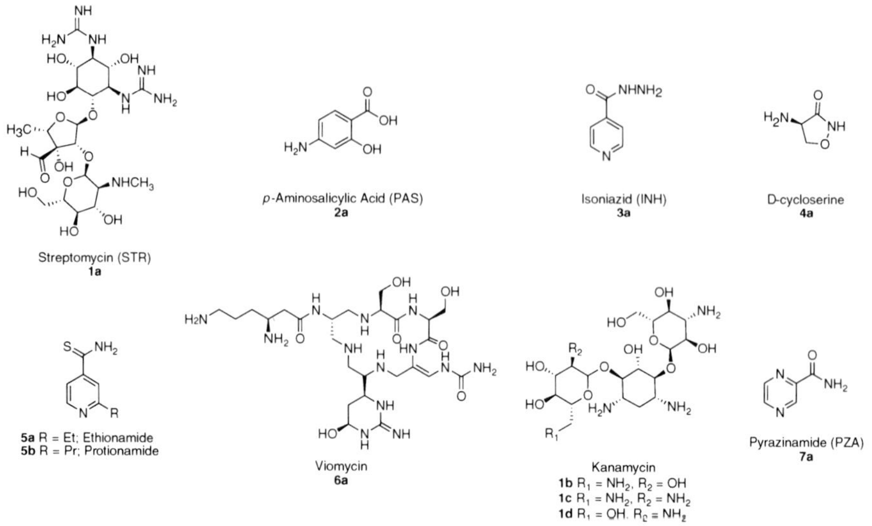

Streptomycin (STR) **1a**

p-Aminosalicylic Acid (PAS) **2a**

Isoniazid (INH) **3a**

D-cycloserine **4a**

5a R = Et; Ethionamide
5b R = Pr; Protionamide

Viomycin **6a**

Kanamycin
1b $R_1 = NH_2$, $R_2 = OH$
1c $R_1 = NH_2$, $R_2 = NH_2$
1d $R_1 = OH$, $R_2 = NH_2$

Pyrazinamide (PZA) **7a**

Fig. 1 TB drugs introduced in the 1940s and 1950s

resistance [11, 12]. PAS was also shown to be useful in pulmonary TB patients as monotherapy, but development of resistance occurred and patients tolerated the drug poorly, mainly due to gastrointestinal side effects and occasional hepatitis [13]. A BMRC trial in subjects with pulmonary TB found that STR with or without PAS was as effective, or even slightly more effective, than PAS alone, but that the combination with PAS greatly reduced the development of drug resistance [10]. By the end of the 1940s, the standard of care was combined therapy with STR and PAS, typically given for 12–24 months.

The 1950s were significant because of the discovery and initial use of isonizaid (INH, Fig. 1; **3a**). There were several trials to optimize treatment combinations of INH with STR and PAS. Although INH was generally well tolerated in patients, some experienced rash or hepatitis with this drug. INH treatment led to rapid improvement over the first month of therapy, but recrudescence of disease was common due to acquired resistance [9, 14, 15]. Drug combination studies showed that INH with STR was superior to INH with PAS as measured by radiographic, microbiologic, and clinical improvement. In addition, such studies showed that STR was more effective than PAS in preventing the emergence of INH resistance [9, 16]. The triple drug combination of INH, STR, and PAS was found to be better than therapy with INH and PAS combined and achieved 98% sputum culture conversion at 6 months compared to 84% with INH and PAS alone [17]. These early studies also underscored the importance of extended treatment durations with chemotherapy of less than 1 year, 1 year, or 15 months associated with an 8%, 1.4%, or 0% relapse rate, respectively [17].

Other drugs discovered during the 1950s include cycloserine (Fig. 1; **4a**) [18], ethionamide (Fig. 1; **5a**), and the closely related prothionamide (Fig. 1; **5b**) [19], viomycin (Fig. 1; **6a**) [20], kanamycin (Fig. 1; **1b-d**) [21], and pyrazinamide (PZA, Fig. 1; **7a**) [22]. At the time of their discovery, these drugs were thought to be inferior to INH, PAS, and STR and were used only in patients with disease refractory to standard therapy [23].

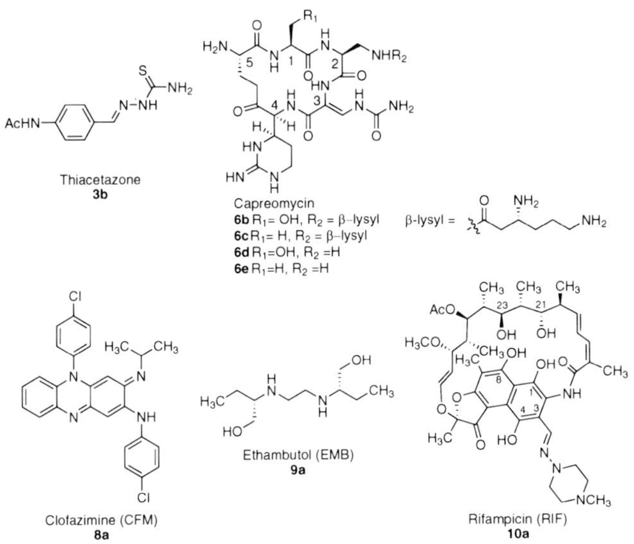

Thiacetazone
3b

Capreomycin
6b R_1= OH, R_2 = β-lysyl
6c R_1= H, R_2 = β-lysyl
6d R_1=OH, R_2 =H
6e R_1=H, R_2 =H

Clofazimine (CFM)
8a

Ethambutol (EMB)
9a

Rifampicin (RIF)
10a

Fig. 2 TB drugs introduced in the 1960s and 1970s

In the 1960s, care shifted from sanatoria or hospitals to the home after a landmark study in Madras, India, which showed that care in the home was equally efficacious to treatment in a sanatorium or hospital [24]. Other TB drugs introduced during the 1960s include thiacetazone (Fig. 2; **3b**) [25], capreomycin (Fig. 2; **6b–e**) [26], and clofazimine (CFM, Fig. 2; **8a**)[27]. These early studies highlight one of the key problems with second-line agents that persists today: tolerability. In the words of one set of authors of these trials "...the patients considered the cure worse than the disease" [23]. This aspect complicated systematic clinical trials to devise an optimal regimen or to establish the relative efficacy of many of these new agents. Notably, these same agents are used in second-line therapy today, where clinicians confront the same issue. Ethambutol (EMB, Fig. 2; **9a**) supplanted PAS in the standard drug regimen since this drug was better tolerated than PAS and also allowed the treatment regimen to be shortened to 18 months [28, 29].

Rifampicin (RIF, Fig. 2; **10a**), one of the last drugs to be introduced into clinical practice, revolutionized TB therapy [30]. Landmark clinical trials in the 1970s in East Africa and Hong Kong showed that addition of RIF to the standard INH/EMB/STR or INH/STR drug regimens allowed the duration of treatment to be decreased from 18 to 9 months without increasing the relapse rate [31, 32]. Renewed interest in PZA was sparked by reports that PZA was more effective than STR in reducing organ burdens in MTb-infected mice when combined with INH [33, 34]. Clinical trials at the end of the 1970s and in the 1980s investigated the use of PZA in various combinations and treatment durations with STR, INH, RIF, EMB, and thiacetazone. PZA was instrumental in allowing shortening of TB chemotherapy to 6 months [35–37]. Although thiacetazone was initially used in chemotherapy instead of RIF due to the high cost of RIF, it was later omitted because of life-threatening

Stevens–Johnson syndrome in those with HIV coinfection. STR was also largely replaced by EMB to avoid the requirement of intravenous administration of STR. The culmination of these studies was the introduction of modern "short-course" chemotherapy for drug susceptible TB where PZA forms part of the drug regimen for the first 2 months (the "intensive phase") in combination with INH, EMB, and RIF, but then is not included in a subsequent continuation phase of 4 months of additional treatment with INH and RIF (sometimes in combination with EMB) to obtain a cure rate greater than 95% [31]. With a new, highly effective treatment regimen, the world celebrated the end of the "White Plague" and quickly turned its attention elsewhere.

One could argue, with some justification, that the resulting collapse of research efforts into developing new antitubercular agents in the 1970s and 1980s happened too soon. We still have little idea why the substitution of EMB for PAS allowed the regimen to be shortened from 24 to 18 months, a poor understanding of why adding RIF allowed the regimen to be shortened to 9 months, and no information at all as to why adding PZA allowed treatment to be further truncated to 6 months. Perhaps more importantly, the consequences of widespread global programs of drug treatment in less-controlled environments than the clinical trials supervised by the BMRC were poorly understood. In retrospect, the trial conclusions in the developed world have been borne out; widespread TB epidemics are a thing of the past, and small outbreaks in the USA or Europe are the subjects of alarmed news headlines. Meanwhile, in the developing world, the rise of drug resistance through improper drug usage, poor compliance, and lack of government commitment to eradication programs began in earnest.

1.3 The Emergence of Drug-Resistant TB

As the history of clinical development of TB drugs shows, to limit the risk of developing resistance developing to every new agent, extensive combination therapies were prescribed. The World Health Organization developed strategies to try to avoid the acquisition of resistance, primarily in the form of the "directly observed therapy, short course" (DOTS) which involved implementing a system to monitor patients' ingestion of pills and recording compliance and treatment completion [38, 39]. Central to the DOTS strategy is government commitment to TB control programs, diagnosis of smear-positive TB cases, observed treatment, ensured drug supply, and standardized reporting. While DOTS can be effective and is recommended by the World Health Organization, it is programmatically difficult and expensive. The natural sequence of events then, despite the introduction of both short-course chemotherapy and DOTS, was that treatment became marked by high relapse rates, and the 1990s marked a period of increasingly resistant TB ranging from mono- to multidrug-resistant tuberculosis (MDR-TB). The phrase "MDR-TB" was coined in the 1990s to refer specifically to isolates that had developed resistance to INH and RIF (according to conventional wisdom the two most important drugs in determining outcome). MDR-TB developed initially by acquisition of resistance during standard treatment as a result of poor compliance or improper chemotherapy with subsequent amplification of resistant populations in treated patients. The development of MDR-TB-infected patients ultimately led to transmission of drug-resistant MTb, first within institutions and hospitals and later in the community [40–43].

Treatment of MDR-TB and higher degrees of resistance has required reintroduction of second-line drugs with unproven efficacy in untested combinations as well as the use of broad-spectrum agents developed for other indications such as fluoroquinolones that have incidental activity against MTb. The treatment of MDR-TB relies upon a backbone of an injectable agent (kanamycin, capreomycin, or amikacin; see Sect. 2.8) [21, 26] and a fluoroquinolone (see Sect. 3.4) [44–48]. The choice of injectable agent and fluoroquinolone for patient treatment is based on drug-sensitivity results from the sputum-borne strain of the patient in question and prior treatment history. Drugs from the first-line agents (EMB, PZA, INH, RIF, and STR) are administered if the strain is sensitive to any of these and combined with second-line drugs (amikacin, kanamycin, capreomycin, viomycin, enviomycin, fluoroquinolones, ethionamide, prothionamide, cycloserine, PAS) with a goal of having five active drugs based on drug-sensitivity results. In cases where extensive resistance does not allow five drugs to be selected from the first- and second-line agents, agents can be selected from non-WHO approved lists of third-line agents (rifabutin, macrolides such as clarithromycin, augmentin, imipenem, clofazimine, linezolid, thioacetazone, and thioridazine have all been reported for such cases). There is minimal data to support the use of the third-line agents [49, 50] with the exception of linezolid (see Sect. 3.3 for further discussion of oxazolidinones) where its use is limited by toxicity and expense [51, 52]. Even in the case of linezolid, the available data are anecdotal and not prospectively collected, but ongoing clinical studies are likely to provide data supporting its use [53, 54]. The duration of treatment for MDR-TB is based on data from the 1950s where 1–2 years of treatment was best at preventing relapse [55]. Treatment is divided into a 6–9-month intensive phase that includes the injectable agent followed by a continuation phase of up to 18 months for a total of 24–30 months of treatment. The injectable agent is stopped to reduce the potential for nephro- and ototoxicity associated with these agents. In the case of uncomplicated MDR-TB, cure rates of greater than 80% have been reported [56].

Following the turn of the century, treatment of MDR-TB with poorly active second- and third-line agents inevitably gave rise to the emergence of extensively drug-resistant tuberculosis (XDR-TB) which has been defined based on the loss of the two components of the MDR treatment backbone perceived to be most important, the fluoroquinolones and the injectable agents [57]. For patients with XDR-TB lucky enough to have access to drug-susceptibility testing and the full suite of second- and third line agents, cure rates now range from 30 to 75% [58–62]. Even more disturbing, there are now reports of totally drug-resistant TB (TDR-TB) for which no chemotherapeutic options remain [63, 64]. The end result of the widespread use of drugs to treat ever-increasingly resistant strains of the organism has been the looming threat of a return to the pre-chemotherapy era. As these strains have evolved, natural selection will restore whatever fitness costs they incur by acquiring drug resistance, and, ultimately, these strains will emerge in the developed world again. The pace of our discovery efforts has been too slow; we are now approaching a situation where we will have lost all of the achievements of the past. We will need to develop entirely new regimens, and we urgently need to consider the mistakes that were made 40 years ago and devise a strategy to avoid repeating them.

1.4 Special Challenges in TB Drug Development

There are four widely accepted primary objectives for improving TB therapy: (1) shortening and simplifying the treatment for active, drug-sensitive TB, (2) improving treatment efficacy, safety, and duration for drug-resistant disease, (3) improving the safety of co-therapy for TB patients coinfected with HIV, and (4) establishing an effective therapy for latent, persistent TB. The solutions achieving each of these goals will have different features, but there are some overarching issues that complicate each area of concern.

One of the most pressing issues in improving TB chemotherapy involves the use of RIF (see Sect. 2.1 for further discussion of development of RIF as a chemotherapeutic agent), the most effective drug in reducing patient bacillary burdens in the first-line regimen for drug-susceptible TB. Recent studies indicate that treatment outcomes worsen with reduced durations and intermittent use of RIF [65]. Reducing RIF treatment to 1–2 months results in increased rates of relapse and acquired resistance compared to established regimens using RIF for a 6-month period. Additionally, intermittent weekly or twice weekly administrations may promote relapse and acquired resistance. Although RIF is an essential drug in first-line therapy for establishing a positive treatment outcome, use of RIF in combination with various drugs is problematic because of drug–drug interactions as a consequence of RIF's powerful induction of many cytochrome P450 (Cyp) enzymes. These enzymes metabolically inactivate other drugs thereby reducing effective serum concentrations and exposure. RIF particularly induces cytochrome P450 (CYP) 3A4, the most abundant enzyme found in the liver and the gut, which metabolizes drugs and toxins [66]. RIF is also associated with upregulation of membrane transporters (P-glycoproteins) that regulate transport of substances across membranes, which often function as cellular efflux pumps thereby limiting bioavailability of drugs [67]. HIV-coinfected patients receiving antiretrovirals whose serum levels are known to be affected by RIF induction of CYP3A4 are sometimes provided rifabutin as a substitute for RIF. Rifabutin has reduced CYP3A4 induction and thus simplifies co-therapy; however, current clinical evaluation does not fully support substitution of RIF with rifabutin [68]. Therefore, any new agent introduced for drug-susceptible disease will likely have to be introduced in combination with RIF and against a background of strong induction of CYP 3A4.

A further complication in TB drug metabolism is malabsorption [69]. Patients presenting with TB are often malnourished; weight loss is a hallmark of the disease.

Sometimes this is linked to advanced HIV disease where patients are malnourished or have diarrhea, but it is also a common complication for patients with diabetes mellitus (DM), another frequent comorbidity of TB patients [69–71].

A particular challenge in the development of new TB drugs is the heterogeneity of TB pathology [2]. TB shows not only the differences in clinical manifestation but also the underlying host and pathogen physiology, which poses particular challenges to antitubercular drug development. Human TB patients harbor a variety of granulomas resulting in different microenvironments to which the resident MTb is exposed. Thus, the metabolism of MTb in each lesion is likely to be different. The presence of discrete populations of MTb in the human host possessing different susceptibilities to antitubercular drugs might explain the combined activity of frontline chemotherapy [72]. One theory (supported by virtually no hard evidence) proposes that rapidly growing bacilli are cleared by drugs such as INH, while sporadically replicating intracellular organisms are killed more efficiently by drugs such as RIF and the more slowly dividing bacteria within acidic environments are selectively sensitive to PZA [72]. Furthermore, it is the eradication of the slowly growing and non-replicating bacilli which requires such an extended duration of chemotherapy. PZA is often put forward as a paradigm for a drug that has the highest capacity to reduce treatment duration. Importantly though, PZA has little or no in vitro activity (except under conditions where the bacteria is acid-stressed), and the precise mechanism of action of this drug remains unclear [72]. The antitubercular activity of PZA was discovered only because it was applied directly to infected mice, an impractical strategy for evaluating large numbers of compounds. Strictly in vitro, a variety of different growth conditions have been shown to result in alterations in the susceptibility of MTb to different drugs; for example, stationary phase [73, 74], anoxia [75], and nutrient deprivation [76, 77] all provide models of treatment-refractory disease, yet none of these has been validated as meaningful in predicting clinical efficacy.

Finally, although serum pharmacokinetic data are widely available for many TB drugs in use and development, TB is not a systemic bacteremia and tissue concentration studies are scarce. Drug penetration is likely to be limited due to tissue damage from disease and the loss of vasculature making primary sites of infection difficult to saturate [78]. A truly effective compound must not only be able to penetrate the bacterial cell wall, but also be able to reach the bacteria within a fibrous, necrotic, or cavitary lesion that may harbor the persistent organisms [78, 79].

2 The Development of Commonly Used First-Line and Second-Line Agents for TB Therapy

2.1 Rifamycins

The rifamycin antibacterials represent one of the most effective and widely used classes of therapeutic compounds used in modern TB treatment. The first of the rifamycins, rifampicin (RIF, Fig. 3; **10a**), was introduced into TB chemotherapy in

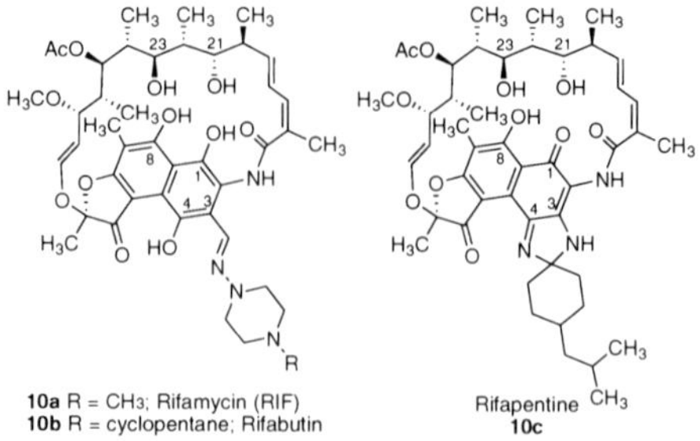

Fig. 3 Clinically used rifamycins

10a R = CH_3; Rifamycin (RIF)
10b R = cyclopentane; Rifabutin

Rifapentine **10c**

the 1960s after extensive structure–activity relationship (SAR) studies performed on rifamycin B, the natural product produced by *Amycolatopsis mediterranei*, from which the rifamycins were derived [80]. This isolated natural product was only active when delivered intravenously, and attaining oral bioavailability of rifampicin required a considerable effort because of the complex chemistry of this scaffold. Shortly thereafter, other rifamycin derivatives, rifabutin (Fig. 3; **10b**) and rifapentine (Fig. 3; **10c**), were developed and currently serve as alternatives to RIF. The rifamycin-derived antituberculous agents all share a general structure characterized by a naphthalene core that is spanned by a 19-atom polyketide bridge. SAR studies have established the role of the aliphatic bridge in stabilizing the overall conformation of the molecule and positioning the C(1) and C(8) phenols and the C(21) and C(23) hydroxyl groups for interaction with their bacterial target, RNA polymerase [81]. As such, modification of the phenol or hydroxyl groups in these positions abolishes antibacterial activity of the molecule. Conversely, modifications made at the C(3) and C(4) positions have been the focus of many efforts to improve the oral bioavailability of the rifamycins, since the C(3) appendages do not appear to interfere with rifamycin-RNA polymerase binding [82].

The primary mode of action of the rifamycins involves disruption of RNA transcription through binding of the drug to bacterial DNA-dependent RNA polymerase [83]. Accordingly, resistance to the rifamycins occurs primarily through point mutations acquired in the RNA polymerase β-subunit gene, *rpoB* [83]. Resistance may also occur through ADP-ribosylation of the alcohol at position C(21) [84].

The most common adverse effects associated with rifamycin therapy are mild influenza-like symptoms, hepatotoxicity, and altered liver function. Additionally, due to the furanonapthoquinone chromophore within the rifamycin structure, bodily fluids (e.g., sweat, tears, or urine) may take on an orange-red color. As discussed in Sect. 1.4, rifamycins may also have adverse interactions with other coadministered drugs, in particular antiretroviral drugs (ARDs). Of the three aforementioned rifamycins, RIF is the most potent inducer of CYP3A and rifabutin is the least [85], making it the preferred rifamycin derivate for treating HIV-TB coinfected patients [86].

2.2 Isoniazid

INH (Fig. 4; **3a**) is an analog developed from the antitubercular drug thiacetazone (Fig. 4; **3b**) which had been used effectively in TB patients in the 1940s but was associated with toxic side effects [87]. In an attempt to improve thiacetazone (**3b**), the phenyl ring was replaced with a pyridine ring based on the observation that nicotinamide (Fig. 4; **3c**) had a growth inhibitory effect on MTb. The isonicotinaldehyde thiosemicarbazone (Fig. 4; **3d**) proved to be more active than thiacetazone, which inspired evaluation of other intermediates in the synthesis, leading to the discovery of isonicotinic acid hydrazide (INH, **3a**) the best antitubercular agent developed to date.

Hundreds of derivatives of INH have been synthesized since its original discovery, but none improved on the activity of INH. *N*-acetyl-INH, an INH metabolite produced in humans, is inactive, although *N*-alkyl derivatives such as iproniazid (Fig. 4; **3e**) and hydrazones such as verazide (Fig. 4; **3f**) show in vivo efficacy although the active metabolite in MTb is INH which is released by in vivo hydrolysis [88–91].

The minimum inhibitory concentration (MIC) of INH is 0.2 μM against rapidly growing MTb, with lower activity against slowly growing MTb and practically no in vitro activity against anaerobically adapted bacteria [92]. INH is a prodrug that is activated by the KatG catalase to an isonicotinoyl radical that reacts with nicotinamide-containing molecules such as NAD(P) to yield acyclic isonicotinoyl-NAD(P) adducts and their cyclic hemiamidals. The INH-NAD adduct is a potent inhibitor of the NADH-dependent enoyl-ACP reductase, InhA, involved in mycolic acid biosynthesis [93–95]. Mutations in *katG* or *inhA* confer the majority of resistance, but other resistant isolates show mutations at targets that use pyrimidine nucleotides (which are structurally similar to adducts formed during INH activation) [96]. Isoniazid is well tolerated although side effects as a result of hepatic enzyme abnormalities resulting in hepatitis occur (especially in older patients). Also, peripheral neuritis can occur but is easily prevented by pyridoxine administration.

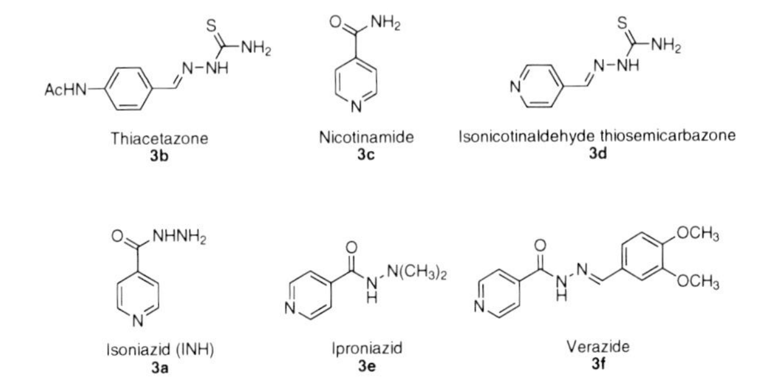

Fig. 4 Thiacetazone derivatives leading to isoniazid (INH)

Fig. 5 Thioisonicotinamide Derivatives

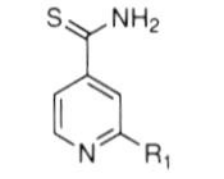

5a R_1 = Et; Ethionamide
5b R_1 = Pr; Protionamide
5c R_1 = H; Thioisonicotinamide

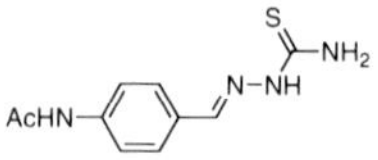

Thiacetazone
3b

2.3 *Thioisonicotinamides and Thiosemicarbazones*

The thioisonicotinamides, ethionamide (Fig. 5; **5a**) and prothionamide (**5b**) were discovered during efforts to improve on the MTb inhibitory activity of nicotinamide. Thioisonicotinamide (Fig. 5; **5c**) showed better in vivo efficacy than in vitro efficacy [97] which prompted further SAR studies on this series resulting in the observation that 2-alkyl derivatives were more active than the parent nicotinamide with 2-ethyl and 2-propyl derivatives showing the best activity. Ethionamide (**5a**) is a prodrug that is activated by *S*-oxidation by a monoxygenase (EtaA) to a 4-pyridylmethane radical intermediate that, similar to the active radical produced from INH by KatG (discussed in Sect. 2.2), reacts with NAD(P) to form a tight-binding inhibitor of InhA [98, 99]. The sulfoxide, the major metabolite produced in humans, is also active against MTb. The thioisonicotinamides have unpleasant gastrointestinal side effects.

Thiacetazone (Fig. 5; **3b**) was discovered to have antitubercular activity in the 1940s and was used as an antitubercular agent despite its toxic side effects [100, 101]. Thiacetazone, similar to the thioisonicotinamides, is activated by EthA resulting in a reactive intermediate that inhibits mycolic acid oxygenation as well as cyclopropanation [102, 103]. Thiacetazone causes gastrointestinal disturbances and, particularly in HIV-infected patients, can cause severe life-threatening skin reactions known as Stevens–Johnson syndrome [104].

2.4 *Pyrazinamide*

PZA (Fig. 6; **7a**) was developed based on reports describing the antitubercular activity of vitamin B3 (niacin) [105]. It is unlikely that PZA would be discovered in modern drug discovery programs since it has no activity against MTb under normal in vitro growth conditions although it has good activity in infected animals [106, 107].

Initial SAR studies [106–109] were performed by in vivo assays of derivatives of nicotinamide (**3c**) and PZA in infected mice. The presence of a pyrazine heterocycle with a carboxamide at the C(2) position was essential for activity. Modification of the carboxamide to tetrazole, nitrile, hydrazide, or carboxylic acid (Fig. 6; **7b–e**) leads to completely inactive compounds in vivo. Substitutions on the amide nitrogen with either a methyl (Fig. 6; **7f**) or an acetyl group (Fig. 6; **7g**) were detrimental

Fig. 6 Derivatives of PZA screened in the murine model of TB

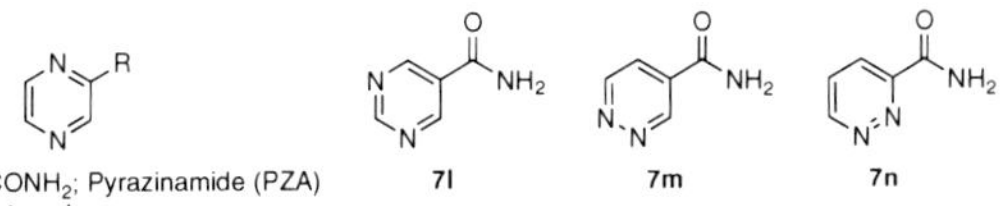

7a R = $CONH_2$; Pyrazinamide (PZA)
7b R = tetrazole
7c R = CN
7d R = $CONHNH_2$
7e R = CO_2H
7f R = $CONHCH_3$
7g R = CONHAc
7h R = CO_2Pr
7i R = $CONHCH_2NEt_2$
7j R = CONH-*N*-(morpholinomethyl)
7k R = $CSNH_2$

Fig. 7 Cycloserine

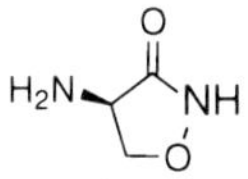

D-cycloserine
4a

to activity. Pyrazinoic acid (**7e**) is considered to be the active metabolite from PZA; hence, various ester derivatives (e.g., **7h**) were synthesized and found to be active in vitro but inactive in vivo probably due to premature hydrolysis or poor solubility. However, more stable aminomethylene prodrugs (**7i** and **7j**) did not show improvement in activity presumably because they were not substrates for the amidase. The thioamide (**7k**), pyrimidine nucleus (**7l**), and the pyridazine nucleus (**7m,n**) were inactive or weakly active. Thus, PZA is the minimum pharmacophore; further substitutions on the amide or changes to the pyrazine ring are detrimental to activity.

Pyrazinamide likely kills MTb by intracellular acidification following hydrolysis by MTb nicotinamidase/pyrazinamidase [110], although inhibition of fatty acid synthase has also been proposed as a mechanism [111–113]. PZA increases serum uric acid concentrations thereby causing nongouty arthralgia and, when used in combination with INH and/or RIF, often causes some hepatotoxicity.

2.5 *Cycloserine*

D-cycloserine (Fig. 7; **4a**) is an antibiotic produced by *Streptomyces* sp. that is currently used in second-line TB therapy [114]. The isoxazolidinone is the pharmacophore of D-cycloserine, and attempts to modify it with additional substituents have been unsuccessful since *N*-substitution prevents the tautomerization which is necessary for its activity, and replacement of the heteroatoms on the isoxazolidinone ring leads to dramatic loss of activity [115, 116]. In addition, the stereochemistry is essential since the L-isomer is inactive [116].

D-cycloserine prevents D-alanine incorporation into the bacterial cell wall peptidoglycan by forming an irreversible isoxazole-pyridoxal adduct in the enzyme alanine racemase, which converts L-alanine to D-alanine [117]. Although the major target in MTb is alanine racemase, D-cycloserine also inhibits the D-alanine–D-alanine ligase involved in synthesis of the terminal D-alanine–D-alanine of the peptidoglycan UDP-*N*-acetylmuramyl-pentapeptide [118]. The MIC of this antibiotic against MTb is about 100 μM. Because of the side effects observed with D-cycloserine (CNS effects and hypersensitivity), it is often given at more frequent but lower doses in TB patients as second-line therapy.

2.6 *Para*-Aminosalicylic Acid

The success of early clinical trials of PAS (Fig. 8; **2a**) in TB patients [119] prompted synthesis of analogs to enhance the activity of the parent compound. These analogs showed that PAS exhibits very specific SAR [120].

The mechanism of action of PAS is not fully understood, although folate biosynthesis has been proposed as the target, since inactivation of thymidylate synthase confers resistance [121]. PAS is generally poorly tolerated in patients due to gastrointestinal disturbances often leading to discontinuation of PAS administration.

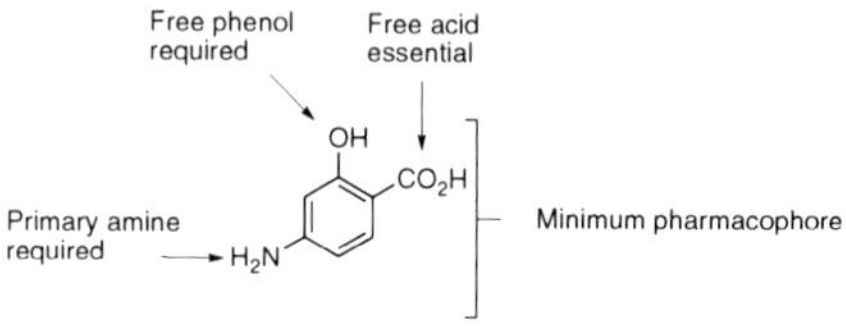

p-Aminosalicylic Acid (PAS)
2a

Fig. 8 SAR of *p*-aminosalicylic acid (PAS)

2.7 *Capreomycin*

Capreomycin is synthesized by *Saccharothrix mutabilis* subsp. *capreolusa* as a mixture of four related cyclic pentapeptides, capreomycins IA (Fig. 9; **6b**), IB (Fig. 9; **6c**), IIA (**6d**), and IIB (**6e**). The peptide backbone is made up of 15 unnatural

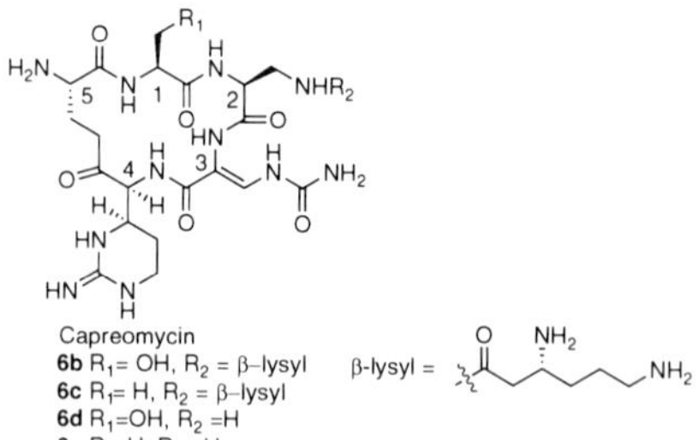

Capreomycin
6b R_1= OH, R_2 = β-lysyl
6c R_1= H, R_2 = β-lysyl
6d R_1=OH, R_2 =H
6e R_1=H, R_2 =H

Fig. 9 Capreomycin

amino acids and either L-serine or L-alanine, known as Capreomycin A and B, respectively.

A few limited SAR studies [122, 123] have shown that both alanine and serine at position 1 have antitubercular activity, that small ureido modifications such as *N*-methyl groups (but not *N*,*N*-dimethyl) are acceptable, whereas *N*-aryl ureido substitution increases general antibacterial activity. Capreomycin inhibits protein synthesis by binding at the interface between helix 44 of the 30S subunit and helix 69 of the 50S subunit of the bacterial ribosome [124]. Like the aminoglycosides with which it is often confused, capreomycin has both nephro- and ototoxic side effects.

2.8 Aminoglycosides

Streptomycin (STR, Fig. 10; **1a**), kanamycin (KM, Fig. 10; **1b–d**), and amikacin (AK, Fig. 10; **1e**) (Fig. 8) comprise the main aminoglycosides used in TB chemotherapy. As discussed in Sect. 1.2, the aminoglycosides are still widely used in modern TB drug regimens although mainly as second-line agents. The general structure of the aminoglycosides is characterized by an aminocyclitol ring connected to one or more amino sugars by a glycosidic linkage. The second generation aminoglycosides KM and AK were largely developed to circumvent resistance mechanisms in other bacteria, not specifically for MTb; hence, their SAR will not be discussed.

This class of antitubercular compounds primarily acts by binding to the 16S rRNA of the bacterial 30S ribosomal subunit, which interferes with protein synthesis and ultimately leads to cell death [125]. As such, resistance mechanisms observed in clinical isolates have principally been the acquisitions of mutations in the 16S rRNA gene (*rrs*) and in genes that encode for proteins that interact with the 16S rRNA in the region where the drug binds [125–129]. Alternative resistance mechanisms that have been reported include drug efflux and inactivation by aminoglycoside-modifying enzymes, but there is little evidence to suggest these are clinically relevant [130–132]. Common adverse effects associated with aminoglycoside therapy include nephro- and ototoxicity [133, 134].

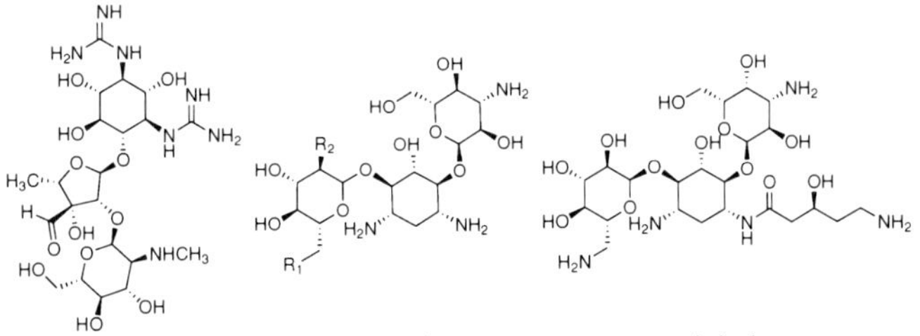

Streptomycin **1a**

Kanamycin
1b $R_1 = NH_2$, $R_2 = OH$
1c $R_1 = NH_2$, $R_2 = NH_2$
1d $R_1 = OH$, $R_2 = NH_2$

Amikacin **1e**

Fig. 10 Aminoglycoside antibiotics

3 Classes of Compounds in Clinical Development

3.1 Nitroimidazoles

3.1.1 History

Nitroimidazoles are a class of compounds with growing importance in the field of tuberculosis chemotherapy. Nitroimidazoles show better activity against obligate anaerobes than aerobic organisms because their bactericidal activity requires a bioreduction of the aromatic nitro group whose reduction potential lies beyond the reach of eukaryotic aerobic redox systems [135, 136]. 2-Nitroimidazole derivatives modified at the 1- and 5-positions were among the first series (Fig. 11) of this class reported to display antimycobacterial activity [137].

Further improvement in antimicrobial activity was gained by lowering the reduction potential by changing from 2-nitro- to 5-nitroimidazole derivatives. A notable example in this class is metronidazole (Fig. 12; **11a**), which was the lead compound from a screen of over 200 derivatives of azomycin (2-nitroimidazole) for antitrichomonal activity at the French pharmaceutical company Rhône-Poulenc in the mid-1950s [136, 138]. Metronidazole (**11a**), which is bactericidal against anaerobic non-replicating Mtb in vitro and in hypoxic granulomas in vivo (as well as other anaerobic bacteria and protozoa) [79, 136], has been in clinical use for four decades and is listed in the essential drug list by the WHO [139]. In 1989, Ciba Geigy India was the first to report antitubercular activity from a series of bicyclic 4- and 5-nitroimidazole [2, 1-*b*]oxazoles. Their lead compound CGI-17341

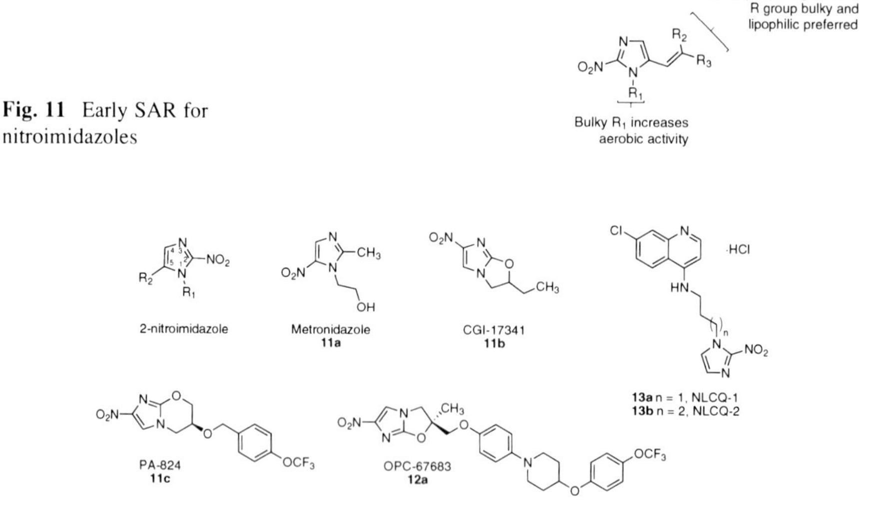

Fig. 11 Early SAR for nitroimidazoles

Fig. 12 Nitroimizadoles with antitubercular activity

(Fig. 12; **11b**) was active against drug susceptible and MDR-TB (MIC of 3.3 μM) [140] and showed dose dependency in a mouse model but was not further developed due to its mutagenicity in the bacterial Ames assay.

The bicyclic 5-nitroimidazole [2,1-*b*]oxazole series showed much lower activity than its 4-nitro counterpart [141]. A decade later, PA-824 (Fig. 12; **11c**), the lead compound from a series of more than 300 nitroimidazooxazine derivatives [142], and OPC-67683 (Fig. 12; **12a**), the lead compound from a series of nitroimidazooxazole derivatives [143], were discovered by PathoGenesis (now Novartis) and Otsuka Pharmaceutical Co. Ltd, respectively. Both compounds showed increased activity against MTb with potential to decrease the current treatment duration.

Most recently, two nitroimidazo-chloroquine derivatives NLCQ-1 (Fig. 12; **13a**) and NLCQ-2 (Fig. 12; **13b**), which are also prodrugs requiring bioreductive activation, have been reported to show a twofold increase in bactericidal activity against non-replicating MTb compared to PA-824 [144].

3.1.2 SAR of Nitroimidazooxazines

PA-824 shows bactericidal activity against drug susceptible (MIC range 0.04-0.8 μM) and resistant (MIC range 0.08–1.5 μM) MTb strains [142]. SAR studies show that the key features responsible for aerobic activity are the nitro group at the 4-position of the imidazole ring (Table 1, Entry 1), the conformationally rigid bicyclic system (Table 1, Entry 3) and the lipophilic tail at the 6-position of the oxazine ring (Table 1, Entries 4 and 5) [145–147]. Antitubercular activity was seen with a biaryl linker (*para* > *meta* > *ortho*), but these compounds exhibited poor solubility in most cases [146]. Heterobiaryl analogs improved solubility over biaryl linkers, and varying lengths of hydrophobic regions at the 6-position of the oxazine

Table 1 SAR of PA-824 [145–147]

Entry	Compound name	Structure	MIC against H37Rv (μM)
1	**11d**	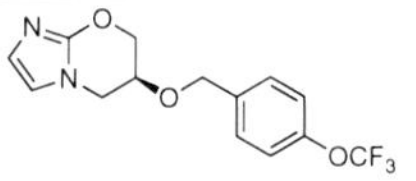	>160
2	**11e**	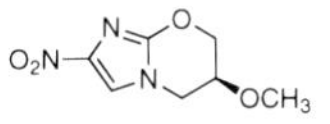	>125
3	**11f**	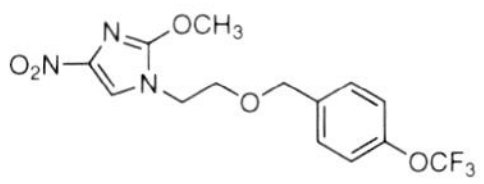	6.25
4	**11g**	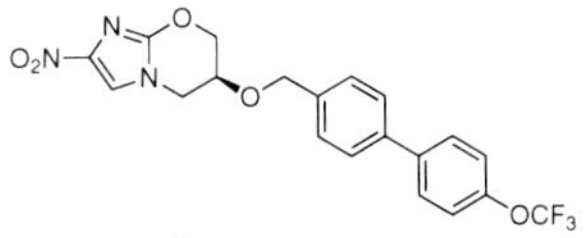	0.04
5	**11h**	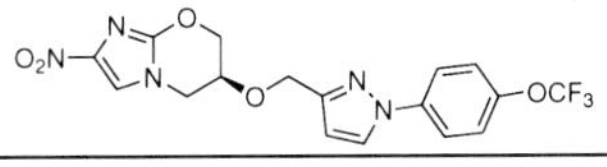	0.05

ring were well tolerated [148] indicating the presence of a large hydrophobic pocket in the active site of Rv3547 (see below for further discussion of mode of action).

The oxygen at 2-position of the nitroimidazole ring could be substituted with nitrogen or sulfur with equipotent aerobic activity, but acylation of the nitrogen, oxidation of sulfur, or replacement of oxygen by a methylene lead to decreased activity (Table 2) [148]. The *S*-enantiomer is more than 100-fold more active than the *R*-enantiomer. Replacement of the benzylic ether at the 6-position with an amine marginally increased activity and improved water solubility [148]. Overall SARs for PA-824 are summarized in Fig. 13.

3.1.3 Biology of Nitroimidazooxazines

Deazaflavin (F_{420} cofactor)-dependent nitroreductase (Ddn) Rv3547 is responsible for the reductive activation of the pro-drug PA-824 (**11c**), generating a reactive nitrogen species (likely ˙NO), production of which correlates with the cidal activity toward anaerobic non-replicating MTb [149, 150]. PA-824 has been shown to inhibit cell wall lipid and protein biosynthesis in a dose-dependent manner

Table 2 SAR on heteroatoms of oxazine ring [148, 149]

Entry	Compound name	X	Y	MIC against H37Rv (μM)
1	PA-824 (**11c**)	O	O	0.80
2	**11i**	CH_2	O	25
3	**11j**	NH	O	0.8
4	**11k**	NAc	O	6.25
5	**11l**	S	O	0.8
6	**11m**	SO_2	O	>100
7	**11n**	O	NH	0.31
8	**11o**	O	$NHCO_2$	0.05
9	**11p**	O	$NH(CH_2)_2$	0.08
10	**11q**	O	$NH(CH_2)_4$	0.08

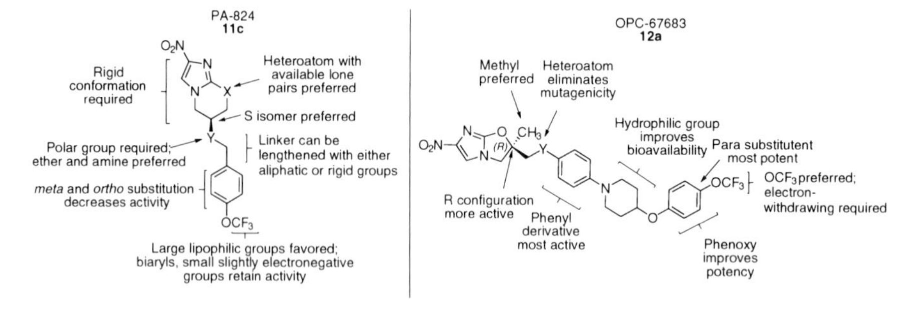

Fig. 13 SAR for PA-824 and OPC-67683 series

[142, 151]. There is poor correlation between aerobic and anaerobic activity, and transcriptional profiling analysis suggests that both respiratory inhibition and inhibition of cell wall biosynthesis are related to the activity of PA-824 [152]. Studies in mice found the daily minimal effective dose and minimal bactericidal dose of PA-824 to be 12.5 mg/kg and 100 mg/kg, respectively, and it exhibited potent bactericidal activity in both the initial and continuation phases of treatment [153]. A combination of PA-824, moxifloxacin, and PZA was shown to cure mice more rapidly than standard regimen of RIF/INH/PZA [154].

3.1.4 Clinical Use of Nitroimidazooxazines

PA-824 has oral bioavailability (subdose proportional) and a relatively long half-life (16–20 h in humans) consistent with once a day regimen [155]. Clinical studies showed that even though PA-824 inhibits excretion of creatinine at high dosage, it did not affect the glomerular filtration rate, thereby limiting concerns about nephrotoxicity [156]. It is non-mutagenic, shows no cross-resistance with current drugs, and can be coadministered with antiretroviral agents. Phase IIa clinical studies on patients with newly diagnosed, uncomplicated, smear-positive, pulmonary tuberculosis[157] ascertained PA-824 to be safe and well tolerated for 2 weeks with daily dosing varying from 200 to 1,200 mg/day in which time-frame decrease of bacillary burdens in sputum was observed [158].

3.1.5 SAR of Nitroimidazooxazoles

OPC-67683 (Fig. 12; **12a**) shows potent antitubercular activity against both replicating and non-replicating bacteria and is equipotent against drug-resistant MTb [143, 159]. Derivatives of these 6-nitro-2,3-dihydroimidazo [2,1-*b*]oxazoles were not mutagenic in contrast to the structurally similar CGI-17341 (Fig. 12; **11b**), with heteroatoms at the side chain of 2-position contributing to the absence of mutagenicity. Addition of a methyl group at the 2-position was found to improve activity, and the absolute stereochemistry was found to be critical with the *R*-enantiomer (MIC of 180 nM) being 60-fold more active than the *S*-enantiomer (Table 3). Subsequent development of *R*-enantiomers of 6-nitro-2,3-dihydroimidazo [2,1-*b*] oxazole culminated in identification of lead compound OPC-67683 [159]. Figure 13 compares the SAR for both the oxazine and the oxazole series of antitubercular nitroimidazoles.

3.1.6 Biology of Nitroimidazooxazoles

OPC-67683 has an MIC of 20 nM, which is more potent than any other nitroimidazole and does not show cross-resistance with currently used antitubercular drugs. It is active against intracellular MTb in a dose-dependent fashion with OPC-67683 being superior to INH and as effective as RIF [160]. In MTb-infected mice,

Table 3 SAR of OPC-67683 [159]

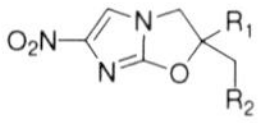

R_1	R_2	Configuration	MIC (μM)
H	OPh	Racemic	2.98
CH_3	OPh	(*R*)	0.18
CH_3	OPh	(*S*)	11.37

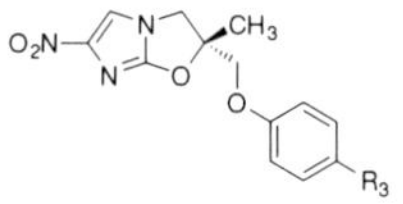

R_3	MIC against MTb strains (μM)		
	H37Rv	H37Rv INH resistant	H37Rv RIF resistant
H	0.18	0.18	0.18
Cl	0.08	0.04	0.02
CF_3	0.58	0.58	0.29
OCF_3	0.56	1.09	0.56
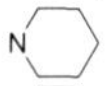 N	2.18	1.09	1.09
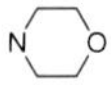 N O	2.16	1.08	1.08
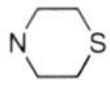 N S	2.07	1.04	0.53

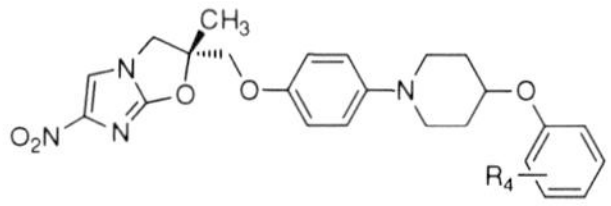

R_4	MIC against MTb strains (μM)		
	H37Rv	H37Rv INH resistant	H37Rv RIF resistant
H	0.87	0.87	0.44
p-Cl	0.10	0.10	0.05
p-F	0.83	0.83	0.43
p-OCF_3	0.01	0.01	0.01
o-OCF_3	0.73	0.73	0.37
m-OCF_3	0.04	0.04	0.04

a combination of OPC-67683 (2.5 mg/kg) with RIF and PZA showed faster rate of MTb clearance from organs than a standard regimen of RIF, PZA, ETB, and INH [143]. OPC-67683 is a prodrug that is likely activated by the same nitroreductase as PA-824 (Ddn/F_{420} reductase) [141, 143]. Similar to PA-824, it is an inhibitor of methoxy- and keto-mycolic acid synthesis, which is essential for biosynthesis of the cell wall [161].

3.1.7 Clinical Use of Nitroimidazooxazoles

OPC-67683 shows a half-life in mouse plasma of 7.6 h with a C_{max} of 0.55 μM (6 h, 2.5 mg/kg) [143]. OPC-67683 is not metabolized by liver microsome enzymes, making it suitable for coadministration with drugs that induce cytochrome P450 enzymes. It is absorbed better with a high fat diet and up to 400 mg/day can be tolerated safely without adverse side effects by healthy volunteers [162]. OPC-67683 is less effective than PA-824 in reducing sputum-borne acid-fast bacilli in the first 4 days of treatment in patients with pulmonary TB. This drug is under development and is currently in the phase II clinical trials for use against MDR-TB [163].

3.2 Diarylquinolines

3.2.1 History

The diarylquinoline TMC207 (Fig. 14; **14a**), first reported in 2005, is the first known antitubercular compound in the diarylquinoline class [164] (for additional details, see [165, 166]). Tibotec, a subsidiary of Johnson and Johnson, has reported the vast majority of research and development on TMC207, although recent efforts by Chattopadhyaya and coworkers have contributed new related compounds [167, 168]. TMC207 was one of the lead compounds discovered in a high-throughput screen for compounds with activity against *Mycobacterium smegmatis* (a nonpathogenic rapid-growing mycobacterium), which were subsequently evaluated against MTb [164].

3.2.2 SAR of TMC207

Correct absolute and relative configuration of the two stereocenters of TMC207 (Fig. 14; **14a**), which have been assigned by NMR and X-ray crystallographic analysis [169, 170], are required for activity [171, 172]. Sterically undemanding functional groups can be substituted for the bromine on the quinoline ring without significant loss of activity, although a bromine atom appears to be preferred. The naphthyl substituent can be replaced with other electron-poor aryl groups and still maintain good activity against MTb. Based on initial reports, the dimethyl-substituted tertiary amine appears to be required for activity, with the replacement of one methyl substituent with a proton or ethyl substituent resulting in a decrease in activity [171]. However, more recent reports suggest that the *N*-monodesmethyl

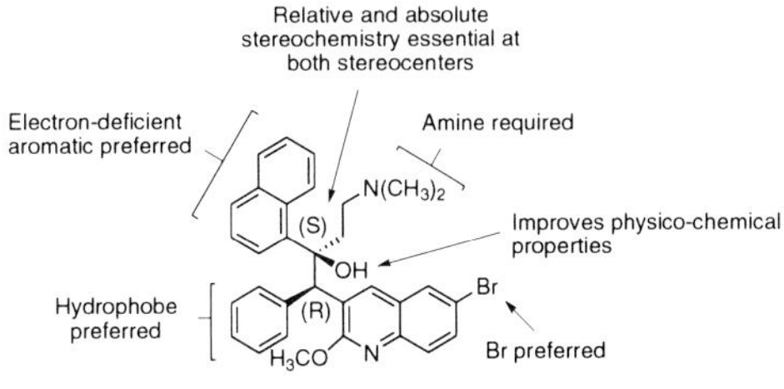

Fig. 14 SAR of TMC207

metabolite of TMC207 produced by oxidation by CYP3A4, a cytochrome P450 that is potently induced by RIF, maintains significant antitubercular activity [173].

3.2.3 Biology

TMC207 is highly specific for mycobacteria [172]. Both H37Rv and clinical isolates show MICs in the range of 54–217 nM. TMC207 targets the c subunit of ATP synthase (*atpE* gene), a mechanism of action distinct from fluoroquinolones and other quinoline derivatives [164, 174, 175]. Docking studies have suggested the tertiary amine of TMC207 serves as an arginine mimic, allowing the compound to disrupt the proton transport chain of ATP synthase [168, 176]. Point mutations in *atpE* confer resistance; these mutations occur at a rate of one in 10^7 to 10^8, similar to the bacterial mutant frequency of rifampicin resistance [164].

Initial in vivo studies showed encouraging results. Treatment of MTb-infected mice exclusively with TMC207 at 25 mg/kg was as effective as triple combination therapy of RIF/INH/PZA [164]. Also in mice, the addition of TMC207 to standard MDR-TB regimens showed an improved cure rate over the standard regimen alone [177]. TMC207 acts synergistically with PZA in mice [178]. In guinea pigs, treatment with TMC207 for 6 weeks resulted in almost complete eradication of MTb bacilli from lesions [179]. Furthermore, TMC207 has also been shown to be bactericidal in vitro against non-replicating MTb, suggesting that TMC207 might prove therapeutically effective against latent tuberculosis [180].

Also, a once-weekly schedule of administration of TMC207/rifapentine/PZA tested in mice was more active than the standard regimen of RIF/INH/PZA given daily [181]. Because TMC207 has a long half-life in humans (44–64 h in plasma) [164], once-weekly tuberculosis treatments might one day be possible. However, metabolism of TMC207 is enhanced by the presence of RIF, suggesting that coadministration of these drugs might not be straightforward [182].

3.2.4 Clinical Use

In humans, C_{max} is reached in 4–5 h [164, 173, 183], and a daily dose of 400 mg administered daily for 7 days results in a C_{max} of 10 μM [183]. A steady-state concentration of 1 μM, which appears to be required for bactericidal activity [183], can be maintained with a dosing schedule of 400 mg daily for 2 weeks followed by reduced doses of 200 mg three times weekly [173]. Adverse events occurred at a low rate and side effects were considered mild to moderate [164, 173, 183].

In preliminary clinical trials, TMC207 showed significant bactericidal activity after 4 days of a 7-day trial treating previously untreated TB patients, although onset of bactericidal activity was delayed in comparison to RIF and INH [173]. In 2009, the first stage of a phase II trial testing TMC207 in combination with a standard, five-drug, second-line antituberculosis regimen in MDR patients showed that after 8 weeks of treatment, 48% of study participants receiving the TMC207 regimen converted to negative sputum culture, compared with 9% of those on the standard regimen [183–185]. Additional trials are ongoing in MDR patients [185], and TMC207 is undergoing further development for drug-susceptible TB [186].

3.3 Oxazolidinones

3.3.1 History

Oxazolidinones are a new structural class of synthetic antibacterial drugs. Reports of structurally novel anti-infectives by DuPont (Fig. 15; **15a,b**) in the mid-1980s [187] drew the interest of researchers at the former Pharmacia and Upjohn Inc. (now Pfizer) [188–190]. Two lead compounds, eperezolid (Fig. 15; **15c**) and linezolid (LZD, Fig. 15; **15d**) [191], proved to be exceedingly effective wide-spectrum drugs, although LZD was better tolerated in clinical trials. Further development of the oxazolidinone scaffold has yielded PNU-100480 (**15e**) [191], a linezolid analog currently in phase II clinical trials [192, 193], as well as DA-7218 (Fig. 15; **15g**) and its metabolite DA-7157 (Fig. 15; **15h**), which are in preclinical development (Dong-A Pharmaceuticals, Ltd.) [194]. Ranbaxy Laboratories Limited (acquired in 2008 by Daiichi Sankyo Company) has also made contributions in the form of RBx-7644 (**15i**) and its more potent analog RBx-8700 (**15j**) [195], which are in preclinical development. Additionally, AstraZeneca has developed two oxazolidinones, AZD2563 (**15k**) [196] (discontinued at preclinical stage) and AZD5847 (structure not yet available) which is starting phase II clinical trials [197, 198].

3.3.2 Structure–Activity Relationship

Because the oxazolidinones were not developed specifically to treat TB, their SARs have been developed mostly against a number of Gram-positive and Gram-negative bacteria, and little is known about TB-specific SAR. DuPont was the first to publish

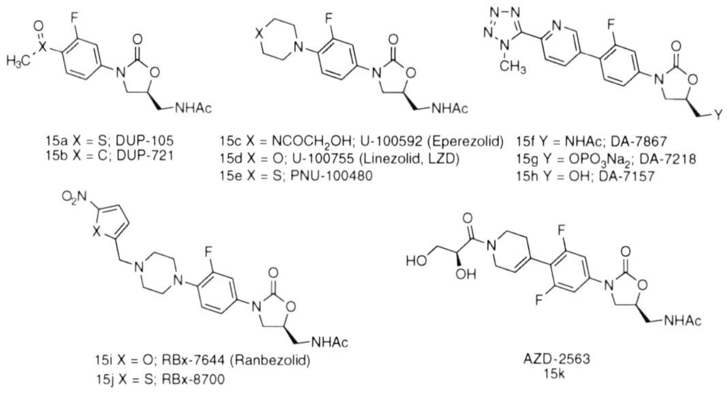

15a X = S; DUP-105
15b X = C; DUP-721

15c X = $NCOCH_2OH$; U-100592 (Eperezolid)
15d X = O; U-100755 (Linezolid, LZD)
15e X = S; PNU-100480

15f Y = NHAc; DA-7867
15g Y = OPO_3Na_2; DA-7218
15h Y = OH; DA-7157

15i X = O; RBx-7644 (Ranbezolid)
15j X = S; RBx-8700

AZD-2563
15k

Fig. 15 Oxazolidinones

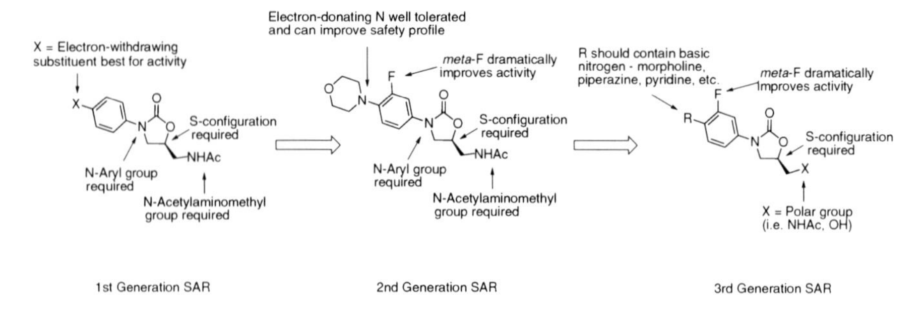

Fig. 16 Evolution of SAR in oxazolidinone class of antibacterials

Table 4 MIC of various oxazolidinone candidates against TB

Drug candidate	MIC against H37Rv (μM) [Ref.]	MIC against RIF or INH-resistant clinical isolates (μM) [Ref.]	MIC against RIF and INH-resistant clinical isolates (μM) [Ref.]
Linezolid (**15d**)	0.74 [204]; 17 [205]	1.40 [204]; 13 [205]	1.24 [204]; 2 [205]
DA-7867 (**15f**)	–	0.15 [206]	0.15 [206]
DA-7157 (**15h**)	–	–	0.25 [207]
DA-7218 (**15g**)	–	–	0.25–1 [207]
RBx-8700 (**15j**)	–	0.09 [205]	0.34

their conclusions about the structural motifs required for antibacterial activity (Fig. 16) [199–201]. These relationships were further refined during the development of eperezolid and LZD [202]. Finally, the development of the RBx and DA compounds has expanded the limits of the functional groups that display antitubercular activity [203].

Activities of the oxazolidinones against TB are shown in Table 4. LZD (Fig 15; **15d**) has an MIC against first-line susceptible TB strains of 1.55 μM [204]. For the DA class of compounds, which contain a triazole as the basic side chain, DA-7867 (Fig 15; **15f**) proved to be poorly water soluble; hence, a water-soluble prodrug DA-7128 (Fig 15; **15g**, which is metabolized to DA-7157, Fig 15; **15h**) was developed. Interestingly, against MTb, prodrug DA-7128 performed similar to its (usually more active) metabolite against MTb, giving an MIC of 0.25 μM [207].

3.3.3 Biology

The oxazolidinones inhibit bacterial protein synthesis by binding to the bacterial 23S rRNA of the 50S subunit, [208, 209] which blocks the interaction between charged tRNAs at the P site and the A site (Fig. 16) [210]. Specifically, LZD disrupts initiation of protein synthesis by inhibiting peptide bond formation between the carboxyl terminus of the *N*-formylmethionine–tRNA complex residue bound at the P site and the amino terminus of the amino acid-tRNA bound at the A site [210]. Crystal structures show LZD bound near the A site of the 50S ribosomal subunit at the 23S rRNA in such a way that peptide bond formation should be inhibited [211].

LZD-resistant tuberculosis has been observed in both following in vitro selection [212] and clinical strains (occurring only rarely) [213] and may arise through an active efflux system [214]. Many reports have shown that LZD is effective against MDR-TB both in vitro [215, 216] (MIC < 24 μM) and in vivo [51, 217, 218].

3.3.4 Clinical Use

For DA-7867 (**15f**), oral bioavailability in rats is 70.8%, with 8.3% not absorbed and 21.8% eliminated by intestinal first-pass metabolism [194]. LZD (**15d**) has nearly 100% bioavailability (regardless of whether or not it is taken with food [219]), and its half-life of 5.40 ± 2.06 h [220] allows for a 12-h dosing schedule [221]. In healthy human subjects, steady-state plasma concentrations of 63 ± 17 μM are obtained at T_{max} of 1.03 ± 0.62 h [220]. LZD's major metabolites (Fig. 17; **15l**, **15m**) are formed via oxidation alpha to the morpholine ring heteroatoms followed by ring opening [220]. It is metabolized through hepatic oxidation (and thus should not affect drugs metabolized by cytochrome P450 enzymes); hence, doses do not have to be altered for patients with renal or hepatic impairment [221]. LZD does not show suppressed antibiotic activity when coadministered with other antibiotics [221] and even shows synergistic activity with fluoroquinolones and RIF [222]. PNU-100480 was studied in healthy volunteers (phase I clinical trials)[192, 193] and appeared to be well tolerated at doses of 1,000 mg/day [216]. Additionally, a whole blood assay against MTb showed PNU-100480 to be more effective

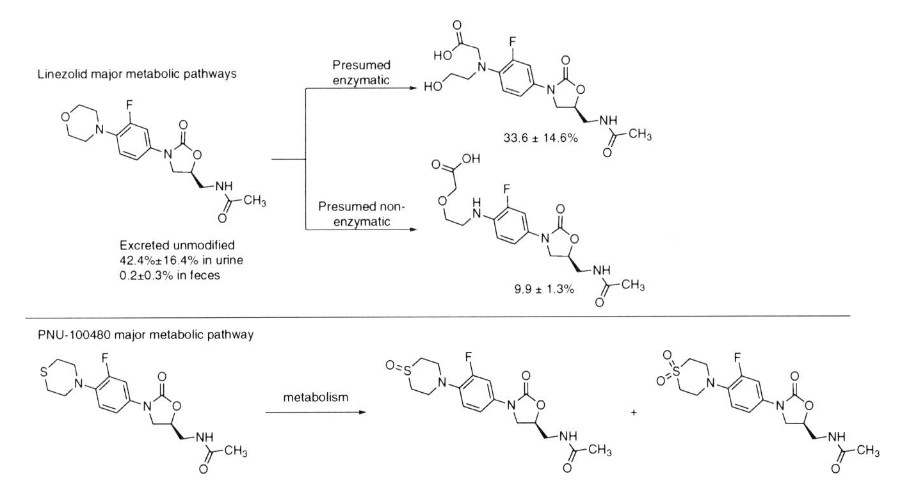

Fig. 17 Metabolism of LZD (*upper*) [220] and PNU-100480 (*lower*)

than LZD, although doses of PNU-100480 used were higher (300 mg linezolid vs 1,000 mg PNU-100480 dosed daily until steady-state plasma concentrations were achieved) [216]. PNU-100480 forms metabolites by the oxidation of the sulfur atom (Fig. 17; **15n,o**).

Side effects reported during phase III clinical trials of linezolid were generally not severe (however, the duration of exposure in such trials has been notably shorter than those used in MTb chemotherapy) [221]. More than half of the patients experienced digestive side effects (including constipation, diarrhea, vomiting, nausea), rash, headache, insomnia, or dizziness [221]. Hematological side effects including thrombocytopenia, anemia, leucopenia, or pancytopenia [221], although rare, warrant monitoring for longer treatment durations [223]. LZD can cause peripheral and optic neuropathy [221], and lactic acidosis has been reported in patients on longer treatment courses [221].

The largest (184 patients) retrospective analysis of patients empirically treated using LZD in a multidrug regimen for MDR- and XDR-TB patients in a multidrug regimen showed an overall 59% cure rate for the entire cohort, with an 87% cure rate in cases with definitive outcomes [224]. The use of LZD was also associated with a favorable outcome in a retrospective analysis of 176 XDR-TB-infected patients [62]. No prospective controlled data are available at this point although two trials are currently underway [53, 54].

3.4 *Fluoroquinolones*

3.4.1 History

The fluoroquinolones are a synthetic class of antibacterial drugs discovered by the Sterling-Winthrop Institute in 1962 as an impurity during synthesis of the antimalarial compound chloroquine [225]. This byproduct, nalidixic acid (Fig. 18; **16a**), was approved by the FDA in 1963 to treat Gram-negative urinary tract infections.

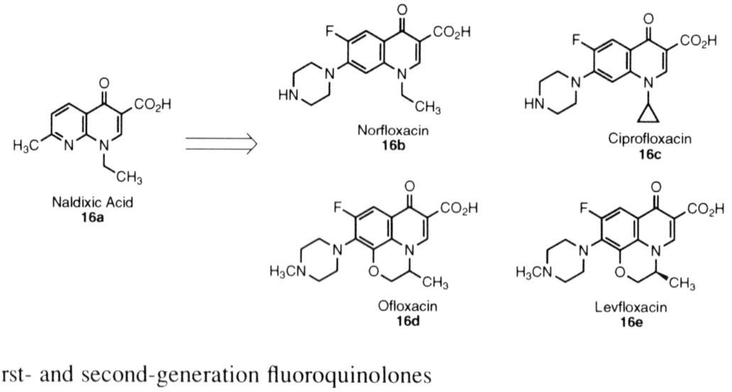

Fig. 18 First- and second-generation fluoroquinolones

However, despite its good bioavailability and straightforward synthesis, nalidixic acid has had limited clinical use due to a poor pharmacokinetic profile and narrow antibacterial spectrum [226]. Interest in the quinolones was renewed in 1980 with the discovery of the first reported antibacterial fluoroquinolone, norfloxacin (Fig. 18; **16b**), by the Dainippon Pharmaceutical Company [227]. Norfloxacin showed broad spectrum antibacterial activity 1,000-fold greater than nalidixic acid [228, 229] as well as improved pharmacokinetic properties, with a longer half-life and improved solubility [228–230]. Norfloxacin and several other second-generation fluoroquinolones such as ciprofloxacin (Fig. 18; **16c**) (first reported in 1982 by Bayer [231]), ofloxacin (Fig. 18; **16d**) (first reported in 1983 by Daiichi Pharmaceutical Co., Ltd., now Daiichi Sankyo Co., Ltd. [232]), and levofloxacin (Fig. 18; **16e**), which is the isolated *S*-isomer of racemic mixture ofloxacin (also developed by Daiichi Pharmaceutical Co., Ltd. [232]), have proven relatively safe and remain among the most frequently prescribed drugs [226].

Following the discovery of norfloxacin (**16b**), SARs for the fluoroquinolone core were studied in detail. This led to the development of a number of analogs with broader antibacterial activity, better solubility, and longer serum half-lives [226, 229]. Among the third and fourth generations of fluoroquinolones, moxifloxacin (Fig. 18; **16f**) (developed in 1991 by Bayer [233]), which has a bulky hydrophobic modification at C(7), has been the most successful. Unfortunately, several third and fourth generation agents have been restricted or withdrawn due to severe adverse effects (Fig. 19) including temafloxacin (**16g**), grepafloxacin (**16h**), trovafloxacin (**16i**), and clinafloxacin (**16j**) [226, 234, 235].

Many new fluoroquinolones are in development such as gemifloxacin (Fig. 20; **16n**), patented in 1998 by LG Life Sciences Ltd. [236], and sitafloxacin (Fig. 20; **16o**) (first reported in 1994 by Daiichi Seiyaku Co. [237]) which show activity against a panel of respiratory pathogens [229]. Sitafloxacin is currently in clinical development;

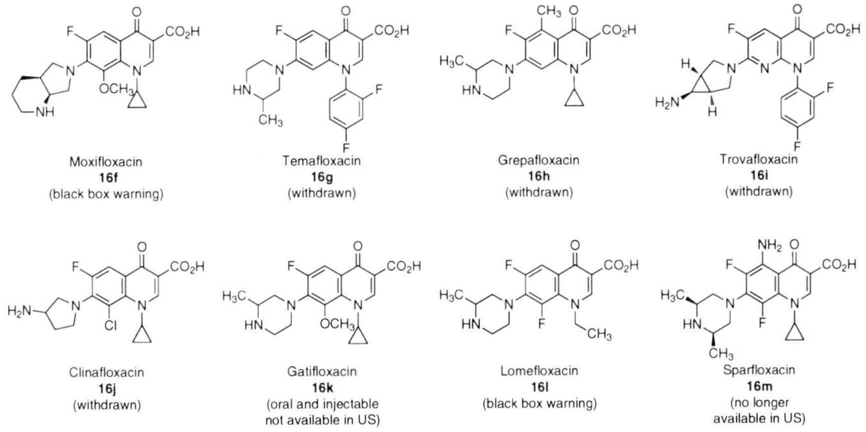

Moxifloxacin **16f** (black box warning)
Temafloxacin **16g** (withdrawn)
Grepafloxacin **16h** (withdrawn)
Trovafloxacin **16i** (withdrawn)
Clinafloxacin **16j** (withdrawn)
Gatifloxacin **16k** (oral and injectable not available in US)
Lomefloxacin **16l** (black box warning)
Sparfloxacin **16m** (no longer available in US)

Fig. 19 Third- and fourth-generation fluoroquinolones

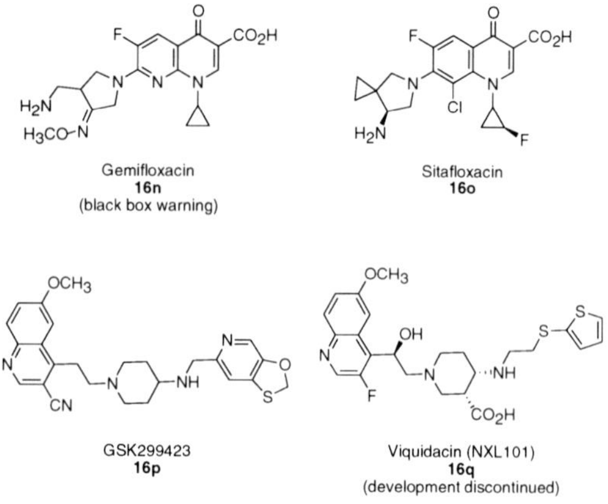

Fig. 20 New DNA gyrase inhibitors

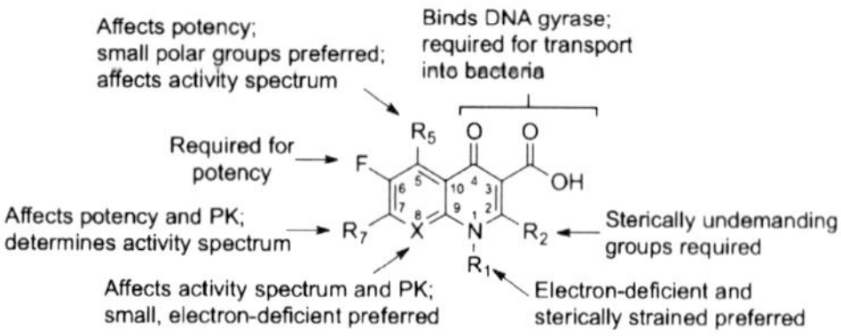

Fig. 21 Fluoroquinolone SAR (adapted from [229, 242, 243])

gemifloxacin is a clinically prescribed drug. Recently, novel bacterial topoisomerase inhibitors (NBTIs) with modes of action similar to the fluoroquinolones have been reported, including GSK 299423 (Fig. 20; **16p**) [238], NXL101 (Fig. 20; **16q**) [239], and a series of tetrahydroindazole compounds [240, 241]. While these new compounds have shown good in vitro activity against a spectrum of both Gram-positive and Gram-negative microbes including strains resistant to fluoroquinolones, it remains to be seen whether they will also exhibit activity toward MTb.

3.4.2 Structure–Activity Relationships

While the SAR of the fluoroquinolones has not been analyzed specifically for mycobacteria, it is reasonable to assume that many of the relationships found in other types of bacteria will be applicable to MTb (Fig. 21). Modifications at N(1) control potency, with electron-poor and sterically strained cyclopropyl being optimal, followed by 2,4-difluorophenyl and *t*-butyl [242]. This substituent also controls Gram-negative and Gram-positive activities, and a 2,4-difluorophenyl group increases activity against anaerobes. The C(2) position is near the DNA gyrase-binding site, and thus a sterically undemanding hydrogen atom at R_2 is optimal [244]. The dicarbonyl moiety is required for binding to DNA gyrase and thus is critical for activity. Modifications at C(5) control in vitro potency with the most active groups being small electron-rich groups such as -NH_2, -OH, and -CH_3 [242]. Additionally, C(5) modifications affect activity against both Gram-negative and Gram-positive organisms. The fluorine atom at C(6) (for which the class is named) enhances DNA gyrase inhibition [226, 244] and can increase the MIC of the compound 100-fold over that of other substitutions [242]. The most active substituents at C(7) have been five- and six-membered nitrogen heterocycles, with pyrrolidines increasing activity against Gram-negative bacteria and piperazines affecting potency against Gram-positive organisms. The C(8) position controls absorption and half-life, and optimal modifications for in vivo efficacy include groups that create an electron-deficient *pi* system, i.e., N, CF, and CCl [245]. Several modifications that create a N(1) to C(8) bridge have also been successful, i.e., ofloxacin (Fig. 18; **16d**) and levofloxacin (Fig. 18; **16e**), which both display significant gyrase inhibition [244].

3.4.3 Biology

The fluoroquinolones alter DNA topology and block replication by inhibiting two essential bacterial enzymes, DNA gyrase (topoisomerase II) and topoisomerase IV. DNA gyrase, encoded by *gyrA* and *gyrB*, maintains the levels of supercoiled DNA required for efficient replication and is the primary target for the fluoroquinolones in most Gram-negative bacteria [246]. Topoisomerase IV, encoded by *parC* and *parE*, is responsible for decatenation of DNA following replication and is the major target of the fluoroquinolones in many Gram-positive bacteria [229, 247]. Mycobacteria are unique in that genome sequence analyses have failed to identify DNA topoisomerase IV [229]. Thus, *gyrA* and *gyrB* are likely the only targets of the fluoroquinolones in MTb.

The MIC for numerous fluoroquinolones has been determined for both wild type (H37Rv) and clinical isolates of MTb. The MIC values against H37Rv for the clinically relevant fluoroquinolones are displayed in Table 5 and range from 0.1 to 5 μM.

3.4.4 Clinical Use

The fluoroquinolones have several pharmacokinetic features that have proven valuable in treating tuberculosis. For example, the oral bioavailability for many of the fluoroquinolones is good, ranging anywhere from 70 to 100%, with levels in

Table 5 MIC data for fluoroquinolones commonly used in treatment of MTb

Fluoroquinolone	MIC (μM)	References
Ciprofloxacin (**16c**)	1.51	[248]
Gatifloxacin (**16k**)	1.25	[249]
Levofloxacin (**16e**)	1.25	[249]
Lomefloxacin (**16l**)	5	[249]
Moxifloxacin (**16f**)	0.16	[249]
Ofloxacin (**16d**)	2.5	[249]
Sparfloxacin (**16m**)	0.08	[249]

the blood peaking soon after administration [248, 250–253]. Moreover, the fluoroquinolones are cell permeable and widely distributed throughout the body, which is important for killing intracellular bacteria and treating disseminated disease [250]. For the most part, the later generation fluoroquinolones have longer serum half-lives, but these vary extensively, from 5.37 h for ciprofloxacin to 18.3 h for sparfloxacin [245]. Finally, most fluoroquinolones are cleared via the kidneys, but liver metabolism and elimination by a combination of routes do occur for several of the compounds [250].

Generally, the fluoroquinolones are well tolerated, causing mild side effects that tend to be self-limiting and rarely require discontinuation or regimen changes [235, 250] (Fig. 22, Table 6). The most frequent adverse events reported include gastrointestinal upset, disturbances of the CNS, and skin reactions [226, 234]. A number of more serious side effects have been documented with fluoroquinolones use as well. In particular, the fluoroquinolones have been associated with tendonitis and tendon rupture due to collagen damage, which in 2008 prompted a black box warning for all currently available drugs within this class [234]. Phototoxicity due to the generation of reactive oxygen species and inflammatory responses to sunlight is also commonly reported [226].

While all fluoroquinolones may cause photosensitivity, there is considerable variation within the class due to structural differences [234]. For example, the presence of halogen atoms at C(5) or C(8) and a bulky side chain or methyl group at C(5) show the highest potential for this effect [226, 242]. Moreover, fluoroquinolones can cause QTc interval prolongation by blocking voltage-gated potassium channels, which has been

Table 6 Notable side effects of selected fluoroquinolones

Fluoroquinolone	Adverse effects	Implications	References
Ciprofloxacin (**16c**)	Tendonitis/tendon rupture	Black box warning; 2008	[234]
Clinafloxacin	Phototoxicity, hypoglycemia	Development stopped	[243]
Gatifloxacin (**16k**)	Dysglycemia	Oral and injectable formulations no longer available in USA; 2006	[234]
Grepafloxacin (**16h**)	Cardiotoxicity	Withdrawn; 1999	[226, 243]
Levofloxacin (**16e**)	Tendonitis/tendon rupture	Black box warning; 2008	[234]
Lomefloxacin (**16l**)	Phototoxicity, CNS effects	Black box warning; 2008	[234, 243]
Moxifloxacin (**16f**)	QTc interval prolongation, tendonitis/tendon rupture	Black box warning; 2008	[234]
Ofloxacin (**16d**)	Tendonitis and tendon rupture	Black box warning; 2008	[234]
Sparfloxacin (**16m**)	Phototoxicity, QTc interval prolongation	No longer available in USA	[243]
Temafloxacin (**16g**)	Severe hemolytic reactions, clotting abnormalities, renal failure	Withdrawn; 1992	[226, 243]
Trovafloxacin (**16i**)	Hepatotoxicity	Withdrawn/limited use; 1999	[226, 234]

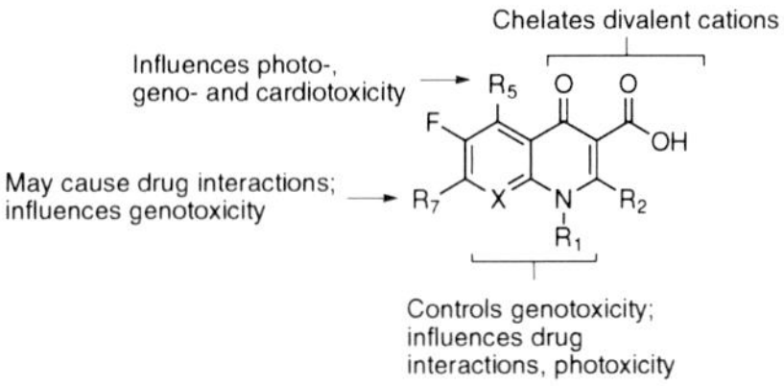

Fig. 22 Structure–toxicity relationships of the fluoroquinolones (adapted from [234, 235, 243])

associated with torsades de pointes syndrome, severe arrhythmia, cardiotoxicity, and death. However, the severity varies according to structural differences and the dose administered [226, 234]. Other adverse effects attributed to fluoroquinolone use include: hepatotoxicity, kidney and liver dysfunction, and dysglycemia [226, 235].

As discussed in Sect. 1.3, patients with MDR-TB receive one of several fluoroquinolones used as second-line agents in the treatment of TB, namely gatifloxacin, levofloxacin, moxifloxacin, or ofloxacin [254, 255]. Based on murine model studies [256–258], the most active fluoroquinolones are: moxifloxacin = gatifloxacin > levofloxacin > ofloxacin [259]. In addition to the aforementioned fluoroquinolones, several clinical studies have investigated the efficacy of sparfloxacin and lomefloxacin [260]. While sparfloxacin appears efficacious for treating MDR-TB, a role for lomefloxacin in tuberculosis therapy is unclear [260]. Based on data from the mouse studies, moxifloxacin and gatifloxacin are currently in phase III clinical trials to determine whether they can shorten duration of therapy [259, 261, 262]. Thus far, clinical trials of fluoroquinolones based on extrapolation of murine results of reduced therapy duration have failed to show similar effects in humans as in mice.

3.5 Ethylenediamines

3.5.1 History

N,*N'*-diisopropylethylenediamine (Fig. 23; **9b**)was the first compound in this series developed in early 1950s against MTb [263]. Structural modification of the lead compound led to the discovery of ethambutol (EMB, Fig. 23; **9a**) [263–266]. Despite its modest potency, EMB is a first-line drug for the treatment of TB.

3.5.2 Structure–Activity Relationship

Initial studies of structural modifications of EMB concluded that the size and nature of the alkyl group on the ethylenediamine nitrogens were critical for activity. These studies confirmed that small α-branched alkyl groups were more effective than alkyl chains branched at positions other than α and that a longer alkyl chain was detrimental to activity [264]. Alterations in the linker region of the molecule were deleterious since any lengthening, incorporation of heteroatoms, or branching of the ethylene

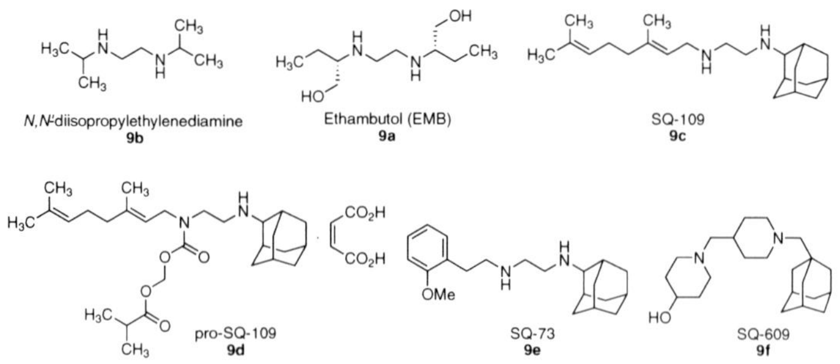

Fig. 23 Ethylenediamines

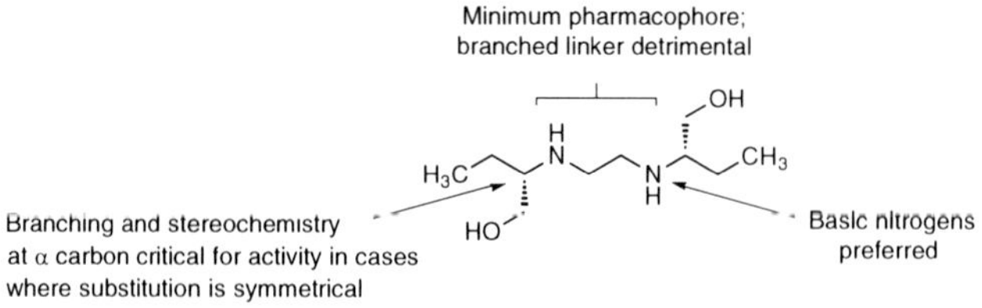

Fig. 24 SAR of ethylenediamines

linker led to reduced activity. In addition, aryldiamines and cycloalkylamines were far less effective than the parent compound. These studies confirmed that the ethylenediamine unit is the minimum pharmacophore required for antitubercular activity. Any change in the basicity of either amino group led to decreased antimycobacterial activity, with the exception of substitution of the amine with an amide that retained partial activity in some analogs [267]. Due to the lack of crystallographic information about the membrane-bound arabinosyltransferase enzyme, which is the presumed target of EMB [268, 269], a thorough study was undertaken to use combinatorial chemistry to develop a comprehensive SAR. A library of 63,238 asymmetric diamines was screened against MTb [270] of which 25 were either more effective or had comparable activity to the parent compound. The most effective compound, SQ-109 (Fig. 23; **9c**), was chosen for development based on its activity and pharmacokinetic properties. A summary of the SAR of the ethylenediamines is shown in Fig. 24.

3.5.3 Biology

Although it was initially assumed that SQ-109 (**9c**) and EMB (**9a**) would share the same arabinosyltransferase target, which catalyzes the transfer of arabinosyl residues to the cell wall arabinogalactan polymer, SQ-109 retained potency against EMB resistant strains. In addition, transcriptional profiling studies and analyses of cell wall-linked sugar residues indicated that MTb responds differently to these compounds, suggesting that SQ-109 acts on a different target than EMB [271, 272].

3.5.4 Clinical Use

Pharmacokinetic profile of SQ-109 after a single dose administration shows C_{max} after intravenous and oral administration as 1,038 and 135 ng/mL, respectively. The $t_{1/2}$ for the drug after i.v. and oral administration were 3.5 and 5.2 h, respectively. SQ-109 displayed a large volume of distribution into various tissues. SQ-109 levels in most tissues after a single administration were significantly higher than that in blood. The highest concentration of SQ-109 was present in lung (>MIC), which was at least 120-fold (p.o.) and 180-fold (i.v.) higher than that in plasma with the next ranked tissues being spleen and kidney [273, 274]. SQ-109 is highly unstable to human microsomes as evidenced by its oxidation, epoxidation, and *N*-demethylation and has been shown to have poor oral bioavailability, presumably due to its poor solubility and first pass metabolism [275]. In a continued effort to enhance the efficacy of SQ-109, carbamate analogs (Fig. 23; **9d**), which act as prodrugs of the parent compound, have recently been synthesized. Carbamate-based esterase-sensitive drug conjugates have been used to create prodrugs of both amines and amidates [276–278]. These carbamates are stable in microsomal assays, but are substrates for plasma esterases. When administered orally, these prodrugs can bypass first pass metabolism in the liver. The bioavailability studies of the new analog **9d**, when compared with SQ-109 in a rat model, showed significant improvement [279]. After oral dosing of 13 mg/kg of SQ-109 or **9d**, bioavailability of free SQ-109 from pro-SQ-109 **9d** was 91.4% compared to 21.9% from SQ-109 [280]. This study also showed that the concentration of SQ-109 after oral administration is higher in lungs than in liver, spleen, and plasma [279], which may be beneficial for a pathogen predominantly associated with lung disease. SQ-109 has currently completed phase Ia clinical trials [280].

3.5.5 Other Diamine Derivatives

Another compound, SQ-73 (Fig. 23; **9e**), having a moderate MIC at 12.5 μM but a better therapeutic index (for macrophage toxicity) of 6.4, was studied further as this compound exhibited better activity in macrophages. In vivo studies with SQ-73 exhibited moderate tissue distribution [271]. A structurally related dipiperidine class of compounds was also recently reported, with the most effective compound from this series exhibiting an MIC of 6.25 μM against MTb [281]. After further optimization and analysis of the dipiperidine library, the compound SQ-609 (Fig. 23; **9f**) was selected as the most promising in the class. This compound has moderate in vitro cytotoxicity in cultured mammalian cells and a suitable therapeutic window. SQ-609 has shown efficacy against intracellular MTb, good aqueous solubility, and oral bioavailability. In murine studies, SQ-73 (5 mg/kg), SQ-109

(10 mg/kg), and SQ-609 (10 mg/kg) all exhibited activity similar to INH (25 mg/kg) after 3 weeks of treatment [281].

4 Series in Preclinical Development

4.1 Benzothiazinones

4.1.1 History

The nitro-benzothiazinone (BTZ) class was originally derived from a series of sulfur-containing heterocycles to develop antibacterial and antifungal agent [282]. BTZ-043 (Fig. 25; **17a**), the most promising compound among the benzothiazinones, shows high antitubercular activity in vitro, in macrophages, and in the murine model of chronic TB [283].

4.1.2 Structure–Activity Relationship

From SAR studies, the sulfur atom and the nitro group at positions 1 and 8, respectively, play a critical role in bactericidal activity. When the nitro is replaced with either an amine or a hydroxylamine at position 8, the resulting analogs show a 500 to 5,000-fold decreased activity [283]. More than 30 different BTZ derivatives showed MICs of less than 116 nM against MTb. Electron-withdrawing group such as CN, CF_3, and Cl at the R_1 position and 1,4-dioxa-8-azaspiro[4.5] decane groups with methyl substituents at R_2 show promising activity against MTb.

4.1.3 Biology

The BTZ class of compounds is thought to inhibit decaprenylphosphoryl-β-D-ribose 2′-epimerase, hereby preventing the conversion of decaprenylphosphoryl ribose (DPR) into decaprenylphosphoryl arabinose (DPA), which is a substrate for

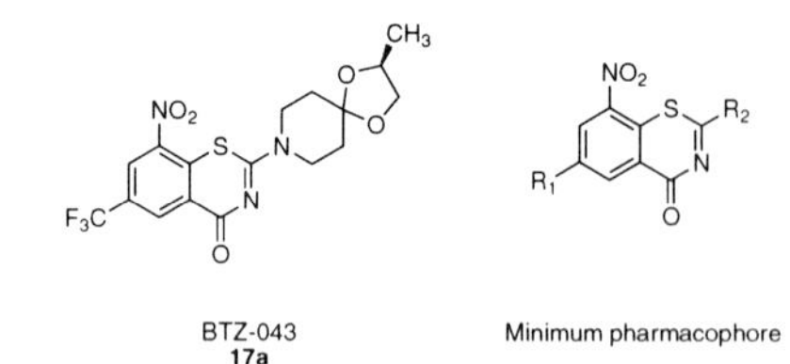

Fig. 25 Structure of BTZ-043 and its pharmacophore

the arabinosyltransferases of mycobacterial cell wall synthesis [284]. The MIC of BTZ-043 against H37Rv is 2.3 nM [283]. Despite the 100-fold better in vitro activity of BTZ-043 against MTb than frontline agents such as INH, its in vivo effect during treatment of chronically infected mice was comparable to that of INH and RIF.

In mice, BTZ-043 has a $t_{1/2}$ greater than 2 h, a $C_{\max}$ of 2 mg/mL, and an AUC of 4.6 h·mg/mL [283]. It is also relatively stable to degradation by human liver microsomes and shows less than 20% inhibition of various cytochrome P450 enzymes. BTZ-043 showed high activity against clinical isolates of MTb including MDR and XDR strains [285]. This compound is in preclinical development and will soon enter phase I clinical trials.

4.2 Nucleosides

4.2.1 History

Nucleoside analogs are a class of drugs typically used in the treatment of infectious diseases and cancer. The requirement for drugs that have activity against MDR-TB and XDR-TB makes the nucleoside analogs particularly attractive, since they have unique mechanisms of action from currently used antitubercular drugs. Among the nucleoside analogs currently under investigation, the capuramycin and caprazamycin classes of antibacterial antibiotics have the most potent activity [286]. Caprazamycin (Fig. 26; **18a–g**) and capuramycin (Fig. 27; **19a**) are natural products originally isolated from the culture broth of *Streptomyces griseus* 447-S3 [287] and culture broth of *Streptomyces* sp. MK730-62F2 [288] and show in vitro activity against drug-resistant MTb strains.

4.2.2 Structure–Activity Relationship

Capuramycin analog SQ-641 (Fig. 27; **19b**) has shown moderate activity against MTb. From SAR studies, the uridine unit and the protic amide are essential for

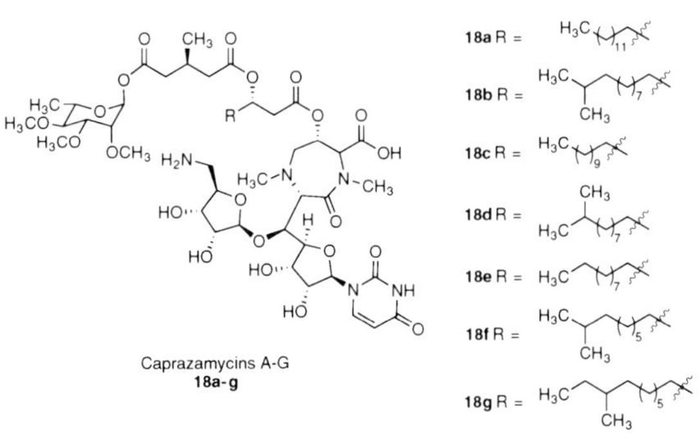

Caprazamycins A-G
18a-g

Fig. 26 Caprazamycins A–G

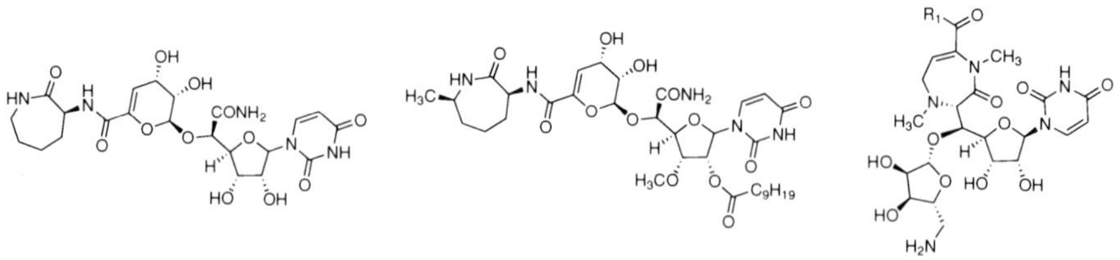

Capuramycin
19a

SQ-641
19b

19c $R_1 = COR_2$, $R_2 = OC_{13}H_{27}$
19d $R_1 = COR_2$, $R_2 = OC_{18}H_{37}$
19e $R_1 = COR_2$, $R_2 =$ NHalkyl
$R_1 =$

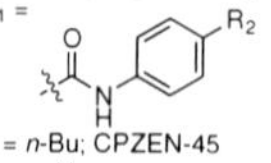

19f $R_2 =$ *n*-Bu; CPZEN-45
19g $R_2 =$ *n*-Hex
19h $R_2 = OC_6H_{13}$

Fig. 27 Capuramycin and analogs SQ-641 and CPZEN-45

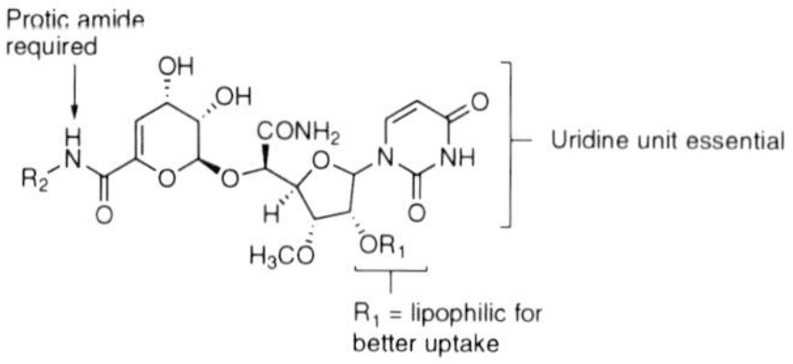

Fig. 28 SAR for capuramycins

bactericidal activity (Fig. 28) [289]. At the R_2 position in Fig. 28, lipophilic groups, including medium size alkyl chains, phenethyl, and phenyl-type substitutions, retained moderate activity but benzyl-type substitution showed decreased activity. When different lipophilic groups were placed at the R_1 position, installation of a decanoate substituent showed the largest increase in whole cell activity compared to shorter alkanoate chains, likely due to the increased lipophilicity, which (see Sect. 5.1) increases intracellular uptake into MTb.

CPZEN-45 (Fig. 27; **19f**), a caprazamycin analog, is shown in Fig. 27. SAR studies revealed that the uridine and the aminoribose are crucial for antibacterial activity [290]. Initially, installation of ester substituents at R_1 with R_2 alkyl chains showed that tridecane ($C_{13}H_{27}$, Fig. 27; **19c**) and octadecane ($C_{18}H_{37}$, Fig. 27; **19d**) esters showed equipotent activity, whereas a 21-carbon chain with unsaturation at C(18) showed decreased potency. Next, the effect of the amide substituent R_2 (**19e**) was investigated, which showed that the potency generally increased up to a 21 carbon alkyl chain and exhibited decreased potency with even longer alkyl substituents. Finally, anilinoamide substituents with *n*-butyl (CPZEN-45, Fig. 27; **19f**), *n*-hexyl (Fig. 27; **19g**), and hexyloxy (Fig. 27; **19h**) showed the most potent activity against MTb. Highly lipophilic molecules are, of course, not good candidates for lead optimization programs; thus a considerable amount of work is still required to discover better candidates from CPZEN-45 (**19f**) as leads for drug development.

4.2.3 Biology

Translocase I (encoded by *mraY*) is an essential enzyme involved in the biosynthesis of peptidoglycans, which makes it an attractive target due to its unique presence in bacteria. Caprazamycin (Fig. 26; **18a–g**) and capuramycin (Fig. 27; **19a**) inhibit Translocase I with an IC_{50} of 18 nM and 90 nM, respectively [289]. The lead compounds of the series are SQ-641 (Fig. 27; **19b**), which has an MIC of 0.67–1.35 μM against drug-susceptible MTb and 0.081–2.71 μM against MDR-TB [291], and CPZEN-45 (Fig. 27; **19e**), which has an MIC of 2.26 and 9.07 μM against drug-susceptible and MDR-TB, respectively.

SQ-641 shows promising efficacy in the murine model of TB infection and exhibits strong synergistic effects with EMB, STR, and SQ-109 (see Sect. 2.5) [286]. CPZEN-45 also exhibits no significant toxicity and a novel mechanism of action making these nucleoside compounds attractive candidates for TB drug development.

4.3 *Macrolides*

4.3.1 History

In the early 1950s, the first-generation prototypical macrolide, erythromycin (EM, Fig. 29; **20a**), was discovered. It is a natural antibiotic isolated from *Saccharopolyspora erythrea* [292, 293]. Erythromycin consists of a 14-membered lactone ring with two attached sugar groups: L-cladinose at the C(3) position and desosamine at the C(5) position [292, 293]. EM shows antibacterial activity against Gram-positive bacteria, but no activity has been observed against MTb [292, 293].

4.3.2 Structure–Activity Relationship

In an effort to increase potency against MTb, a series of EM analogs was synthesized with modifications at the 2, 3, 6, 9, 11, and 12 positions of the 14-membered lactone ring, as well as at the 4′ position of cladinose and the 2″ position of

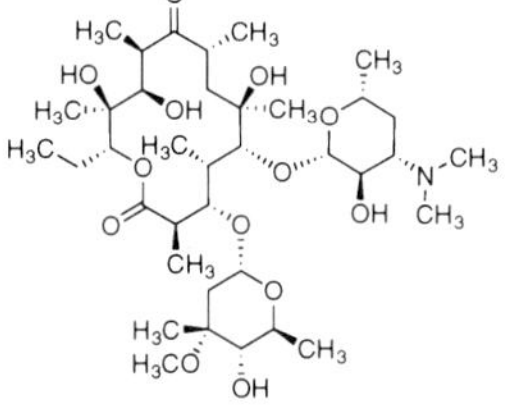

Erythromycin (EM)
20a

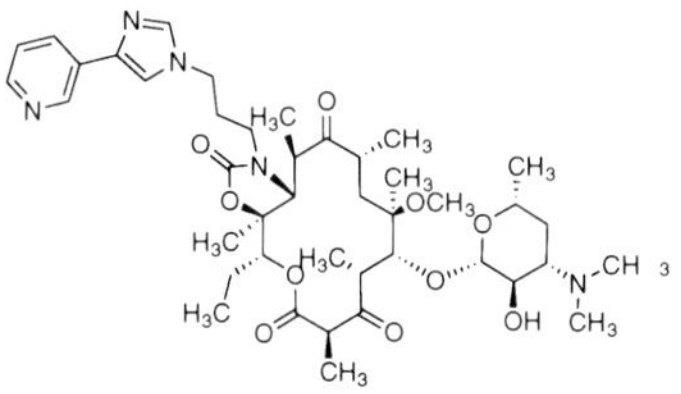

Telithromycin
20b

Fig. 29 Erythromycin (EM), 20a, and telithromycin, 20b

desosamine [292–294] (Fig. 30). Specific modifications on the lactone ring such as 6-substitution, 11, 12-carbamate, 11, 12-carbazate, and 9-oxime substitutions enhance potency [294]. Substitution of fluorine at position 2 in ketolides appears to improve both potency and selectivity (i.e., cytotoxicity vs activity against MTb) [293, 294]. The C(6) substituent is critical for activity of the ketolides [293], as it affords acid stability by preventing internal hemiketalization with the 3-keto group [292]. In general, ketolides are less potent than the corresponding cladinose-containing compounds for all substituents on the 6-position [293]. Among 9-oxime-substituted ketolides and macrolides, there is a correlation between the lipophilicity of the substituent on the 9-position (defined as calculated logP) and the potency [293], with some C(9) oximes showing submicromolar MIC against MTb [294].

The substituent at 11, 12 position appears to significantly affect potency. A variety of aryl-substituted 11, 12-carbamate and carbazate macrolides and ketolides demonstrated low or submicromolar MICs [294]. Also, the aryl substituent may be involved in determining cytotoxicity [294]. The substituted 11, 12-carbazate compounds demonstrated significant dose-dependent inhibition of MTb growth in mice, with a 10–20-fold reduction of colony forming units (CFUs) in lung tissue [294].

To further enhance lipophilicity, the 2′ and 4″-positions on desosamine and cladinose rings, respectively, have been modified via esterification, which generally improved potency and sometimes decreased CYP3A4 inhibition (more commonly in the cladinose-containing macrolides; see Sect. 4.3.3 for discussion of CYP3A4 inhibition) [293], although the substituent on the 9-position is generally more important than modifications on 2′ and 4″ positions [293].

4.3.3 Biology

Macrolides bind reversibly to the 50S subunit of 70S bacterial ribosomes, which inhibits protein synthesis [293, 295]. Although macrolides are effective for other bacterial infections, including some mycobacteria, they have not demonstrated significant efficacy against MTb [293, 294]. Ribosome methylation is the most

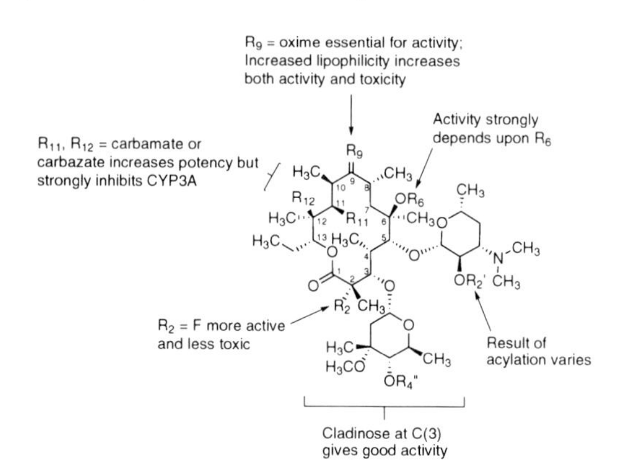

Fig. 30 General SAR for macrolides (adapted from [293])

widespread mechanism involved with macrolide resistance in MTb, and the gene *erm*MT plays an essential role [294–296]. Therefore, the current development goal for macrolides has been primarily to overcome bacterial resistances resulting from methylation of the rRNA and drug efflux [294, 295]. Furthermore, since macrolides are well-known inhibitors of CYP3A4, a cytochrome P450 enzyme, developing compounds with decreased inhibition against CYP3A4 is critical.

In addition to low efficacy against MTb, the pharmacokinetics of EM are somewhat unsatisfactory, as it is unstable to gastric acid and displays a short serum half-life (~1.4 h) [292, 293]. The second-generation macrolides such as clarithromycin and roxithromycin improved both properties [293], but clarithromycin showed only weak activity against MTb either in vitro or in vivo (200 mg/kg dose for low-dose aerosol infection mouse models), suggesting that second generation macrolides cannot be expected to offer significant antimicrobial clinical benefits for TB [293, 294]. Further improvements focused on replacement of the L-cladinose substituent, as it is associated with both drug efflux (one mechanism for development of macrolide resistance) and metabolic instability of the macrolides [293]. This third generation of macrolides replaced the cladinose ring with a ketone moiety (ketolides), leading to more metabolically stable drugs [292–294]. However, it appears that C(3) cladinose is important for antitubercular potency of macrolides, which remain more potent than either ketolides or other substituents such as 3-OH and 3-carbamoyloxy groups [293, 294]. Telithromycin (Fig. 29; **20b**), the first clinically approved ketolide, has been developed for use against respiratory pathogens but is not active against MTb [292–294].

Currently, there have been some improvements in in vitro activity with macrolides against MTb, but to date no promising drug candidate has emerged. Preclinical work in this area is ongoing.

4.4 β-Lactams

4.4.1 History

Since the discovery of 6-amino penicillanic acid (penicillin) in 1929 [297], β-lactams have been one of the most successfully used classes of antibiotics. They are irreversible inhibitors of peptidoglycan-cross-linking enzymes, D, D-transpeptidases and D, D-carboxypeptidases [298]. β-lactams are rarely used in chemotherapy of TB, however, because of the limited permeability of the mycobacterial cell envelope, expression of inactivating enzymes (β-lactamases), and involvement of β-lactam-insensitive targets in peptidoglycan transpeptidation.

4.4.2 Structure–Activity Relationship

The pharmacophore of the β-lactams is a highly reactive four-membered azetidinone ring which is generally fused to a five- or six-membered ring (Table 7). There

Table 7 Basic structures of clinically relevant β-lactams and their pharmacologic properties

β-lactam class	Pharmacophore	Spectrum of activity	Inactivation	Adverse effects
Penams		Gram-positives	Classical β-lactamases, carbapenemases	Diarrhea, hypersensitivity, anaphylaxis, pseudomembranous colitis, yeast infections
Cephems		Gram-negatives	Extended-spectrum β-lactamases, carbapenemases, cephalosporinases	Diarrhea, hypersensitivity, pseudomembranous colitis, yeast infections
Carbapenems		Gram-positives, Gram-negatives, anaerobes	Renal dehydropeptidase, carbapenemases	Diarrhea, anaphylaxis, pseudomembranous colitis, nephrotoxicity, neurotoxicity
Mono-bactams		Select Gram-negatives	Extended-spectrum β-lactamases, carbapenemases	Diarrhea, pseudomembranous colitis, yeast infections

is an absolute requirement for the β-lactam ring and a carboxylic acid on the fused ring (or an electron withdrawing moiety such as the sulfonyl as in monobactams). An amide α to the β-lactam ring is preferred. Conformationally, this core resembles the acyl-D-alanyl-D-alanine moiety of the natural substrate. The serine nucleophile in the enzyme active site attacks the electrophilic carbonyl of the β-lactam amide leading to ring opening and irreversible acylation of the enzyme.

4.4.3 Biology

β-lactams are commonly used in combination with β-lactamase inhibitors such as clavulanic acid (CA), sulbactam, and tazobactam which themselves are β-lactams. A variety of β-lactams have been tested for in vitro efficacy against MTb (see Table 8 for examples). However, despite activity of some β-lactams against MTb in vitro, especially in the presence of β-lactamase inhibitors, none have to date shown good efficacy in vivo. Amoxicillin (Table 8; **21a**)/CA was found to be ineffective in mice [302], and imipenem (Table 8; **21f**) showed only a 16-fold reduction in bacterial burden in lungs of infected mice over 4 weeks of treatment [50]. Furthermore, only a modest decrease in viable numbers was seen in sputum of patients receiving amoxicillin(Table 8; **21a**)/CA or ampicillin(Table 8; **21b**)/sulbactam monotherapy [303]. The poor in vivo efficacy may be due to the intracellular environment of MTb, making it difficult for drugs to penetrate the phagosomal compartment. Additionally, the bacterial physiology in vivo may

Table 8 Biological activity of select β-lactams against MTb H37Rv in the presence or absence of clavulanic acid [299–301]

Compound	Structure	MIC (μM)	
		−CA	+CA
Amoxicillin (**21a**)	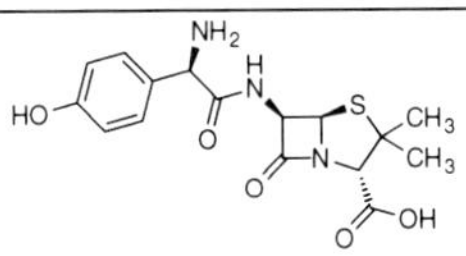	>250	0.5–3
Ampicillin (**21b**)	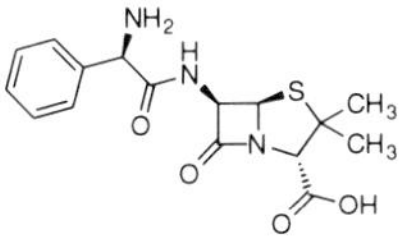	>25	13
Ceftriaxone (**21c**)	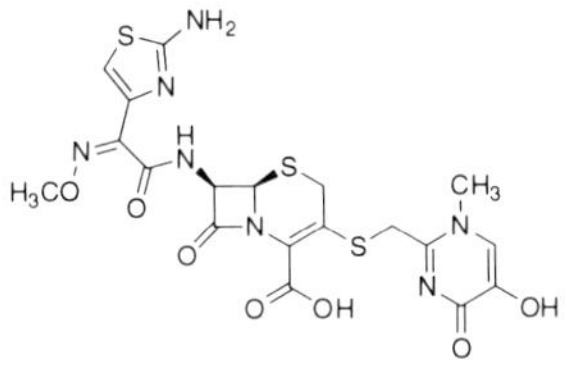	>230	7–28
Cephalothin (**21d**)	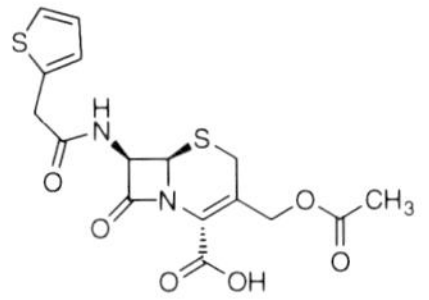	>40	25
Meropenem (**21e**)	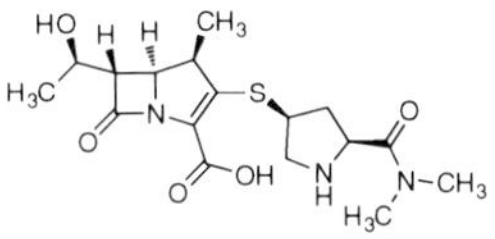	6.5	0.8
Imipenem (**21f**)	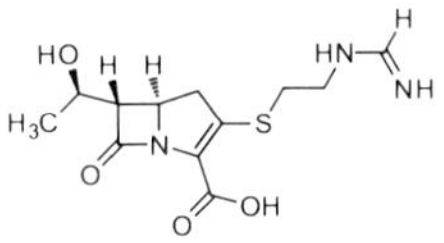	4	0.5
Aztreonam (**21g**)	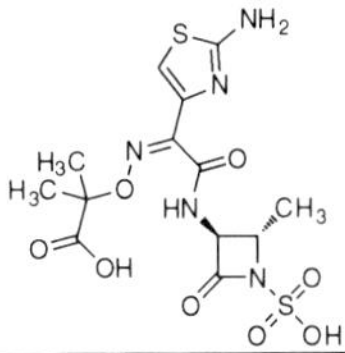	>147	>147

be different making it less responsive to β-lactam therapy, although the recent demonstration of activity of meropenem/CA against non-replicating persistent MTb has raised the possibility of use of this carbapenem against TB [299, 304].

CP-5484 (a carbapenem with activity against MRSA) [305] is currently in preclinical development for use against tuberculosis. Additionally, the meropenem/CA combination has shown potent activity against strains of MTb [299] and is currently being investigated for possible clinical use.

4.5 Rhiminophenazines

4.5.1 History

Clofazimine (CFM, Fig. 31; **8a**), a member of the riminophenazine class of compounds, was originally developed as an antitubercular drug in 1950s, but inconsistent outcomes in animal studies (effective in mouse models but less activity against in monkeys and guinea pigs) halted development [306]. Recently, CFM was reinvestigated, and several analogs of CFM that are active against MDR-TB were reported [307].

4.5.2 Structure–Activity Relationship

It has been shown that the imino group is essential for activity. Substitution with electron-withdrawing groups such as -Cl or -CF_3 at the R_1 and R_2 positions results in higher antituberculosis activity, but increased lipophilicity particularly of R_3 substituents can exacerbate the accumulation already observed with CFM in fat tissues and cells of the reticuloendothelial system [27]. Installation of a tetramethyl-piperidine in the R_3 position showed higher activity than that of ethyl or isopropyl.

4.5.3 Biology

B-4157 (Fig. 32; **8b**) exhibited promising in vitro activity against H37Rv and MDR strains of MTb with MIC range of ≤114 nM to 228 nM. Tetramethylpiperidine-substituted riminophenazines (such as B-4169, Fig. 32; **8c**) showed MICs of 42.4–169 nM [308]. A generalized membrane disrupting effect, interference with potassium transport, and generation of reactive oxygen intermediates have been suggested as the mechanism of action for riminophenazines [27, 309, 310], but detailed information is still not clear. The very low mutation frequency suggests that rhiminophenazines may affect multiple aspects of metabolism [27].

The 2,2,6,6-tetramethylpiperidine-substituted riminophenazines such as B4169 have superior activity to CFM against MTb growing in macrophages and are also

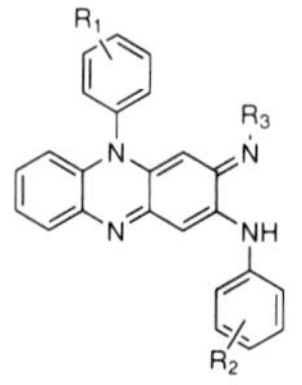

8a R_1 = 4-Cl, R_2 = 4-Cl, R_3 = ^{i}Pr; Clofazimine (CFM)
8b R_1 = 4-CF_3, R_2 = 4-CF_3, R_3 = Et; B4157
8c R_1 = 3,4,5-Cl R_2 = 3,4,5-Cl, R_3 = 4-(2,2,6,6-tetramethylpiperidine); B4169

Fig. 31 Antitubercular rhiminophenazines

less toxic in animal models [308]. The B-4157 analog has similar activity to CFM in infected mice, and at a dose of 20 mg/kg was as effective as similar doses of RIF or INH in long-term monotherapy in infected animals [307]. The in vivo potency may be due to the long half-life of CFM in tissues (>70 days after repeated dosing of human patients) [27]. The absoption, distribution, metabolism, and excretion (ADME) of CFM analogs have not yet been reported. CFM is 45–62% orally bioavailable in humans, reaches serum concentrations of 0.7–1 mg/mL, and is metabolized by the liver through dehalogenation or deamination followed by glucuronidation or by hydroxylation along with glucuronidation [27]. Riminophenazines are currently in the preclinical development state, but their high potency makes them attractive drug candidates.

4.6 Pyrroles

4.6.1 History

Naturally occurring pyrrolnitrin (Fig. 32; **22a**) and its analogs were tested against MTb, and the most effective exhibited an MIC of 3.9 μM [311]. However, most of the compounds from this series were cytotoxic, presumably because of the nitro group. Structural optimization of pyrrolnitrin and other azole analogs led to the discovery of the more potent pyrrole, BM-212 (Fig. 33; **22b**), exhibiting MIC values of 1.68 μM against MTb [312]. BM-212 (**22b**) was also found to be effective against strains resistant to EMB, INH, amikacin, STR, RIF, and rifabutin, as well as against MTb growing within a human monocyte cell line.

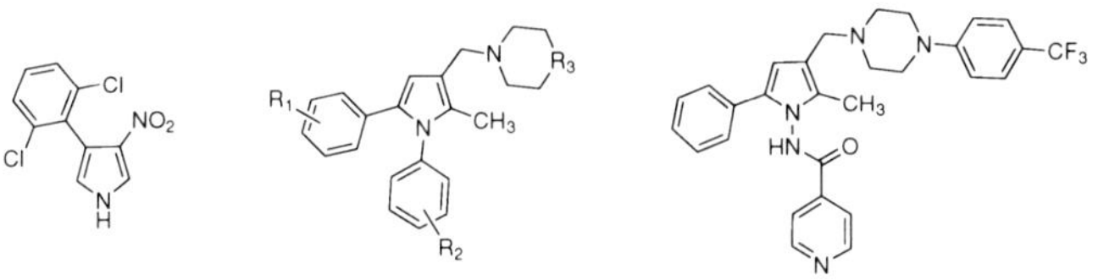
Cl
NO2
Cl
N
H
R1
N
R3
N
CH3
R2
N
N
CF3
N
CH3
HN
O
N

Pyrrolnitrin lead analog **22a**

22b R_1 = 4-Cl, R_2 = 4-Cl, R_3 = NCH_3; BM-212
22c R_1 = 4-F, R_2 = 4-F, R_3 = S
22d R_1 = 4-F, R_2 = 4-F, R_3 = S
22e R_1,R_2 = H, R_3 = S
22f R_1 = 4-F, R_2 = 2-F, R_3 = S
22g R_1 = 4-CH_3, R_2 = 4-F, R_3 = S

LL-3858 **22h**

Fig. 32 Pyrrole-based antitubercular compounds

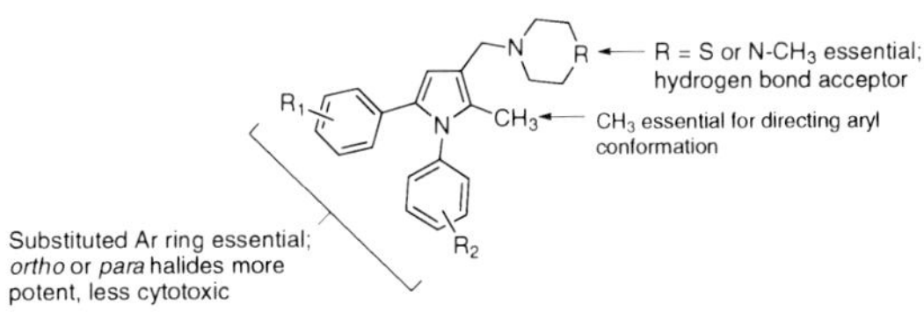
R = S or N-CH3 essential; hydrogen bond acceptor
CH3 essential for directing aryl conformation
Substituted Ar ring essential; *ortho* or *para* halides more potent, less cytotoxic
R1
N
R
N
CH3
R2

Fig. 33 Pyrrole SAR

4.6.2 Structure–Activity Relationship

Using BM-212 (**22b**) as a lead compound, systematic structural optimization led to the discovery of improved analogs, with similar or better activity in the range of 0.5–2 μM and an improved therapeutic index (ratio of cytotoxicity to in vitro activity against MTb) of 16–160 [313–316].

Based on whole cell biological activity, SARs could be deduced, with aromatic groups at N(1) and C(5) and a methyl group at C(2) as essential features (Fig. 33). Additionally, methylene-linked thiomorpholine or *N*-methylpiperazine substituents at C(3) act as hydrogen bond acceptors to improve activity (Fig. 34; **22b–g**) [316, 317]. Thus, a 1,2,3,5-tetrasubstituted pyrrole is the pharmacophore essential for antitubercular activity (Fig. 32). The 2-methyl group is not involved in any pharmacophoric interaction but influences the conformation of the substituents at positions 1 and 3 of the pyrrole ring [316, 317].

4.6.3 Biology

Another pyrrole analog, LL-3858 (Fig. 34; **22h**) (first reported by Lupin Limited in 2004), a pyrrole derivative, also complies with this pharmacophore model and is currently in phase IIa clinical trials in India. This compound has exhibited MIC values in the range of 0.05–0.1 μM, against MTb. LL-3858 (**22h**) has been reported to sterilize the lungs and spleen of infected mice after 12 weeks of treatment, none of which relapsed after 2 months of therapy termination [318].

4.7 *Deazapteridines*

4.7.1 History

Tetrahydrofolate (reduced dihydrofolate) is a key cofactor for the synthesis of many biomolecules, and inhibition of dihydrofolate reductase (DHFR) leads to cell death [319]. Bacterial DHFR is sufficiently different from human DHFR to serve as a novel drug target [320]. To this end, the deazapteridines were designed as inhibitors of mycobacterial DHFR by researchers at the Southern Research Institute [321].

4.7.2 Structure–Activity Relationship

MIC assays against MTb and cytotoxicity assays using Vero cells were used to compare selectivity for mycobacterial DHFR. Based on the limited number of structures reported, it appears that a smaller R_1 group is better tolerated (compounds SRI-8117 vs SRI-8922 or SRI-8229 vs SRI-8911; Table 9), and either a secondary or tertiary amine is tolerated (compounds SRI-8710 vs SRI-8117 and SRI-8687 vs SRI-8686) [321]. 2,5-substituted electron-rich aromatics are preferred (data not shown). Modeling studies of these small molecules binding to MTb DHFR

Table 9 MIC vs MTb and IC_{50} vs Vero cells [321]

Compound name	R_1	R_2	R_3	MIC vs H37Rv (μM)	IC_{50} vs Vero cells (μM)
SRI-8117 (**23a**)	CH_3	2,5-$(CH_3O)_2$Ph	H	37	2,106
SRI-8922 (**23b**)	CH_2CH_3	2,5-$(CH_3O)_2$Ph	H	>35	ND
SRI-8710 (**23c**)	CH_3	2,5-$(CH_3O)_2$Ph	CH_3	8.8	200
SRI-8686 (**23d**)	CH_3	2,5-$(CH_3CH_2O)_2$Ph	H	>34	ND
SRI-8687 (**23e**)	CH_3	2,5-$(CH_3CH_2O)_2$Ph	CH_3	>32	ND
SRI-8202 (**23f**)	CH_3	2-CH_3-5-CH_3OPh	H	19	1,421
SRI-8229 (**23g**)	CH_3	2-CH_3O-5-CH_3Ph	H	19	231
SRI-8911 (**23h**)	CH_2CH_3	2-CH_3O-5-CH_3Ph	H	>37	ND
SRI-8228 (**23i**)	CH_3	2-CH_3O-5-CF_3OPh	H	4.0	2.6

suggest that the 2-ethoxy or 2-methoxy group acts as a hydrogen bond donor [322, 323]. Additionally, modeling shows that for human DHFR, the cleft to which the deazapteridines bind is lined with hydrophobic residues, whereas the analogous MTb DHFR cleft is larger and more accessible to solvent [322, 323].

4.7.3 Biology

Subsequent publications refined the pharmacophore using members of the *Mycobacterium avium* complex (MAC), which established that SRI-8686 had the highest IC_{50} ratio for MAC DHFR vs human DHFR (0.84 nM vs 2,300 nM, a 2,700-fold selectivity). SRI-8117 showed similar selectivity (1.1 nM vs 1,000 nM, 900-fold selectivity) [324]. Other MTb DHFR inhibitors are in early stages of development [325]. Although the MIC assays suggest that this series may be worth developing, the on-target effect of these compounds in MTb still needs to be verified.

5 Critical Issues in TB Drug Development

5.1 Cell Penetration

The complex, lipid-rich envelope of MTb acts as a permeation barrier to a broad range of therapeutic agents and has likely contributed to both the fitness and the success of the pathogen (Fig. 33). The plasma membrane (PM) forms the innermost region of the cell envelope and is a typical lipid bilayer, structurally and functionally similar to the PM of other eubacteria. External to the PM is the peptidoglycan sacculus. This contains repeating units of *N*-acetylglucosamine and

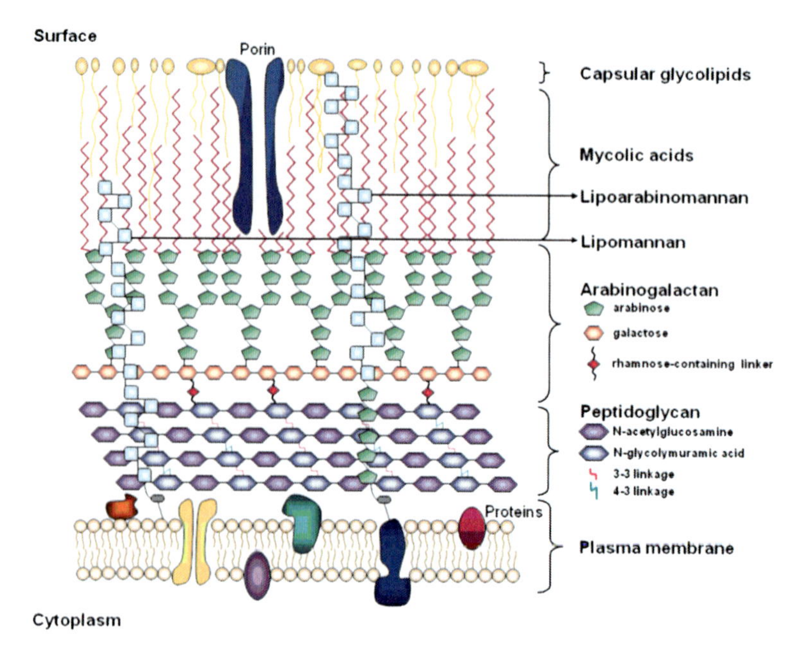

Fig. 34 Schematic of the mycobacterial cell envelope

N-glycolymuramic acid with stem peptides joined mostly through 3–3 cross-links rather than the 4–3 linkages commonly found in other bacteria [326, 327].

Peptidoglycan serves as a scaffold for arabinogalactan, a polymer of D-arabinose and D-galactose, which are covalently attached by β-1,5 linkages [328]. The arabinogalactan chain bridges between peptidoglycan and a thick layer of mycolic acids where the galactose portion of the polymer is connected to peptidoglycan by a unique glycosyl–phosphoryl bridge, and the arabinose moieties are ester-linked to four mycolic acid residues [329, 330]. The mycolic acids that are largely responsible for the impermeability of the mycobacterial cell wall consist of branched 2-alkyl-3-hydroxy fatty acids 70–90 carbon atoms in length [330]. In contrast to other bacteria, the MTb membrane also contains a number of lipids with unusual structures including: phosphatidylinositol mannosides, lipomannans and lipoarabinomannans, trehalose-6,6′-dimycolate, sulfolipids, phthicerol dimycocerosates, and phenolic glycolipids [331]. A loosely attached capsule defines the outermost layer of MTb and is composed primarily of glucans, arabinomannans, and mannans, with a small number of lipids and proteins decorating the structure as well.

Historically, the intrinsic resistance of mycobacteria to antibiotics has been attributed to the low permeability of the cell wall. For example, perturbations in the cell envelope caused by detergents, mutations, or through inhibition of cell wall polymer biosynthesis increases the susceptibility of mycobacteria to various classes of antibiotics, including aminoglycosides, EMB, and RIF [332–335]. Moreover, data indicate that the outer membrane of *Mycobacterium chelonei* is 1,000-fold less permeable to hydrophilic molecules than *Escherichia coli* and tenfold less permeable than the notoriously impermeable *Pseudomonas aeruginosa* [331, 336, 337]. Instead, hydrophilic molecules such as the cephalosporins likely penetrate the envelope using water-filled porins located on the outer leaflet of the cell wall [338]. While typical lipid bilayers are highly permeable to lipophilic molecules, permeability is inversely correlated with membrane fluidity [331]. When a cell wall is erected from lipids containing long, hydrocarbon chains with few double bonds or cyclopropane groups, the result is membrane rigidity [331, 339]. The mycolic acids within the MTb cell wall are unique in this regard, and accordingly, the inner leaflet displays extremely low fluidity [340]. In addition, lipids with more than one fatty acid chain attached to a single head group, similar to the mycolyl-arabinogalactan found in the cell wall of MTb, decrease membrane fluidity further [331]. Data predict that the lipophilic antibiotics such as fluoroquinolones, macrolides, rifamycins, and tetracyclines do penetrate the bacteria, but this likely occurs via the lipid bilayer rather than the inefficient porins found on the outer leaflet [331]. In support of this model, the more hydrophobic agents within an antibacterial class tend to be more effective against mycobacteria [331].

5.2 Animal Models for Evaluation

As most vertebrates can be infected with a mycobacterial pathogen, it is no surprise that there are a wide range of TB animal models that to different extents recapitulate the characteristics of human disease. A summary of some of the current experimental animal models of TB chemotherapy, their typical uses, and the comparative compound requirement for use in drug efficacy studies are shown in Table 10 and briefly summarized here.

Zebrafish infected with *Mycobacterium marinum* are gaining popularity as a model of TB as the costs and space requirements are quite modest and experimental work with infected fish or their embryos does not require the biological safely laboratory level 3 containment of any work with virulent strains of the MTb complex (*M. africanum, M. bovis, M. tuberculosis*, and *M. caprae*). *M. marinum* is a natural pathogen of fish that causes necrotizing lesions within a nearly transparent host where the progress of infection can be imaged with high resolution microscopy [341] allowing real-time data collection. This model is contributing to our understanding of TB pathogenesis and may become useful for drug screening in the near future [342, 343, 361].

The inbred mouse has been used most extensively in TB studies and can be infected by a variety of routes including i.v., intranasal inoculation, and by aerosol exposure. Many strains of mice, each with different genetic backgrounds for investigating certain immunological parameters, can be reliably maintained for many months in a state of chronic infection, and used for a variety of different readouts (Table 10). While the mouse model is often used in studies of MTb strains

Table 10 Characteristics of experimental animal models of tuberculosis chemotherapy

Model	Pathology	Utility: limitations	Compound requirement (g) [mass of animal]	References
Zebrafish	Necrotizing granuloma in embryo	Early granuloma formation, high-throughput: non-pharmacodynamic	0.025 [100 mL dish]	[341–343]
Mice	Pulmonary pneumonia and aggregates of leukocytes, spleen, and liver permissive for growth	Standard model for drug efficacy: very limited pathology	0.35 [25 g]	[79, 344–347]
Rats	Similar to mice	Accepted PK/PD model: little additional information cf. mice	4.6 [0.3 kg]	[348, 349]
Guinea pigs	Pulmonary, splenic, and liver lesions; 50% solid, 50% necrotizing granulomas	Small size with more similarity to human pathology: non-cavitating	12.6 [0.9 kg]	[79, 350–353]
Rabbits	Limited infection with Mtb strains; pulmonary solid and necrotizing lesions, cavities; extensive secondary lesions (*M. bovis* only)	Pathology including pulmonary, CNS system, ocular sites: highly sensitive to GI distress from many agents	40 [3.5 kg]	[354–357]
Nonhuman primates	Diverse presentation of lesions in multiple organs	Pathology accurately reproduces human, PK similar: expensive, dosing can be difficult	56 [5 kg]	[358–360]

to define the contribution of various mycobacterial genes to virulence or in studies of immunological responses to MTb infection, it does not reproduce the lung pathology observed in the human [290, 344]. However, the poor sterilizing activity of INH and the treatment-shortening effect of treating with INH, RIF, and PZA containing regimens in the mouse model are superficially similar to their efficacy in human studies; hence, it is argued that the mouse model is predictive of human relapse rates [345]. Since the mouse does not develop the latent disease that characterizes the spectrum of human TB but rather develops a chronic disease marked by very high bacterial burdens and progressive destruction of lung tissue, it has been argued that this animal model is unsuitable for testing drugs under development for latent disease [362]. The development of the gamma interferon gene-disrupted C57BL6 mouse has shortened initial in vivo drug evaluation to about 2 weeks, but may only reflect drug bioavailability as the host immune system is crippled and unable to control bacterial replication, resulting in fulminant disseminated infection [363]. The discovery of the *sst1* susceptibility locus in the C3HeB/FeJ mouse has provided an immune competent mouse model (C3H.B6-*sst1* mice) with lesions demonstrating central necrosis which more closely resemble those observed in diseased human lungs [346]. Experiments to benchmark standard tuberculosis chemotherapy studies to determine relapse rates in this model are underway [364]. The "Cornell" model, where chemotherapy is used to completely sterilize mouse tissues of MTb bacilli, or studies where a combination of BCG vaccination and drug treatment are used to sterilize tissues require extended durations of chemotherapy (>8 months) and have been argued to reflect the length and sterilizing activity of TB drug regimens in humans [365–367]. Results of these studies have been used to determine optimal drug combinations and drug exposures in human clinical trials to identify treatment-shortening regimens for human clinical trials [368–370].

Actual data substantiating the predictive ability of the mouse model to determine the length of treatment are very sparse. Anatomically, one major predictor of relapse in patients is the presence of cavities, a pathologic feature not represented in any mouse model [371]. Recent trials of shorter therapies based on substituting moxifloxacin for various components of the standard frontline regimen have failed to recapitulate the therapy-shortening observed in mice models [372]. Unfortunately, this evidence seems to be widely disregarded in the rush to introduce new regimens based exclusively on their comparative efficacy in mice.

A larger rodent, the Wistar or Sprague Drawley rat, is often used in medical research and is widely accepted as a model for toxicology studies. The rat model allows dual comparison of toxicology and efficacy in the same animal and, as a result, has been used in several TB drug development studies. In one of these studies, RIF was reported to give a good dose–response curve for bacillary clearance, but other TB chemotherapeutics gave less encouraging results, [348] suggesting that this model may not be a suitable model for evaluation of antitubercular drugs.

The guinea pig, used by Robert Koch in the late 1800s to demonstrate that the MTb bacillus was the etiological agent of TB, was one of the earliest animal models in mycobacterial research. Even with very low aerosolized infectious doses, these animals experience rapid progression to granulomatous non-cavitary disease that is ultimately fatal [350]. Because of its extreme susceptibility to MTb infection, the guinea pig is used extensively in vaccine studies, but to a lesser extent in drug efficacy studies because of its relatively large size [352, 353]. Due to its high susceptibility to MTb infection, it is also gaining popularity for virulence testing of gene knockout mutants of MTb, since large differences in organ burdens are generally observed with attenuated strains of MTb in the guinea pig [179, 351]. Recently, the sterilizing activity of a species-specific, human-equivalent dosage of a INH, RIF, and PZA regimen given in a 2-week intensive phase followed by a 6-month continuation phase of biweekly dosing (the so-called Denver regimen) was tested in both the guinea pig and the BALB/c mouse [369]. The guinea pigs were found to respond to treatment more quickly and have lower relapse rates in the 6 months following treatment as compared to similarly treated mice, although the guinea pigs experienced gastrointestinal toxicities of unknown origin making the model more challenging.

The New Zealand White rabbit, a relatively resistant animal to TB infection, has often been used to study the development of lung pathology including necrotizing lesions and cavities after either i.v. or aerosol infection with *M. bovis*, MTb, and even *M. avium* strains [354, 357, 373, 374]. It has been used for studies of extra-pulmonary dissemination and/or growth of MTb especially in studies of pathogenesis of the central nervous system as a model of human TB meningitis [355, 375]. It is also used for testing indwelling venous lines and ports containing drugs which are technically not feasible on smaller animals such as mice, but less often used for oral drug efficacy studies due to the large amount of compound needed for efficacy testing in rabbits (see Table 10) [376]. On the other hand, it is particularly advantageous that oral dosing and PK/PD blood collection are possible without anesthesia in this species. It has been reported that latent disease can be achieved in the rabbit model with certain MTb strains although additional validation is required [377].

Nonhuman primates (NHP), especially cynomolgus and rhesus macaques, have a long history in TB research for both vaccine and drug testing, but the advent of more restrictive laboratory practices and the requirement for BSL-3 housing has made the model prohibitively expensive and thus less utilized [344, 378]. Like other species, aerosol infection or direct installation of the bacilli into the lung is the usual route of infection. These monkeys reproduce the spectrum of disease observed in humans including pulmonary, extrapulmonary, and latent tuberculosis infection (LTBI) as well as many of the different types of granulomas observed in human patients [379]. Low dose infection of the cynomolgus macaque is associated with the induction of LTBI with roughly 60% of animals showing no signs of disease after skin test conversion [359]. The genetic similarity between NHP and humans has allowed use of the same immunological reagents such as TNF-α blockers to elicit reactivation disease from LTBI and other reagents to query host immune function for vaccine studies as those that have been used in human clinical trials [380]. A disadvantage with these relatively large NHPs is the amount of GMP or minimally GLP compound needed (Table 10) and the requirement for anesthesia for most manipulations including dosing to achieve reliable administration in drug studies. For these and other regulatory reasons, NHP work is usually reserved for proof of concept studies to establish the link between findings in prior (lower) animal studies and anticipated outcome in humans or as the final stage in a preclinical drug development pipeline before seeking approval for investigational studies in humans.

5.3 *Pharmacological Models for Antitubercular Drugs*

An evaluation of the pharmacological properties of new TB drugs is essential for effective treatment and overall cure of disease. Currently, most TB drugs are evaluated through preclinical animal models to establish appropriate dosing levels that promote optimal bacterial killing with limited toxicity. In vivo efficacy for TB drugs is not solely dependent on plasma concentration but more significantly dependent on tissue concentrations near and within lesions [78]. These concentrations must remain above MIC levels for an effective period of time to eradicate bacilli (described as Time > MIC). Lower levels are associated with the development of resistance and relapse of disease. However, due to the duration of treatment, maintaining high levels of drugs in combination often poses severe issues of toxicities and tolerabilities for the patient [381]. Thus, pharmacological evaluations for TB drugs depend on optimizing treatment to an often narrowly constrained therapeutic window.

Effective therapy with any drug is dependent on the relationship between pharmacokinetic and pharmacodynamics (PK/PD) parameters [382, 383]. Pharmacokinetics (PK) defines the ADME properties of the compound, while pharmacodynamics (PD) reveals the correlation between the serum concentration and the biological effect, efficacy or toxicity. For most drugs, the primary measurement under evaluation is plasma concentration. However, it is more relevant to assess drug levels within the infected lesion. Many efforts have attempted to define key parameters for TB drugs such as C_{max}/MIC (ratio of peak serum drug levels to MIC), AUC > MIC (overall drug exposure over the dosing interval must be greater than MIC), and Time > MIC (time period at which the drug remains in the blood must be greater than MIC per dosing interval) which are all established from plasma concentrations [78, 384]. Typically, high C_{max}/MIC ratios can offer sterilizing activity as well as limit adaptive resistance or the selection of resistant subpopulations whereas for TB, AUC > MIC and T > MIC are thought to be most relevant for both to maintain long-term exposure above MIC with limited dosing. These parameters are typically measured from blood, yet it is presumed that the primary driving factor for efficacy in TB therapy involves lesion penetration at effective concentrations [78, 290]. It is this factor that will eradicate persistent bacilli and circumvent the development of resistance.

There are several confounding factors that play a role in antitubercular drug efficacy in vivo. As with most drugs, stability and bioavailability with limited metabolism are important. Delivery to the site of infection at active concentrations increases the overall efficacy of any drug. However, active pulmonary tuberculosis is a chronic complex disease with a diverse spectrum of lesions within the lung. Predominantly caseous lesions are central to the "life cycle" of a pulmonary TB infection, as they eventually erode into air passages to allow bacilli to reach blood vessels and permit dissemination. These granulomas are generally poorly vascularized, hypoxic, lipid-rich, and often fortified with fibrous tissue forming an impenetrable fortress for the TB bacilli [290]. The ability of a drug to penetrate these lesions and kill bacilli is most critical to eradicate the bacteria and prevent disseminated or extensive disease.

Currently, there are various models designed to assess the pharmacological activity of TB drugs. Traditional models focus on determining drug levels in the blood from preclinical and clinical animal and human evaluation [385]. These studies are used to optimize dosing and evaluate tolerance. Dose fractionation models in animals can determine relevant PK indices with a strong correlation with PD effects. This type of experiment can elucidate important information for clinical development and optimal dosing strategies to prioritize compounds through drug development [258, 386, 387]. Recent in vitro models have been designed to mimic human PK (half-lives and dosing schedules) to assess the development of resistance [388]. The data obtained from these models help to identify drug-exposure breakpoints required for maximal bactericidal activity and the suppression of drug resistance. Current lesion penetration studies involving tissues from animals and human resections are providing important information in regard to drug levels found in the various lesion types enabling a better understanding to drug efficacy and therapeutic response [290].

Finally, the use of therapeutic drug monitoring (TDM) has been a useful tool for assessing drug levels during treatment in the clinical setting [71]. The use of TDM in tuberculosis treatment can allow physicians the ability to adjust dosing to provide an efficacious therapeutic concentration throughout the extensive duration of treatment [71]. Patients who most benefit are those with complications which may alter drug exposure, such as those on co-therapy for HIV for which there are known drug–drug interactions, those with diabetes mellitus with typical delayed absorption or malabsorption concerns, those with renal failure undergoing dialysis, and those experiencing hepatic dysfunction (see Sect. 1.4) [69]. TDM in combination with bacteriological and clinical data can be a useful tool to assess treatment and ensure as successful outcome [71].

Understanding pharmacological activities of TB drugs is essential not only for addressing drug levels for effective sterilizing activity and optimizing dosing strategies, but more importantly, they are also useful in limiting the development of acquired resistance. PK/PD for TB agents is relevant to understanding important phenomena associated with TB. Efforts are ongoing to develop PK/PD analyses which will effectively predict success or failure of new antituberculosis drugs and combination regimens.

5.4 Clinical Development Methodologies

The goal of TB chemotherapy is to cure clinical symptoms, prevent the development of resistance, and to prevent relapse. Since TB treatment entails 6–24 months of chemotherapy depending on drug susceptibility patterns, early markers that predict durable cure would greatly facilitate and speed the evaluation of new drugs. Currently, it requires 2 years of follow-up after termination of chemotherapy to capture more than 90% of relapses, thus necessitating trials lasting up to 4 years during evaluation of new therapies. The FDA recognizes the need to develop new drugs for TB and the associated need to look for early predictors of durable cure, making regulatory approval processes for such trials logistically easier.

Biomarkers that predict durable cure could include pathogen-specific measurements, determination of host responses to the pathogen or nonspecific disease-associated responses, and imaging. Unfortunately, no biomarker of any nature has been validated as of this time. Microbiologic markers traditionally used to evaluate TB chemotherapy include: 2-month culture conversion (in which sputum samples taken at regular intervals during chemotherapy are evaluated for eradication of culturable MTb at two months posttreatment), days to positivity (the time required to obtain mycobacterial growth from sputum in liquid culture), and serial counts of CFUs of MTb from sputum samples on agar combined with scoring acid-fast bacilli in sputum during treatment. Two-month culture positivity as surrogate marker for relapse was first recognized and used as a surrogate endpoint to predict treatment efficacy during the BMRC studies where culture negativity at 2 months was associated with cure and lack of subsequent relapse and is widely used for current evaluation of TB chemotherapeutic trials [389–392].

While other culture-dependent methods show promise, they do not always predict 2-month culture conversion and there is no data to correlate findings with risk of relapse [390, 393]. In a large ongoing clinical trial (REMoxTB), where moxifloxacin is evaluated as a replacement for INH or EMB, various culture-dependent and -independent methods will be assessed against the primary endpoint of relapse within 2 years of treatment termination, which may give some insight into the utility of other biomarkers [394].

Culture-dependent methods suffer from their long turnaround time, often several weeks, due to the slow growth of mycobacteria such as MTb. Potential culture-independent pathogen-derived biomarkers include detection of MTb DNA [395], lipoarabinomannan [396, 397] in urine, MTb mRNA in sputum [393], MTb characteristic volatile organic compounds in breath [398], and several other pathogen-derived biomarkers currently under evaluation.

Host biomarkers of disease include interferon-gamma release assays in response to MTb antigens [399, 400], measurements of non-MTb-specific host responses such as C-reactive protein, serum interleukin-2, neopterin and procalcitonin which still require a definitive TB diagnosis by other means [401–403]. These may especially be useful as clinical trial endpoints in smear-negative, paucibacillary, extra-pulmonary, and pediatric TB where microbiologic culture is less reliable.

Imaging biomarkers such as high resolution computed tomography (HRCT) have been shown to be helpful in diagnosing active TB and distinguishing it from latent disease [404], and may be useful in detecting the early stages of disease in TB contacts [405, 406] but is also used to evaluate treatment response. Positron emission tomography (PET) using 2-fluorodeoxyglucose has been used to visualize inflammatory regions in pulmonary mycobacteriosis caused by MTb and *M. avium* but is even less discriminating than HRCT in distinguishing between TB and nontuberculous mycobacterial infections [407].

Traditionally, clinical trials for new TB drugs include a 7–14-day assessment of early bactericidal activity (EBA) of a drug given as monotherapy (often at different doses). Daily sputum microbiology is performed and bacterial counts quantified. Unfortunately, reproducibility of such microbiological assays is questionable, and EBA rarely gives a definitive answer. In most cases, ambiguous EBA results are ignored and a longer study (2 months or more depending on the available safety and toxicity data) of the new drug or placebo in combination with current TB drugs followed by completion of chemotherapy with regular extended phase regimens. These studies are followed up with expensive phase III trials that give the new drug as part of combined therapy (6–24 months) with prolonged follow-up after treatment completion [369]. The caveat to using EBA as a predictor of successful treatment is that drugs such as RIF and PZA, which have a poor EBA but have good activity against MTb populations in certain lesions, would score as poor drugs [408]. PZA is the essential drug that allowed the shortening of treatment of drug-sensitive TB from 9 to 6 months [409], indicating that EBA studies may not be a good way to cull ineffective drugs.

Phase II trials typically evaluate sputum conversion at 2 months as a predictor of durable cure (cure without relapse), but the utility of 2-month culture conversion as a biomarker was derived from studies of drug-sensitive TB and may not apply to MDR-TB. In addition, evaluation of new drugs is typically compared to the standard of care, which can achieve a 95% cure rate in clinical settings. Thus, the sample size of such a trial must be large to detect either non-inferiority or superiority of new agents. With MDR- and XDR-TB, the cure rates are much lower, which may make it easier to detect the effect of the investigational drug, but such study populations are much more heterogeneous due to different degrees and spectra of drug susceptibility results, different background drug regimens, and a wide range of disease severity and treatment history all of which may confound results. Some of these problems can be dealt with by randomized stratification of strong covariates, for example by stratifying for fluoroquinolone sensitivity [259].

In summary, EBA studies, while having the advantage of seeing the effect of a single drug on MTb in lesions that are the source of sputum-borne bacilli, are clearly not a useful predictor of drug efficacy. Studies of combination therapy using 2-month culture conversion as an end point have the advantage of a partially validated end point with a weak correlation for drug-sensitive disease but should be interpreted with caution in drug-resistant disease. Imaging modalities such as

HRCT and PET are attractive endpoints that do not require culture and evaluated disease at the relevant site of infection but, like host-derived biomarkers, require validation for predicting treatment efficacy.

6 Concluding Remarks

TB drug discovery remains a challenging enterprise at every level. There is an urgent need to validate criteria for lead compounds with promise to reduce treatment duration, treat drug-resistant disease, and provide utility for prophylaxis of subclinical disease. Poorly validated in vitro assays, poorly validated in vivo assays, and a severe lack of predictive animal models all compromise efforts to improve the quality of the TB drug pipeline. Existing SAR around series explored in the 1950s and 1960s provides some starting points, but the chemistry of most of these series is either intractable or linked to enzymatic activation of prodrug and is therefore difficult to optimize systematically. As in most anti-infectives, target-based strategies have failed to lead to viable candidates, and in most cases the reason for this failure has been unexplained. In the current climate, it is no surprise that the momentum is toward improving whole cell screening and attempting to identify leads with novel targets. Biological studies working toward more predictive assays are in progress and very promising. For example, the promising use of titratable promoter elements to systematically knock down genes and mimic the pharmacological action of a drug can truly validate drug action and directly address "target vulnerability" by quantitating how much a target needs to be inhibited to affect cell death [410]. Current efforts to understand the "systems biology" of MTb also have the potential to reshape the landscape and improve our ability to select targets.

The nitroimidazoles and diarylquinolines, series developed specifically for TB, provide some reason for optimism and perhaps a sort of loose road map for how to approach the problems. Both series suffer from critical limitations including extremely poor water solubility and potential hERG problems. In general, one serious problem that persists in new drug discovery programs for TB lies in the lack of broad application of preclinical ADME and toxicology studies in lead optimization. The academic sector lacks an appreciation of the role of such studies and lacks resources dedicated to performing them. Both series may well fail because of these considerations, but at a minimum it has made clinical development programs for these two new classes considerably more complex than if these issues had been addressed systematically in lead optimization. Candidate selection in most cases continues to happen predominantly by MIC and mouse activity, and it is doubtful that the situation will improve more than incrementally without more sophistication in lead optimization programs. Nonetheless, there are ongoing phase II studies with these agents, and it is crucial that clinical development of these compounds is paired with meaningful attempts to understand the relationship of the preclinical studies that led to selection of the candidates with the ultimate clinical properties of these agents. Through such efforts, in vitro and in vivo assays (not to mention phase I and II trials) could perhaps be optimized to be more predictive, and these could be used to guide both backup programs for the existing agents and optimization efforts for new series not yet in play.

Funding sources for TB drug discovery have also expanded dramatically over the last 10 years and now include significant efforts from private foundations (such as the Bill and Melinda Gates' Foundation through its TB Drug Accelerator Program) as well as public institutions such as the US National Institutes of Health and the European Union's Framework Programmes. Many of the activities being funded are taking place at the interface of academic laboratories and commercial pharmaceutical companies using a wide variety of models for cooperation. Although this is precisely where such programs appear to be best placed, there has been a rather steep learning curve as these two cultures are brought together with competing needs. Nonetheless, there is significant cause for excitement at the number and quality of programs that are currently operating. If we can successfully leverage the strengths of both academia and industry, there is hope that we will be able to raise the victory flag in earnest in the long struggle against TB.

References

1. Dye C, Lonnroth K, Jaramillo E, Williams BG, Raviglione M (2009) Trends in tuberculosis incidence and their determinants in 134 countries. Bull World Health Organ 87:683–691
2. Barry CE III, Boshoff HI, Dartois V, Dick T, Ehrt S, Flynn J, Schnappinger D, Wilkinson RJ, Young D (2009) The spectrum of latent tuberculosis: rethinking the biology and intervention strategies. Nat Rev Microbiol 7:845–855
3. Young D, Dye C (2006) The development and impact of tuberculosis vaccines. Cell 124:683–687
4. Kaufmann SH, Hussey G, Lambert PH (2010) New vaccines for tuberculosis. Lancet 375:2110–2119
5. Parwati I, van Crevel R, van Soolingen D (2010) Possible underlying mechanisms for successful emergence of the *Mycobacterium tuberculosis* Beijing genotype strains. Lancet Infect Dis 10:103–111
6. Dye C, Williams BG (2010) The population dynamics and control of tuberculosis. Science 328:856–861
7. Iseman MD (2002) Tuberculosis therapy: past, present and future. Eur Respir J Suppl 36:87s–94s
8. Murray JF (2004) A century of tuberculosis. Am J Respir Crit Care Med 169:1181–1186
9. Thoren M, Hinshaw HC (1952) Therapy of pulmonary tuberculosis with isoniazid alone and in combination with streptomycin and with para-amino-salicylic acid. Stanford Med Bull 10:316–318
10. Marshall G (1948) Streptomycin treatment of pulmonary tuberculosis. Br Med J 2:769–782
11. Youmans GP, Raleigh GW, Youmans AS (1947) The tuberculostatic action of para-amino-salicylic acid. J Bacteriol 54:409–416
12. Graessle OE, Pietrowski JJ (1949) The in vitro effect of para-aminosalicylic acid (PAS) in preventing acquired resistance to streptomycin by *Mycobacterium tuberculosis*. J Bacteriol 57:459–464
13. Dooneief AS, Buchberg A, Steinbach MM (1950) Para-aminosalicylic acid (PAS) in chronic pulmonary tuberculosis. N Engl J Med 242:859–862
14. Bowen DA, Collins DM (1952) Development of drug resistance to isoniazid in cases of pulmonary tuberculosis. Tubercle 33:276–278

15. Joiner CL, Maclean KS, Carroll JD, Marsh K, Collard P, Knox R (1954) Isoniazid and P. A. S. in chronic pulmonary tuberculosis: a warning. Lancet 267:663–666
16. Marshall G (1955) Various combinations of isoniazid with streptomycin or with P.A.S. in the treatment of pulmonary tuberculosis; seventh report to the Medical Research Council by their Tuberculosis Chemotherapy Trials Committee. Br Med J 1:435–445
17. Capon AW (1954) Streptomycin and PAS vs. streptomycin, PAS and isoniazid in the treatment of pulmonary tuberculosis. Can Med Assoc J 70:62–67
18. Hudgins PC, Patnode RA, Cummings MM (1955) The effect of cycloserine on growing and resting tubercle bacilli. Am Rev Tuberc 72:685–686
19. Hutton PW, Tonkin IM (1960) Ethionamide (1314') with streptomycin in acute tuberculosis of recent origin in Uganda Africans: a pilot study. Tubercle 41:253–256
20. Bartz QR, Ehrlich J, Mold JD, Penner MA, Smith RM (1951) Viomycin, a new tuberculostatic antibiotic. Am Rev Tuberc 63:4–6
21. Patnode RA, Hudgins PC (1958) Effect of kanamycin on *Mycobacterium tuberculosis* in vitro. Am Rev Tuberc 78:138–139
22. Kaida K, Sugiyama K (1959) Clinical experience with PZA-INH therapy: report on study of resected specimens following the above therapy in particular. Dis Chest 36:378–388
23. Somner AR, Brace AA (1962) Ethionamide, pyrazinamide and cycloserine used successfully in the treatment of chronic pulmonary tuberculosis. Tubercle 43:345–360
24. Tuberculosis Chemotherapy Center M (1959) A concurrent comparison of home and sanatorium treatment of pulmonary tuberculosis in South India. Bull World Health Organ 21:51–144
25. Bienenstock J, Shaldon S (1963) Thiacetazone in tuberculosis. Lancet 2:817–818
26. Cuthbert J, Bruce LG (1964) Treatment of pulmonary tuberculosis by capreomycin and PAS: a small preliminary trial. Tubercle 45:205–210
27. (2008) Clofazimine. Tuberculosis (Edinb) 88:96–99
28. Doster B, Murray FJ, Newman R, Woolpert SF (1973) Ethambutol in the initial treatment of pulmonary tuberculosis. U.S. Public Health Service tuberculosis therapy trials. Am Rev Respir Dis 107:177–190
29. Lees AW, Tyrrell WF, Smith J, Allan GW (1970) Ethambutol in the retreatment of chronic pulmonary tuberculosis. Br J Dis Chest 64:85–89
30. Fisher L (1971) Rifampin–new and potent drug for TB treatment. Bull Natl Tuberc Respir Dis Assoc 57:11–12
31. Mitchison DA (2005) The diagnosis and therapy of tuberculosis during the past 100 years. Am J Respir Crit Care Med 171:699–706
32. (1974) Controlled clinical trial of four short-course (6-month) regimens of chemotherapy for treatment of pulmonary tuberculosis. Third report. East African-British Medical Research Councils. Lancet 2:237–240
33. McCune RM Jr, Tompsett R (1956) Fate of *Mycobacterium tuberculosis* in mouse tissues as determined by the microbial enumeration technique. I. The persistence of drug-susceptible tubercle bacilli in the tissues despite prolonged antimicrobial therapy. J Exp Med 104:737–762
34. Grosset J (1978) The sterilizing value of rifampicin and pyrazinamide in experimental short-course chemotherapy. Bull Int Union Tuberc 53:5–12
35. (1986) Long-term follow-up of a clinical trial of six-month and four-month regimens of chemotherapy in the treatment of pulmonary tuberculosis. Singapore Tuberculosis Service/British Medical Research Council. Am Rev Respir Dis 133:779–783
36. (1988) Five-year follow-up of a clinical trial of three 6-month regimens of chemotherapy given intermittently in the continuation phase in the treatment of pulmonary tuberculosis. Singapore Tuberculosis Service/British Medical Research Council. Am Rev Respir Dis 137:1147–1150
37. (1983) Controlled clinical trial of 4 short-couse regimens of chemotherapy (three 6-month and one 8-month) for pulmonary tuberculosis. Tubercle 64:153–166

38. (1991) Controlled trial of 2, 4, and 6 months of pyrazinamide in 6-month, three-times-weekly regimens for smear-positive pulmonary tuberculosis, including an assessment of a combined preparation of isoniazid, rifampin, and pyrazinamide. Results at 30 months. Hong Kong Chest Service/British Medical Research Council. Am Rev Respir Dis 143:700–706
39. Bayer R, Wilkinson D (1995) Directly observed therapy for tuberculosis: history of an idea. Lancet 345:1545–1548
40. Edlin BR, Tokars JI, Grieco MH, Crawford JT, Williams J, Sordillo EM, Ong KR, Kilburn JO, Dooley SW, Castro KG et al (1992) An outbreak of multidrug-resistant tuberculosis among hospitalized patients with the acquired immunodeficiency syndrome. N Engl J Med 326:1514–1521
41. Friedman CR, Stoeckle MY, Kreiswirth BN, Johnson WD Jr, Manoach SM, Berger J, Sathianathan K, Hafner A, Riley LW (1995) Transmission of multidrug-resistant tuberculosis in a large urban setting. Am J Respir Crit Care Med 152:355–359
42. Bifani PJ, Plikaytis BB, Kapur V, Stockbauer K, Pan X, Lutfey ML, Moghazeh SL, Eisner W, Daniel TM, Kaplan MH, Crawford JT, Musser JM, Kreiswirth BN (1996) Origin and interstate spread of a New York City multidrug-resistant *Mycobacterium tuberculosis* clone family. JAMA 275:452–457
43. Agerton T, Valway S, Gore B, Pozsik C, Plikaytis B, Woodley C, Onorato I (1997) Transmission of a highly drug-resistant strain (strain W1) of *Mycobacterium tuberculosis*. Community outbreak and nosocomial transmission via a contaminated bronchoscope. JAMA 278:1073–1077
44. (1992) A controlled study of rifabutin and an uncontrolled study of ofloxacin in the retreatment of patients with pulmonary tuberculosis resistant to isoniazid, streptomycin and rifampicin. Hong Kong Chest Service/British Medical Research Council. Tuber Lung Dis 73:59–67
45. Kohno S, Koga H, Kaku M, Maesaki S, Hara K (1992) Prospective comparative study of ofloxacin or ethambutol for the treatment of pulmonary tuberculosis. Chest 102:1815–1818
46. Sahoo RC (1993) Ofloxacin in the retreatment of patients with pulmonary tuberculosis resistant to isoniazid, streptomycin and rifampicin – a south Indian experience. Tuber Lung Dis 74:140–141
47. Kennedy N, Berger L, Curram J, Fox R, Gutmann J, Kisyombe GM, Ngowi FI, Ramsay AR, Saruni AO, Sam N, Tillotson G, Uiso LO, Yates M, Gillespie SH (1996) Randomized controlled trial of a drug regimen that includes ciprofloxacin for the treatment of pulmonary tuberculosis. Clin Infect Dis 22:827–833
48. Moadebi S, Harder CK, Fitzgerald MJ, Elwood KR, Marra F (2007) Fluoroquinolones for the treatment of pulmonary tuberculosis. Drugs 67:2077–2099
49. Bergmann JS, Woods GL (1998) In vitro activity of antimicrobial combinations against clinical isolates of susceptible and resistant *Mycobacterium tuberculosis*. Int J Tuberc Lung Dis 2:621–626
50. Chambers HF, Turner J, Schecter GF, Kawamura M, Hopewell PC (2005) Imipenem for treatment of tuberculosis in mice and humans. Antimicrob Agents Chemother 49:2816–2821
51. Anger HA, Dworkin F, Sharma S, Munsiff SS, Nilsen DM, Ahuja SD (2010) Linezolid use for treatment of multidrug-resistant and extensively drug-resistant tuberculosis, New York City, 2000-06. J Antimicrob Chemother 65:775–783
52. Schecter GF, Scott C, True L, Raftery A, Flood J, Mase S (2010) Linezolid in the treatment of multidrug-resistant tuberculosis. Clin Infect Dis 50:49–55
53. Barry CE (2010) Linezolid Pharmacokinetics (PK) in Multi-Drug Resistant (MDR) and Extensively-Drug Resistant (XDR) Tuberculosis (TB) (S30PK). Clinical Trial NCT00727844. Sponsored by National Institute for Allergy and Infectious Diseases. Accessed from National Library of Medicine and National Institutes of Health (US), ClinicalTrials.gov
54. MacKenzie WR (2010) Linezolid to treat extensively drug-resistant tuberculosis. Clinical Trial NCT00691392. Sponsored by the Centers for Disease Control. Accessed from National Library of Medicine and National Institutes of Health (US), ClinicalTrials.gov

55. Ross JD, Horne NW, Grant IW, Crofton JW (1958) Hospital treatment of pulmonary tuberculosis; a follow-up study of patients admitted to Edinburgh hospitals in 1953. Br Med J 1:237–242
56. Caminero JA, Sotgiu G, Zumla A, Migliori GB (2010) Best drug treatment for multidrug-resistant and extensively drug-resistant tuberculosis. Lancet Infect Dis 10:621–629
57. Centers for Disease Control and Prevention (2006) Emergence of *Mycobacterium tuberculosis* with extensive resistance to second-line drugs–worldwide, 2000–2004. MMWR Morb Mortal Wkly Rep 55:301–305
58. Kwon YS, Kim YH, Suh GY, Chung MP, Kim H, Kwon OJ, Choi YS, Kim K, Kim J, Shim YM, Koh WJ (2008) Treatment outcomes for HIV-uninfected patients with multidrug-resistant and extensively drug-resistant tuberculosis. Clin Infect Dis 47:496–502
59. Jacobson KR, Tierney DB, Jeon CY, Mitnick CD, Murray MB (2010) Treatment outcomes among patients with extensively drug-resistant tuberculosis: systematic review and meta-analysis. Clin Infect Dis 51:6–14
60. Dheda K, Shean K, Zumla A, Badri M, Streicher EM, Page-Shipp L, Willcox P, John MA, Reubenson G, Govindasamy D, Wong M, Padanilam X, Dziwiecki A, van Helden PD, Siwendu S, Jarand J, Menezes CN, Burns A, Victor T, Warren R, Grobusch MP, van der Walt M, Kvasnovsky C (2010) Early treatment outcomes and HIV status of patients with extensively drug-resistant tuberculosis in South Africa: a retrospective cohort study. Lancet 375:1798–1807
61. Migliori GB, Sotgiu G, D'Arcy Richardson M, Centis R, Facchini A, Guenther G, Spanevello A, Lange C (2009) MDR-TB and XDR-TB: drug resistance and treatment outcomes. Eur Respir J 34:778–779
62. Jeon DS, Kim DH, Kang HS, Hwang SH, Min JH, Kim JH, Sung NM, Carroll MW, Park SK (2009) Survival and predictors of outcomes in non-HIV-infected patients with extensively drug-resistant tuberculosis. Int J Tuberc Lung Dis 13:594–600
63. Velayati AA, Farnia P, Masjedi MR, Ibrahim TA, Tabarsi P, Haroun RZ, Kuan HO, Ghanavi J, Varahram M (2009) Totally drug-resistant tuberculosis strains: evidence of adaptation at the cellular level. Eur Respir J 34:1202–1203
64. Velayati AA, Masjedi MR, Farnia P, Tabarsi P, Ghanavi J, Ziazarifi AH, Hoffner SE (2009) Emergence of new forms of totally drug-resistant tuberculosis bacilli: super extensively drug-resistant tuberculosis or totally drug-resistant strains in Iran. Chest 136:420–425
65. Menzies D, Benedetti A, Paydar A, Martin I, Royce S, Pai M, Vernon A, Lienhardt C, Burman W (2009) Effect of duration and intermittency of rifampin on tuberculosis treatment outcomes: a systematic review and meta-analysis. PLoS Med 6:e1000146
66. Zhou SF, Xue CC, Yu XQ, Li C, Wang G (2007) Clinically important drug interactions potentially involving mechanism-based inhibition of cytochrome P450 3A4 and the role of therapeutic drug monitoring. Ther Drug Monit 29:687–710
67. Finch CK, Chrisman CR, Baciewicz AM, Self TH (2002) Rifampin and rifabutin drug interactions: an update. Arch Intern Med 162:985–992
68. Davies G, Cerri S, Richeldi L (2007) Rifabutin for treating pulmonary tuberculosis. Cochrane Database Syst Rev:CD005159
69. Peloquin CA (2001) Pharmacological issues in the treatment of tuberculosis. Ann NY Acad Sci 953:157–164
70. Sen T, Joshi SR, Udwadia ZF (2009) Tuberculosis and diabetes mellitus: merging epidemics. J Assoc Physicians India 57:399–404
71. Peloquin CA (2002) Therapeutic drug monitoring in the treatment of tuberculosis. Drugs 62:2169–2183
72. Mitchison DA (2000) Role of individual drugs in the chemotherapy of tuberculosis. Int J Tuberc Lung Dis 4:796–806
73. Herbert D, Paramasivan CN, Venkatesan P, Kubendiran G, Prabhakar R, Mitchison DA (1996) Bactericidal action of ofloxacin, sulbactam-ampicillin, rifampin, and isoniazid on logarithmic- and stationary-phase cultures of Mycobacterium tuberculosis. Antimicrob Agents Chemother 40:2296–2299

74. Garcia-Tapia A, Rodriguez JC, Ruiz M, Royo G (2004) Action of fluoroquinolones and linezolid on logarithmic- and stationary-phase culture of *Mycobacterium tuberculosis*. Chemotherapy 50:211–213
75. Wayne LG, Sohaskey CD (2001) Nonreplicating persistence of *Mycobacterium tuberculosis*. Annu Rev Microbiol 55:139–163
76. Xie Z, Siddiqi N, Rubin EJ (2005) Differential antibiotic susceptibilities of starved *Mycobacterium tuberculosis* isolates. Antimicrob Agents Chemother 49:4778–4780
77. Betts JC, Lukey PT, Robb LC, McAdam RA, Duncan K (2002) Evaluation of a nutrient starvation model of *Mycobacterium tuberculosis* persistence by gene and protein expression profiling. Mol Microbiol 43:717–731
78. Dartois V, Barry CE (2010) Clinical pharmacology and lesion penetrating properties of second- and third-line antituberculous agents used in the management of multidrug-resistant (MDR) and extensively-drug resistant (XDR) tuberculosis. Curr Clin Pharmacol 5:96–114
79. Via LE, Lin PL, Ray SM, Carrillo J, Allen SS, Eum SY, Taylor K, Klein E, Manjunatha U, Gonzales J, Lee EG, Park SK, Raleigh JA, Cho SN, McMurray DN, Flynn JL, Barry CE III (2008) Tuberculous granulomas are hypoxic in guinea pigs, rabbits, and nonhuman primates. Infect Immun 76:2333–2340
80. Sensi P (1983) History of the development of rifampin. Rev Infect Dis 5(Suppl 3):S402–406
81. Lancini G, Zanichelli W (1977) Structure-activity relationships in rifamycins. In: Perlman D (ed) Structure-activity relationships among the semisynthetic antibiotics. Academic, New York, pp 531–600
82. Campbell EA, Korzheva N, Mustaev A, Murakami K, Nair S, Goldfarb A, Darst SA (2001) Structural mechanism for rifampicin inhibition of bacterial RNA polymerase. Cell 104:901–912
83. Floss HG, Yu TW (2005) Rifamycin-mode of action, resistance, and biosynthesis. Chem Rev 105:621–632
84. Baysarowich J, Koteva K, Hughes DW, Ejim L, Griffiths E, Zhang K, Junop M, Wright GD (2008) Rifamycin antibiotic resistance by ADP-ribosylation: structure and diversity of Arr. Proc Natl Acad Sci USA 105:4886–4891
85. Li AP, Reith MK, Rasmussen A, Gorski JC, Hall SD, Xu L, Kaminski DL, Cheng LK (1997) Primary human hepatocytes as a tool for the evaluation of structure-activity relationship in cytochrome P450 induction potential of xenobiotics: evaluation of rifampin, rifapentine and rifabutin. Chem Biol Interact 107:17–30
86. Narita M, Stambaugh JJ, Hollender ES, Jones D, Pitchenik AE, Ashkin D (2000) Use of rifabutin with protease inhibitors for human immunodeficiency virus-infected patients with tuberculosis. Clin Infect Dis 30:779–783
87. Fox HH (1952) The chemical approach to the control of tuberculosis. Science 116:129–134
88. Kakimoto S, Tone I (1965) Antituberculous compounds. 23. Alkyl- and acylisonicotinic acid hydrazides. J Med Chem 8:868
89. Fox HH (1953) The chemical attack on tuberculosis. Trans NY Acad Sci 15:234–242
90. Rubbo SD, Edgar J, Vaughan G (1957) Chemotherapy of tuberculosis. I. Antituberculous activity of verazide and related hydrazones. Am Rev Tuberc 76:331–345
91. Rubbo SD, Rouch LC, Egan JB, Waddington AL, Tellesson WG (1958) Chemotherapy of tuberculosis. III. Verazide in the treatment of pulmonary tuberculosis. Am Rev Tuberc 78:251–258
92. Wayne LG, Sramek HA (1994) Metronidazole is bactericidal to dormant cells of *Mycobacterium tuberculosis*. Antimicrob Agents Chemother 38:2054–2058
93. Vilcheze C, Jacobs WR Jr (2007) The mechanism of isoniazid killing: clarity through the scope of genetics. Annu Rev Microbiol 61:35–50
94. Rozwarski DA, Grant GA, Barton DH, Jacobs WR Jr, Sacchettini JC (1998) Modification of the NADH of the isoniazid target (InhA) from *Mycobacterium tuberculosis*. Science 279:98–102
95. Rawat R, Whitty A, Tonge PJ (2003) The isoniazid-NAD adduct is a slow, tight-binding inhibitor of InhA, the *Mycobacterium tuberculosis* enoyl reductase: adduct affinity and drug resistance. Proc Natl Acad Sci USA 100:13881–13886

96. Argyrou A, Jin L, Siconilfi-Baez L, Angeletti RH, Blanchard JS (2006) Proteome-wide profiling of isoniazid targets in *Mycobacterium tuberculosis*. Biochemistry 45:13947–13953
97. Rist N, Grumbach F, Libermann D (1959) Experiments on the antituberculous activity of alpha-ethylthioisonicotinamide. Am Rev Tuberc 79:1–5
98. Wang F, Langley R, Gulten G, Dover LG, Besra GS, Jacobs WR Jr, Sacchettini JC (2007) Mechanism of thioamide drug action against tuberculosis and leprosy. J Exp Med 204:73–78
99. DeBarber AE, Mdluli K, Bosman M, Bekker LG, Barry CE III (2000) Ethionamide activation and sensitivity in multidrug-resistant *Mycobacterium tuberculosis*. Proc Natl Acad Sci USA 97:9677–9682
100. Domagk G (1950) Investigations on the antituberculous activity of the thiosemicarbazones in vitro and in vivo. Am Rev Tuberc 61:8–19
101. Behnisch R, Mietzsch F, Schmidt H (1950) Chemical studies on thiosemicarbazones with particular reference to antituberculous activity. Am Rev Tuberc 61:1–7
102. Alahari A, Trivelli X, Guerardel Y, Dover LG, Besra GS, Sacchettini JC, Reynolds RC, Coxon GD, Kremer L (2007) Thiacetazone, an antitubercular drug that inhibits cyclopropanation of cell wall mycolic acids in mycobacteria. PLoS ONE 2:e1343
103. Alahari A, Alibaud L, Trivelli X, Gupta R, Lamichhane G, Reynolds RC, Bishai WR, Guerardel Y, Kremer L (2009) Mycolic acid methyltransferase, MmaA4, is necessary for thiacetazone susceptibility in *Mycobacterium tuberculosis*. Mol Microbiol 71:1263–1277
104. Lawn SD, Frimpong EH, Acheampong JW (1999) Life-threatening cutaneous reactions to thiacetazone-containing antituberculosis treatment in Kumasi, Ghana. West Afr J Med 18:249–253
105. Mc Kenzie D, Malone L (1948) The effect of nicotinic acid amide on experimental tuberculosis of white mice. J Lab Clin Med 33:1249–1253
106. Rogers EF, Leanza WJ, Becker HJ, Matzuk AR, O'Neill RC, Basso AJ, Stein GA, Solotorovsky M, Gregory FJ, Pfister K III (1952) Antitubercular diazine carboxamides. Science 116:253–254
107. Kushner S, Dalalian H, Sanjurjo JL, Bach FL, Safir SR, Smith VK, Williams JH (1952) Experimental chemotherapy of tuberculosis. 2. The synthesis of pyrazinamides and related compounds. J Am Chem Soc 74:3617–3621
108. Felder E, Pitre D, Tiepolo U (1962) N-Morpholinomethyl-pyrazinamide: chemico-physical characteristics and determination in biological fluids. Minerva Med 53:1699–1703
109. Chung WJ, Kornilov A, Brodsky BH, Higgins M, Sanchez T, Heifets LB, Cynamon MH, Welch J (2008) Inhibition of *M. tuberculosis* in vitro in monocytes and in mice by aminomethylene pyrazinamide analogs. Tuberculosis (Edinb) 88:410–419
110. Scorpio A, Zhang Y (1996) Mutations in *pncA*, a gene encoding pyrazinamidase/nicotinamidase, cause resistance to the antituberculous drug pyrazinamide in tubercle bacillus. Nat Med 2:662–667
111. Boshoff HI, Mizrahi V (2000) Expression of *Mycobacterium smegmatis* pyrazinamidase in *Mycobacterium tuberculosis* confers hypersensitivity to pyrazinamide and related amides. J Bacteriol 182:5479–5485
112. Boshoff HI, Mizrahi V, Barry CE III (2002) Effects of pyrazinamide on fatty acid synthesis by whole mycobacterial cells and purified fatty acid synthase I. J Bacteriol 184: 2167–2172
113. Zimhony O, Cox JS, Welch JT, Vilcheze C, Jacobs WR Jr (2000) Pyrazinamide inhibits the eukaryotic-like fatty acid synthetase I (FASI) of *Mycobacterium tuberculosis*. Nat Med 6:1043–1047
114. Epstein IG, Nair KG, Boyd LJ (1956) The treatment of human pulmonary tuberculosis with cycloserine: progress report. Dis Chest 29:241–257
115. Kim MG, Strych U, Krause K, Benedik M, Kohn H (2003) N(2)-substituted D, L-cycloserine derivatives: synthesis and evaluation as alanine racemase inhibitors. J Antibiot (Tokyo) 56:160–168
116. Neuhaus F (1967) D-cycloserine and O-carbamoyl-D-serine. In: Gottlieb D, Shaw P (eds) Antibiotics I (mode of action) pp 40-83. Springer-Verlag, New York

117. Peisach D, Chipman DM, Van Ophem PW, Manning JM, Ringe D (1998) D-cycloserine inactivation of D-amino acid aminotransferase leads to a stable noncovalent protein complex with an aromatic cycloserine-PLP derivative. J Am Chem Soc 120:2268–2274
118. Feng Z, Barletta RG (2003) Roles of *Mycobacterium smegmatis* D-alanine:D-alanine ligase and D-alanine racemase in the mechanisms of action of and resistance to the peptidoglycan inhibitor D-cycloserine. Antimicrob Agents Chemother 47:283–291
119. Lehmann J (1949) The treatment of tuberculosis in Sweden with *para*-aminosalicylic acid: a review. Dis Chest 16:684–703, illust
120. Doub L, Schaefer JJ, Bambas LL, Walker CT (1951) Some derivatives of 4-amino-2-hydroxybenzoic acid (*para*-aminosalicylic acid). J Am Chem Soc 73:903–906
121. Rengarajan J, Sassetti CM, Naroditskaya V, Sloutsky A, Bloom BR, Rubin EJ (2004) The folate pathway is a target for resistance to the drug *para*-aminosalicylic acid (PAS) in mycobacteria. Mol Microbiol 53:275–282
122. Nomoto S, Shiba T (1977) Chemical studies on tuberactinomycin. XIII. Modification of beta-ureidodehydroalanine residue in tuberactinomycin N. J Antibiot (Tokyo) 30: 1008–1011
123. Linde RG, Birsner NC, Chandrasekaran RY, Clancy J, Howe RJ, Lyssikatos JP, MacLelland CP, Magee TV, Petitpas JW, Rainville JP, Su WG, Vu CB, Whipple DA (1997) Cyclic homopentapeptides.3. Synthetic modifications to the capreomycins and tuberactinomycins: compounds with activity against methicillin-resistant Staphylococcus aureus and vancomycin-resistant Enterococci. Bioorg Med Chem Lett 7:1149–1152
124. Stanley RE, Blaha G, Grodzicki RL, Strickler MD, Steitz TA (2010) The structures of the anti-tuberculosis antibiotics viomycin and capreomycin bound to the 70S ribosome. Nat Struct Mol Biol 17:289–293
125. Kotra LP, Haddad J, Mobashery S (2000) Aminoglycosides: perspectives on mechanisms of action and resistance and strategies to counter resistance. Antimicrob Agents Chemother 44:3249–3256
126. Ramaswamy S, Musser JM (1998) Molecular genetic basis of antimicrobial agent resistance in *Mycobacterium tuberculosis*: 1998 update. Tuber Lung Dis 79:3–29
127. Bottger EC, Springer B (2008) Tuberculosis: drug resistance, fitness, and strategies for global control. Eur J Pediatr 167:141–148
128. Suzuki Y, Katsukawa C, Tamaru A, Abe C, Makino M, Mizuguchi Y, Taniguchi H (1998) Detection of kanamycin-resistant *Mycobacterium tuberculosis* by identifying mutations in the 16S rRNA gene. J Clin Microbiol 36:1220–1225
129. Okamoto S, Tamaru A, Nakajima C, Nishimura K, Tanaka Y, Tokuyama S, Suzuki Y, Ochi K (2007) Loss of a conserved 7-methylguanosine modification in 16S rRNA confers low-level streptomycin resistance in bacteria. Mol Microbiol 63:1096–1106
130. Magnet S, Blanchard JS (2005) Molecular insights into aminoglycoside action and resistance. Chem Rev 105:477–498
131. Zaunbrecher MA, Sikes RD Jr, Metchock B, Shinnick TM, Posey JE (2009) Overexpression of the chromosomally encoded aminoglycoside acetyltransferase eis confers kanamycin resistance in *Mycobacterium tuberculosis*. Proc Natl Acad Sci USA 106:20004–20009
132. Ramon-Garcia S, Martin C, De Rossi E, Ainsa JA (2007) Contribution of the Rv2333c efflux pump (the Stp protein) from *Mycobacterium tuberculosis* to intrinsic antibiotic resistance in *Mycobacterium bovis* BCG. J Antimicrob Chemother 59:544–547
133. Peloquin CA, Berning SE, Nitta AT, Simone PM, Goble M, Huitt GA, Iseman MD, Cook JL, Curran-Everett D (2004) Aminoglycoside toxicity: daily versus thrice-weekly dosing for treatment of mycobacterial diseases. Clin Infect Dis 38:1538–1544
134. (2008) Handbook of anti-tuberculosis agents. Introduction. Tuberculosis (Edinb) 88:85–86
135. Lamp KC, Freeman CD, Klutman NE, Lacy MK (1999) Pharmacokinetics and pharmacodynamics of the nitroimidazole antimicrobials. Clin Pharmacokinet 36:353–373
136. Barry CE III, Boshoff HI, Dowd CS (2004) Prospects for clinical introduction of nitroimidazole antibiotics for the treatment of tuberculosis. Curr Pharm Des 10:3239–3262

137. Cavalleri B, Ballotta R, Arioli V, Lancini G (1973) New 5-substituted 1-alkyl-2-nitroimidazoles. J Med Chem 16:557–560
138. Roe FJ (1977) Metronidazole: review of uses and toxicity. J Antimicrob Chemother 3:205–212
139. Bendesky A, Menendez D, Ostrosky-Wegman P (2002) Is metronidazole carcinogenic? Mutat Res 511:133–144
140. Ashtekar DR, Costa-Perira R, Nagrajan K, Vishvanathan N, Bhatt AD, Rittel W (1993) In vitro and in vivo activities of the nitroimidazole CGI 17341 against *Mycobacterium tuberculosis*. Antimicrob Agents Chemother 37:183–186
141. Barry CE III, Blanchard JS (2010) The chemical biology of new drugs in the development for tuberculosis. Curr Opin Chem Biol 14:456–466
142. Stover CK, Warrener P, VanDevanter DR, Sherman DR, Arain TM, Langhorne MH, Anderson SW, Towell JA, Yuan Y, McMurray DN, Kreiswirth BN, Barry CE, Baker WR (2000) A small-molecule nitroimidazopyran drug candidate for the treatment of tuberculosis. Nature 405:962–966
143. Matsumoto M, Hashizume H, Tomishige T, Kawasaki M, Tsubouchi H, Sasaki H, Shimokawa Y, Komatsu M (2006) OPC-67683, a nitro-dihydro-imidazooxazole derivative with promising action against tuberculosis in vitro and in mice. PLoS Med 3:e466
144. Papadopoulou MV, Bloomer WD, McNeil MR (2007) NLCQ-1 and NLCQ-2, two new agents with activity against dormant *Mycobacterium tuberculosis*. Int J Antimicrob Agents 29:724–727
145. Kim P, Zhang L, Manjunatha UH, Singh R, Patel S, Jiricek J, Keller TH, Boshoff HI, Barry CE III, Dowd CS (2009) Structure-activity relationships of antitubercular nitroimidazoles. 1. Structural features associated with aerobic and anaerobic activities of 4- and 5-nitroimidazoles. J Med Chem 52:1317–1328
146. Palmer BD, Thompson AM, Sutherland HS, Blaser A, Kmentova I, Franzblau SG, Wan B, Wang Y, Ma Z, Denny WA (2010) Synthesis and structure-activity studies of biphenyl analogues of the tuberculosis drug (6S)-2-nitro-6-{[4-(trifluoromethoxy)benzyl]oxy}-6,7-dihydro-5H-imidazo[2, 1-*b*][1,3]oxazine (PA-824). J Med Chem 53:282–294
147. Sutherland HS, Blaser A, Kmentova I, Franzblau SG, Wan B, Wang Y, Ma Z, Palmer BD, Denny WA, Thompson AM (2010) Synthesis and structure-activity relationships of antitubercular 2-nitroimidazooxazines bearing heterocyclic side chains. J Med Chem 53:855–866
148. Kim P, Kang S, Boshoff HI, Jiricek J, Collins M, Singh R, Manjunatha UH, Niyomrattanakit P, Zhang L, Goodwin M, Dick T, Keller TH, Dowd CS, Barry CE III (2009) Structure-activity relationships of antitubercular nitroimidazoles. 2. Determinants of aerobic activity and quantitative structure-activity relationships. J Med Chem 52:1329–1344
149. Singh R, Manjunatha U, Boshoff HI, Ha YH, Niyomrattanakit P, Ledwidge R, Dowd CS, Lee IY, Kim P, Zhang L, Kang S, Keller TH, Jiricek J, Barry CE III (2008) PA-824 kills nonreplicating *Mycobacterium tuberculosis* by intracellular NO release. Science 322: 1392–1395
150. Manjunatha UH, Boshoff H, Dowd CS, Zhang L, Albert TJ, Norton JE, Daniels L, Dick T, Pang SS, Barry CE III (2006) Identification of a nitroimidazo-oxazine-specific protein involved in PA-824 resistance in *Mycobacterium tuberculosis*. Proc Natl Acad Sci USA 103:431–436
151. (2008) Pa-824. Tuberculosis (Edinb) 88:134–136
152. Manjunatha U, Boshoff HI, Barry CE (2009) The mechanism of action of PA-824: novel insights from transcriptional profiling. Commun Integr Biol 2:215–218
153. Tyagi S, Nuermberger E, Yoshimatsu T, Williams K, Rosenthal I, Lounis N, Bishai W, Grosset J (2005) Bactericidal activity of the nitroimidazopyran PA-824 in a murine model of tuberculosis. Antimicrob Agents Chemother 49:2289–2293
154. Nuermberger E, Tyagi S, Tasneen R, Williams KN, Almeida D, Rosenthal I, Grosset JH (2008) Powerful bactericidal and sterilizing activity of a regimen containing PA-824, moxifloxacin, and pyrazinamide in a murine model of tuberculosis. Antimicrob Agents Chemother 52:1522–1524

155. Ginsberg AM, Laurenzi MW, Rouse DJ, Whitney KD, Spigelman MK (2009) Safety, tolerability, and pharmacokinetics of PA-824 in healthy subjects. Antimicrob Agents Chemother 53:3720–3725
156. Ginsberg AM, Laurenzi MW, Rouse DJ, Whitney KD, Spigelman MK (2009) Assessment of the effects of the nitroimidazo-oxazine PA-824 on renal function in healthy subjects. Antimicrob Agents Chemother 53:3726–3733
157. Dawson R, Diacon A (2010) A phase IIa trial to evaluate the safety, tolerability, extended early bactericidal activity and pharmacokinetics of 14 days' treatment with four oral doses of PA-824 in adult participants with newly diagnosed, uncomplicated, smear-positive, pulmonary tuberculosis. In: PA-824-CL-007: phase IIa evaluation of early bactericidal activity in pulmonary tuberculosis. Clinical Trial NCT00567840. Sponsored by the Global Alliance for TB Drug Development. Accessed from National Library of Medicine and National Institutes of Health (US), ClinicalTrials.gov
158. Diacon AH, Dawson R, Hanekom M, Narunsky K, Maritz SJ, Venter A, Donald PR, van Niekerk C, Whitney K, Rouse DJ, Laurenzi MW, Ginsberg AM, Spigelman MK (2010) Early bactericidal activity and pharmacokinetics of PA-824 in smear-positive tuberculosis patients. Antimicrob Agents Chemother 54:3402–3407
159. Sasaki H, Haraguchi Y, Itotani M, Kuroda H, Hashizume H, Tomishige T, Kawasaki M, Matsumoto M, Komatsu M, Tsubouchi H (2006) Synthesis and antituberculosis activity of a novel series of optically active 6-nitro-2,3-dihydroimidazo[2,1-*b*]oxazoles. J Med Chem 49:7854–7860
160. Saliu OY, Crismale C, Schwander SK, Wallis RS (2007) Bactericidal activity of OPC-67683 against drug-tolerant *Mycobacterium tuberculosis*. J Antimicrob Chemother 60:994–998
161. van den Boogaard J, Kibiki GS, Kisanga ER, Boeree MJ, Aarnoutse RE (2009) New drugs against tuberculosis: problems, progress, and evaluation of agents in clinical development. Antimicrob Agents Chemother 53:849–862
162. Rivers EC, Mancera RL (2008) New anti-tuberculosis drugs with novel mechanisms of action. Curr Med Chem 15:1956–1967
163. Diacon, AH, Rustomjee, R, Dawson R (2007) A phase 2, multi-center, non-controlled, open-label dose escalation trial to assess the safety, tolerability, pharmacokinetics, and efficacy of orally administered OPC-67683 two times daily to patients with pulmonary multidrug-resistant tuberculosis refractory to conventional treatment. In: Safety and pharmacokinetics (PK) in multidrug-resistant (MDR) refractive tuberculosis. Clinical Trial NCT01131351. Sponsored by Otsuka Pharmaceutical Development and Commercialization, Inc. Accessed from National Library of Medicine and National Institutes of Health (US), ClinicalTrials.gov
164. Andries K, Verhasselt P, Guillemont J, Gohlmann HW, Neefs JM, Winkler H, Van Gestel J, Timmerman P, Zhu M, Lee E, Williams P, de Chaffoy D, Huitric E, Hoffner S, Cambau E, Truffot-Pernot C, Lounis N, Jarlier V (2005) A diarylquinoline drug active on the ATP synthase of *Mycobacterium tuberculosis*. Science 307:223–227
165. Matteelli A, Carvalho AC, Dooley KE, Kritski A (2010) TMC207: the first compound of a new class of potent anti-tuberculosis drugs. Future Microbiol 5:849–858
166. Arjona A (2008) TMC-207 mycobacterial ATP synthase inhibitor treatment of tuberculosis. Drugs Future 33:1018–1024
167. Upadhayaya RS, Vandavasi JK, Kardile RA, Lahore SV, Dixit SS, Deokar HS, Shinde PD, Sarmah MP, Chattopadhyaya J (2010) Novel quinoline and naphthalene derivatives as potent antimycobacterial agents. Eur J Med Chem 45:1854–1867
168. Upadhayaya RS, Vandavasi JK, Vasireddy NR, Sharma V, Dixit SS, Chattopadhyaya J (2009) Design, synthesis, biological evaluation and molecular modelling studies of novel quinoline derivatives against *Mycobacterium tuberculosis*. Bioorg Med Chem 17:2830–2841
169. Gaurrand S, Desjardins S, Meyer C, Bonnet P, Argoullon JM, Oulyadi H, Guillemont J (2006) Conformational analysis of r207910, a new drug candidate for the treatment of tuberculosis, by a combined NMR and molecular modeling approach. Chem Biol Drug Des 68:77–84

170. Petit S, Coquerel G, Meyer C, Guillemont J (2007) Absolute configuration and structural features of R207910, a novel anti-tuberculosis agent. J Mol Struct 837:252–256
171. Andries, KJLM, Van Gestel, JFE (2005) Use of substituted quinoline derivatives for the treatment of drug resistant mycobacterial diseases. Johnson & Johnson: United States. WO2005117875
172. Lounis N, Guillemont J, Veziris N, Koul A, Jarlier V, Andries K (2010) R207910 (TMC207): a new antibiotic for the treatment of tuberculosis. Méd Mal Infect 40:383–390
173. Rustomjee R, Diacon AH, Allen J, Venter A, Reddy C, Patientia RF, Mthiyane TC, De Marez T, van Heeswijk R, Kerstens R, Koul A, De Beule K, Donald PR, McNeeley DF (2008) Early bactericidal activity and pharmacokinetics of the diarylquinoline TMC207 in treatment of pulmonary tuberculosis. Antimicrob Agents Chemother 52:2831–2835
174. Petrella S, Cambau E, Chauffour A, Andries K, Jarlier V, Sougakoff W (2006) Genetic basis for natural and acquired resistance to the diarylquinoline R207910 in mycobacteria. Antimicrob Agents Chemother 50:2853–2856
175. Koul A, Dendouga N, Vergauwen K, Molenberghs B, Vranckx L, Willebrords R, Ristic Z, Lill H, Dorange I, Guillemont J, Bald D, Andries K (2007) Diarylquinolines target subunit c of mycobacterial ATP synthase. Nat Chem Biol 3:323–324
176. de Jonge MR, Koymans LH, Guillemont JE, Koul A, Andries K (2007) A computational model of the inhibition of Mycobacterium tuberculosis ATPase by a new drug candidate R207910. Proteins 67:971–980
177. Lounis N, Veziris N, Chauffour A, Truffot-Pernot C, Andries K, Jarlier V (2006) Combinations of R207910 with drugs used to treat multidrug-resistant tuberculosis have the potential to shorten treatment duration. Antimicrob Agents Chemother 50:3543–3547
178. Ibrahim M, Truffot-Pernot C, Andries K, Jarlier V, Veziris N (2009) Sterilizing activity of R207910 (TMC207)-containing regimens in the murine model of tuberculosis. Am J Respir Crit Care Med 180:553–557
179. Lenaerts AJ, Hoff D, Aly S, Ehlers S, Andries K, Cantarero L, Orme IM, Basaraba RJ (2007) Location of persisting mycobacteria in a Guinea pig model of tuberculosis revealed by r207910. Antimicrob Agents Chemother 51:3338–3345
180. Koul A, Vranckx L, Dendouga N, Balemans W, Van den Wyngaert I, Vergauwen K, Gohlmann HW, Willebrords R, Poncelet A, Guillemont J, Bald D, Andries K (2008) Diarylquinolines are bactericidal for dormant mycobacteria as a result of disturbed ATP homeostasis. J Biol Chem 283:25273–25280
181. Veziris N, Ibrahim M, Lounis N, Chauffour A, Truffot-Pernot C, Andries K, Jarlier V (2009) A once-weekly R207910-containing regimen exceeds activity of the standard daily regimen in murine tuberculosis. Am J Respir Crit Care Med 179:75–79
182. Lounis N, Gevers T, Van Den Berg J, Andries K (2008) Impact of the interaction of R207910 with rifampin on the treatment of tuberculosis studied in the mouse model. Antimicrob Agents Chemother 52:3568–3572
183. Diacon AH, Pym A, Grobusch M, Patientia R, Rustomjee R, Page-Shipp L, Pistorius C, Krause R, Bogoshi M, Churchyard G, Venter A, Allen J, Palomino JC, De Marez T, van Heeswijk RP, Lounis N, Meyvisch P, Verbeeck J, Parys W, de Beule K, Andries K, Mc Neeley DF (2009) The diarylquinoline TMC207 for multidrug-resistant tuberculosis. N Engl J Med 360:2397–2405
184. Barry CE III (2009) Unorthodox approach to the development of a new antituberculosis therapy. N Engl J Med 360:2466–2467
185. Tibotec-Virco Virology BVBA Clinical Trial (2009) A Phase II, Open-Label Trial With TMC207 as Part of a Multi-drug Resistant Tuberculosis (MDR-TB) Treatment Regimen in Subjects with Sputum Smear-positive Pulmonary Infection with MDR-TB. Clinical Trial NCT00910871. Sponsored by Tibotec BVBA. Accessed from National Library of Medicine and National Institutes of Health (US), ClinicalTrials.gov
186. Webb S (2009) Public-private partnership tackles TB challenges in parallel. Nat Rev Drug Discov 8:599–600

187. Gregory WA (1984) *p*-Oxazolidinylbenzene compounds as antibacterial agents. E. I. DuPont de Nemours and Company: United States US4461772
188. Zurenko GE, Yagi BH, Schaadt RD, Allison JW, Kilburn JO, Glickman SE, Hutchinson DK, Barbachyn MR, Brickner SJ (1996) In vitro activities of U-100592 and U-100766, novel oxazolidinone antibacterial agents. Antimicrob Agents Chemother 40:839–845
189. Ford CW, Hamel JC, Wilson DM, Moerman JK, Stapert D, Yancey RJ, Hutchinson DK, Barbachyn MR, Brickner SJ (1996) In vivo activities of U-100592 and U-100766, novel oxazolidinone antimicrobial agents, against experimental bacterial infections. Antimicrob Agents Chemother 40:1508–1513
190. Jones RN, Johnson DM, Erwin ME (1996) In vitro antimicrobial activities and spectra of U-100592 and U-100766, two novel fluorinated oxazolidinones. Antimicrob Agents Chemother 40:720–726
191. Barbachyn MR, Brickner SI, Hutchinson DK (1997) Substituted oxazine and thiazine oxazolidinone antimicrobials. Pharmacia & Upjohn Co.: United States US199713
192. Pfizer, Inc. (2009) A Phase 1, Double-Blind, Randomized, Placebo-Controlled Study To Evaluate The Safety, Tolerability, Pharmacokinetics And Pharmacodynamics Of PNU-100480 (PF-02341272) After Administration Of Multiple Escalating Oral Doses To Healthy Adult Subjects. Clinical Trial NCT 0990990 Sponsored by Pfizer. Accessed from National Library of Medicine and National Institutes of Health (US), ClinicalTrials.gov
193. Pfizer, Inc. (2009) A Phase 1 Study To Evaluate The Safety, Tolerability, And Pharmacokinetics Of PNU-100480 (PF-02341272) After First Time Administration Of Ascending Oral Doses To Healthy Adult Subjects Under Fed And Fasted Conditions. Clinical Trial NCT00871949 Sponsored by Pfizer. Accessed from National Library of Medicine and National Institutes of Health (US), ClinicalTrials.gov
194. Bae SK, Chung WS, Kim EJ, Rhee JK, Kwon JW, Kim WB, Lee MG (2004) Pharmacokinetics of DA-7867, a new oxazolidinone, after intravenous or oral administration to rats: Intestinal first-pass effect. Antimicrob Agents Chemother 48:659–662
195. Rudra S, Yadav A, Rao A, Srinivas A, Pandya M, Bhateia P, Mathur T, Malhotra S, Rattan A, Salman M, Mehta A, Cliffe IA, Das B (2007) Synthesis and antibacterial activity of potent heterocyclic oxazolidinones and the identification of RBx 8700. Bioorg Med Chem Lett 17:6714–6719
196. Wookey A, Turner PJ, Greenhalgh JM, Eastwood M, Clarke J, Sefton C (2004) AZD2563, a novel oxazolidinone: definition of antibacterial spectrum, assessment of bactericidal potential and the impact of miscellaneous factors on activity in vitro. Clin Microbiol Infect 10:247–254
197. Meier PA (2009) A Phase-1, Single Center, Double-blind, Randomized, Placebo-controlled, Parallel-group Study to Assess the Safety, Tolerability and Pharmacokinetics (Including Food Effect) of Ascending Oral Doses of AZD5847 in Healthy Male Subjects and Female Subjects of Non-childbearing Potential. Clinical Trial NCT01037725. Sponsored by AstraZeneca. Accessed from National Library of Medicine and National Institutes of Health (US), ClinicalTrials.gov
198. Shaw A (2010) A Phase I, Single-center, Double-blind, Randomized, Placebo-controlled, Parallel-group, Multiple Ascending Dose Study to Assess the Safety, Tolerability, and Pharmacokinetics of AZD5847 Following Oral Administration to Healthy Male Subjects and Female Subjects of Non-childbearing Potential. Clinical Trial NCT0116258. Sponsored by AstraZeneca. Accessed from National Library of Medicine and National Institutes of Health (US), ClinicalTrials.gov
199. Gregory WA, Brittelli DR, Wang CLJ, Wuonola MA, McRipley RJ, Eustice DC, Eberly VS, Bartholomew PT, Slee AM, Forbes M (1989) Antibacterials – synthesis and structure activity studies of 3-aryl-2-oxooxazolidines.1. The B-group. J Med Chem 32: 1673–1681
200. Gregory WA, Brittelli DR, Wang CLJ, Kezar HS, Carlson RK, Park CH, Corless PF, Miller SJ, Rajagopalan P, Wuonola MA, McRipley RJ, Eberly VS, Slee AM, Forbes M (1990)

Antibacterials – synthesis and structure activity studies of 3-aryl-2-oxooxazolidines. 2. The A group. J Med Chem 33:2569–2578
201. Park CH, Brittelli DR, Wang CLJ, Marsh FD, Gregory WA, Wuonola MA, McRipley RJ, Eberly VS, Slee AM, Forbes M (1992) Antibacterials – synthesis and structure activity studies of 3-aryl-2-oxooxazolidines. 4. Multiply-substituted aryl derivatives. J Med Chem 35:1156–1165
202. Barbachyn MR, Ford CW (2003) Oxazolidinone structure-activity relationships leading to linezolid. Angew Chem Int Ed 42:2010–2023
203. Selvakumar N, Rajulu GG, Reddy KCS, Chary BC, Kumar PK, Madhavi T, Praveena K, Reddy KHP, Takhi M, Mallick A, Amarnath PVS, Kandepu S, Iqbal J (2008) Synthesis, SAR, and antibacterial activity of novel oxazolidinone analogues possessing urea functionality. Bioorg Med Chem Lett 18:856–860
204. Alcala L, Ruiz-Serrano MJ, Turegano CPF, Garcia de Viedma D, Diaz-Infantes M, Marin-Arriaza M, Bouza E (2003) In vitro activities of linezolid against clinical isolates of *Mycobacterium tuberculosis* that are susceptible or resistant to first-line antituberculous drugs. Antimicrob Agents Chemother 47:416–417
205. Rao M, Sood R, Malhotra S, Fatma T, Upadhyay DJ, Rattan A (2006) In vitro bactericidal activity of oxazolidinone, RBx 8700 against *Mycobacterium tuberculosis* and *Mycobacterium avium* complex. J Chemother 18:144–150
206. Sood R, Rao M, Singhal S, Rattan A (2005) Activity of RBx 7644 and RBx 8700, new investigational oxazolidinones, against *Mycobacterium tuberculosis* infected murine macrophages. Int J Antimicrob Agents 25:464–468
207. Vera-Cabrera L, Gonzalez E, Rendon A, Ocampo-Candiani J, Welsh O, Velazquez-Moreno VM, Choi SH, Molina-Torres C (2006) In vitro activities of DA-7157 and DA-7218 against *Mycobacterium tuberculosis* and *Nocardia brasiliensis*. Antimicrob Agents Chemother 50:3170–3172
208. Lin AH, Murray RW, Vidmar TJ, Marotti KR (1997) The oxazolidinone eperezolid binds to the 50S ribosomal subunit and competes with binding of chloramphenicol and lincomycin. Antimicrob Agents Chemother 41:2127–2131
209. Aoki H, Ke LZ, Poppe SM, Poel TJ, Weaver EA, Gadwood RC, Thomas RC, Shinabarger DL, Ganoza MC (2002) Oxazolidinone antibiotics target the P site on *Escherichia coli* ribosomes. Antimicrob Agents Chemother 46:1080–1085
210. Shinabarger DL, Marotti KR, Murray RW, Lin AH, Melchior EP, Swaney SM, Dunyak DS, Demyan WF, Buysse JM (1997) Mechanism of action of oxazolidinones: effects of linezolid and eperezolid on translation reactions. Antimicrob Agents Chemother 41: 2132–2136
211. Wilson DN, Schluenzen F, Harms JM, Starosta AL, Connell SR, Fucini P (2008) The oxazolidinone antibiotics perturb the ribosomal peptidyl-transferase center and effect tRNA positioning. Proc Natl Acad Sci USA 105:13339–13344
212. Hillemann D, Rusch-Gerdes S, Richter E (2008) In vitro-selected linezolid-resistant *Mycobacterium tuberculosis* mutants. Antimicrob Agents Chemother 52:800–801
213. Richter E, Rusch-Gerdes S, Hillemann D (2007) First linezolid-resistant clinical isolates of *Mycobacterium tuberculosis*. Antimicrob Agents Chemother 51:1534–1536
214. Escribano I, Rodriguez C, Llorca B, Garcia-Pachon E, Ruiz M, Royo G (2007) Importance of the efflux pump systems in the resistance of *Mycobacterium tuberculosis* to Fluoroquinolones and linezolid. Chemotherapy 53:397–401
215. Erturan Z, Uzun M (2005) In vitro activity of linezolid against multidrug-resistant *Mycobacterium tuberculosis* isolates. Int J Antimicrob Agents 26:78–80
216. Prammananan T, Chaiprasert A, Leechawengwongs M (2009) In vitro activity of linezolid against multidrug-resistant tuberculosis (MDR-TB) and extensively drug-resistant (XDR)-TB isolates. Int J Antimicrob Agents 33:190–191
217. Schaaf HS, Willemse M, Donald PR (2009) Long-term linezolid treatment in a young child with extensively drug-resistant tuberculosis. Pediatr Infect Dis J 28:748–750

218. von der Lippe B, Sandven P, Brubakk O (2006) Efficacy and safety of linezolid in multidrug resistant tuberculosis (MDR-TB) – a report of ten cases. J Infect 52:92–96
219. Welshman IR, Sisson TA, Jungbluth GL, Stalker DJ, Hopkins NK (2001) Linezolid absolute bioavailability and the effect of food on oral bioavailability. Biopharm Drug Dispos 22:91–97
220. Slatter JG, Stalker DJ, Feenstra KL, Welshman IR, Bruss JB, Sams JP, Johnson MG, Sanders PE, Hauer MJ, Fagerness PE, Stryd RP, Peng GW, Shobe EM (2001) Pharmacokinetics, metabolism, and excretion of linezolid following an oral dose of C-14 linezolid to healthy human subjects. Drug Metab Dispos 29:1136–1145
221. Metaxas EI, Falagas ME (2009) Update on the safety of linezolid. Expert Opin Drug Saf 8:485–491
222. Diaz JCR, Ruiz M, Lopez M, Royo G (2003) Synergic activity of fluoroquinolones and linezolid against *Mycobacterium tuberculosis*. Int J Antimicrob Agents 21:354–356
223. Gerson SL, Kaplan SL, Bruss JB, Le V, Arellano FM, Hafkin B, Kuter DJ (2002) Hematologic effects of linezolid: summary of clinical experience. Antimicrob Agents Chemother 46:2723–2726
224. Eker B, Ortmann J, Migliori GB, Sotgiu G, Muetterlein R, Centis R, Hoffmann H, Kirsten D, Schaberg T, Ruesch-Gerdes S, Lange C, German TG (2008) Multidrug- and extensively drug-resistant tuberculosis, Germany. Emerg Infect Dis 14:1700–1706
225. Lesher GY, Froelich EJ, Gruett MD, Bailey JH, Brundage RP (1962) 1,8-Naphthyridine derivatives. A new class of chemotherapeutic agents. J Med Pharm Chem 91:1063–1065
226. Mitscher LA (2005) Bacterial topoisomerase inhibitors: quinolone and pyridone antibacterial agents. Chem Rev 105:559–592
227. Nakamura S, Nakata K, Katae H, Minami A, Kashimoto S, Yamagishi J, Takase Y, Shimizu M (1983) Activity of AT-2266 compared with those of norfloxacin, pipemidic acid, nalidixic acid, and gentamicin against various experimental infections in mice. Antimicrob Agents Chemother 23:742–749
228. Koga H, Itoh A, Murayama S, Suzue S, Irikura T (1980) Structure-activity relationships of antibacterial 6,7- and 7,8-disubstituted 1-alkyl-1,4-dihydro-4-oxoquinoline-3-carboxylic acids. J Med Chem 23:1358–1363
229. De Souza MV (2005) New fluoroquinolones: a class of potent antibiotics. Mini Rev Med Chem 5:1009–1017
230. Wise R (1984) Norfloxacin – a review of pharmacology and tissue penetration. J Antimicrob Chemother 13(Suppl B):59–64
231. Grohe DK, Zeiler H-j, Metzger KG (1984) 7-amino-1-cyclopropyl-4-oxo-1,4-dihydronaphthyridine-3-carboxylic acids, process for their preparation and pharmaceutical compositions containing them. Bayer AG: Germany. EP0049355
232. Tanaka Y, Hayakawa I, Hiramitsu T (1982) 1,8-cyclic substituted quinoline derivative. Dai Ichi Seiyaku Co Ltd: Japan; JP57203085
233. Grohe KO, Zeiler H-j, Metzger KG (1991) 7-amino-1-cyclopropyl-4-oxo-1,4-dihydroquinoline- and naphthyridine-3-carboxylic acids, processes for their preparation and antibacterial agents containing these compounds. Bayer AG: United States. US5077429
234. Liu HH (2010) Safety profile of the fluoroquinolones: focus on levofloxacin. Drug Saf 33:353–369
235. Mandell L, Tillotson G (2002) Safety of fluoroquinolones: an update. Can J Infect Dis 13:54–61
236. Hong CY, Kim YK, Chang JH, Kim SH, Choi H, Nam D, Kwak J-H, Jeong YN, Oh JI, Kim MY (1998) Quinoline carboxylic acid derivatives having 7-(4-amino-methyl-3-oxime) pyrrolidine substituent and processes for preparing thereof. LG Chemical Ltd, Seoul
237. Hayakawa IC, Atarashi S, Imamura M, Kimura Y (1994) Spiro compound. Daiichi Seiyaku Co., Ltd, Tokyo
238. Bax BD, Chan PF, Eggleston DS, Fosberry A, Gentry DR, Gorrec F, Giordano I, Hann MM, Hennessy A, Hibbs M, Huang J, Jones E, Jones J, Brown KK, Lewis CJ, May EW, Saunders MR, Singh O, Spitzfaden CE, Shen C, Shillings A, Theobald AJ, Wohlkonig A, Pearson ND,

Gwynn MN (2010) Type IIA topoisomerase inhibition by a new class of antibacterial agents. Nature 466:935–940
239. Black MT, Stachyra T, Platel D, Girard AM, Claudon M, Bruneau JM, Miossec C (2008) Mechanism of action of the antibiotic NXL101, a novel nonfluoroquinolone inhibitor of bacterial type II topoisomerases. Antimicrob Agents Chemother 52:3339–3349
240. Gomez L, Hack MD, Wu J, Wiener JJ, Venkatesan H, Santillan A Jr, Pippel DJ, Mani N, Morrow BJ, Motley ST, Shaw KJ, Wolin R, Grice CA, Jones TK (2007) Novel pyrazole derivatives as potent inhibitors of type II topoisomerases. Part 1: synthesis and preliminary SAR analysis. Bioorg Med Chem Lett 17:2723–2727
241. Wiener JJ, Gomez L, Venkatesan H, Santillan A Jr, Allison BD, Schwarz KL, Shinde S, Tang L, Hack MD, Morrow BJ, Motley ST, Goldschmidt RM, Shaw KJ, Jones TK, Grice CA (2007) Tetrahydroindazole inhibitors of bacterial type II topoisomerases. Part 2: SAR development and potency against multidrug-resistant strains. Bioorg Med Chem Lett 17:2718–2722
242. Domagala JM (1994) Structure-activity and structure-side-effect relationships for the quinolone antibacterials. J Antimicrob Chemother 33:685–706
243. Rubinstein E (2001) History of quinolones and their side effects. Chemotherapy 47(Suppl 3):3–8, discussion 44–48
244. Gootz TD, Brighty KE (1996) Fluoroquinolone antibacterials: SAR mechanism of action, resistance, and clinical aspects. Med Res Rev 16:433–486
245. Bolon MK (2009) The newer fluoroquinolones. Infect Dis Clin North Am 23:1027–1051
246. Gellert M, O'Dea MH, Itoh T, Tomizawa J (1976) Novobiocin and coumermycin inhibit DNA supercoiling catalyzed by DNA gyrase. Proc Natl Acad Sci USA 73:4474–4478
247. Takei M, Fukuda H, Kishii R, Hosaka M (2001) Target preference of 15 quinolones against Staphylococcus aureus, based on antibacterial activities and target inhibition. Antimicrob Agents Chemother 45:3544–3547
248. (2008) Moxifloxacin. Tuberculosis (Edinb) 88:127–131
249. Lougheed KE, Taylor DL, Osborne SA, Bryans JS, Buxton RS (2009) New anti-tuberculosis agents amongst known drugs. Tuberculosis (Edinb) 89:364–370
250. Berning SE (2001) The role of fluoroquinolones in tuberculosis today. Drugs 61:9–18
251. (2008) Gatifloxacin. Tuberculosis (Edinb) 88:109–111
252. Wise R, Honeybourne D (1999) Pharmacokinetics and pharmacodynamics of fluoroquinolones in the respiratory tract. Eur Respir J 14:221–229
253. (2008) Levofloxacin. Tuberculosis (Edinb) 88:119–121
254. Lalloo UG, Ambaram A (2010) New antituberculous drugs in development. Curr HIV AIDS Rep 7:143–151
255. Ma Z, Lienhardt C, McIlleron H, Nunn AJ, Wang X (2010) Global tuberculosis drug development pipeline: the need and the reality. Lancet 375:2100–2109
256. Alvirez-Freites EJ, Carter JL, Cynamon MH (2002) In vitro and in vivo activities of gatifloxacin against *Mycobacterium tuberculosis*. Antimicrob Agents Chemother 46: 1022–1025
257. Poissy J, Aubry A, Fernandez C, Lott MC, Chauffour A, Jarlier V, Farinotti R, Veziris N (2010) Should moxifloxacin be used for the treatment of XDR-TB? An answer from the murine model. Antimicrob Agents Chemother 54(11):4765–4771
258. Shandil RK, Jayaram R, Kaur P, Gaonkar S, Suresh BL, Mahesh BN, Jayashree R, Nandi V, Bharath S, Balasubramanian V (2007) Moxifloxacin, ofloxacin, sparfloxacin, and ciprofloxacin against *Mycobacterium tuberculosis*: evaluation of in vitro and pharmacodynamic indices that best predict in vivo efficacy. Antimicrob Agents Chemother 51:576–582
259. Lienhardt C, Vernon A, Raviglione MC (2010) New drugs and new regimens for the treatment of tuberculosis: review of the drug development pipeline and implications for national programmes. Curr Opin Pulm Med 16:186–193
260. Bryskier A, Lowther J (2002) Fluoroquinolones and tuberculosis. Expert Opin Investig Drugs 11:233–258

261. Gillespie SH (2009) A randomized placebo-controlled double blind trial comparing 1) a two month intensive phase of ethambutol, moxifloxacin, rifampicin, pyrazinamide versus the standard regimen (two months ethambutol, isoniazid, rifampicin, pyrazinamide followed by four months isoniazid and rifampicin) for the treatment of adults with pulmonary tuberculosis. In: Controlled comparison of treatment shortening regiments in pulmonary tuberculosis (REMoxTB) Clinical Trial NCT 00864383. Sponsored by University College, London. Accessed from National Library of Medicine and National Institutes of Health (US), ClinicalTrials.gov.
262. Lienhardt, C (2005) A randomized open-label controlled trial of a 4-month gatifloxacin-containing regimen versus standard regiment for the treatment of adult patients with pulmonary tuberculosis. Clinical Trial NCT 00216385. Sponsored by Institut de Recherche pour le Developpement. Accessed from National Library of Medicine and National Institutes of Health (US), ClinicalTrials.gov.
263. Thomas JP, Baughn CO, Wilkinson RG, Shepherd RG (1961) A new synthetic compound with antituberculous activity in mice: ethambutol (dextro-2,2'-(ethylenediimino)-di-l-butanol). Am Rev Respir Dis 83:891–893
264. Shepherd RG, Wilkinson RG (1962) Antituberculous agents. II. *N, N'*-Diisopropylethylenediamine and analogs. J Med Pharm Chem 91:823–835
265. Shepherd RG, Baughn C, Cantrall ML, Goodstein B, Thomas JP, Wilkinson RG (1966) Structure-activity studies leading to ethambutol, a new type of antituberculous compound. Ann NY Acad Sci 135:686–710
266. Wilkinson RG, Shepherd RG (1969) Compositions and method of treating mycobacterium tuberculosis with 2,2'-(ethylenediimino)-di-1-butanols. American Cyanamid Co: United States, US3463861
267. Hausler H, Kawakami RP, Mlaker E, Severn WB, Stutz AE (2001) Ethambutol analogues as potential antimycobacterial agents. Bioorg Med Chem Lett 11:1679–1681
268. Belanger AE, Besra GS, Ford ME, Mikusova K, Belisle JT, Brennan PJ, Inamine JM (1996) The embAB genes of *Mycobacterium avium* encode an arabinosyl transferase involved in cell wall arabinan biosynthesis that is the target for the antimycobacterial drug ethambutol. Proc Natl Acad Sci USA 93:11919–11924
269. Telenti A, Philipp WJ, Sreevatsan S, Bernasconi C, Stockbauer KE, Wieles B, Musser JM, Jacobs WR Jr (1997) The emb operon, a gene cluster of Mycobacterium tuberculosis involved in resistance to ethambutol. Nat Med 3:567–570
270. Lee RE, Protopopova M, Crooks E, Slayden RA, Terrot M, Barry CE III (2003) Combinatorial lead optimization of [1,2]-diamines based on ethambutol as potential antituberculosis preclinical candidates. J Comb Chem 5:172–187
271. Protopopova M, Hanrahan C, Nikonenko B, Samala R, Chen P, Gearhart J, Einck L, Nacy CA (2005) Identification of a new antitubercular drug candidate, SQ109, from a combinatorial library of 1,2-ethylenediamines. J Antimicrob Chemother 56:968–974
272. Boshoff HI, Myers TG, Copp BR, McNeil MR, Wilson MA, Barry CE III (2004) The transcriptional responses of *Mycobacterium tuberculosis* to inhibitors of metabolism: novel insights into drug mechanisms of action. J Biol Chem 279:40174–40184
273. Jia L, Tomaszewski JE, Hanrahan C, Coward L, Noker P, Gorman G, Nikonenko B, Protopopova M (2005) Pharmacodynamics and pharmacokinetics of SQ109, a new diamine-based antitubercular drug. Br J Pharmacol 144:80–87
274. Jia L, Noker PE, Coward L, Gorman GS, Protopopova M, Tomaszewski JE (2006) Interspecies pharmacokinetics and in vitro metabolism of SQ109. Br J Pharmacol 147:476–485
275. Jia L, Tomaszewski JE, Noker PE, Gorman GS, Glaze E, Protopopova M (2005) Simultaneous estimation of pharmacokinetic properties in mice of three anti-tubercular ethambutol analogs obtained from combinatorial lead optimization. J Pharm Biomed Anal 37:793–799
276. Sun X, Zeckner D, Zhang Y, Sachs RK, Current WL, Rodriguez M, Chen SH (2001) Prodrugs of 3-amido bearing pseudomycin analogues: novel antifungal agents. Bioorg Med Chem Lett 11:1881–1884

277. Maryanoff BE, McComsey DF, Costanzo MJ, Yabut SC, Lu T, Player MR, Giardino EC, Damiano BP (2006) Exploration of potential prodrugs of RWJ-445167, an oxyguanidine-based dual inhibitor of thrombin and factor Xa. Chem Biol Drug Des 68:29–36
278. Burkhart DJ, Barthel BL, Post GC, Kalet BT, Nafie JW, Shoemaker RK, Koch TH (2006) Design, synthesis, and preliminary evaluation of doxazolidine carbamates as prodrugs activated by carboxylesterases. J Med Chem 49:7002–7012
279. Meng Q, Luo H, Liu Y, Li W, Zhang W, Yao Q (2009) Synthesis and evaluation of carbamate prodrugs of SQ109 as antituberculosis agents. Bioorg Med Chem Lett 19:2808–2810
280. (2008) Sq109. Tuberculosis (Edinb) 88:159–161
281. Protopopova N, Bogatcheva E (2003) Methods of use and compositions for the diagnosis and treatment of infectious disease. Sequella, Inc.: United States. WO2003096987
282. Makarov V, Riabova OB, Yuschenko A, Urlyapova N, Daudova A, Zipfel PF, Mollmann U (2006) Synthesis and antileprosy activity of some dialkyldithiocarbamates. J Antimicrob Chemother 57:1134–1138
283. Makarov V, Manina G, Mikusova K, Mollmann U, Ryabova O, Saint-Joanis B, Dhar N, Pasca MR, Buroni S, Lucarelli AP, Milano A, De Rossi E, Belanova M, Bobovska A, Dianiskova P, Kordulakova J, Sala C, Fullam E, Schneider P, McKinney JD, Brodin P, Christophe T, Waddell S, Butcher P, Albrethsen J, Rosenkrands I, Brosch R, Nandi V, Bharath S, Gaonkar S, Shandil RK, Balasubramanian V, Balganesh T, Tyagi S, Grosset J, Riccardi G, Cole ST (2009) Benzothiazinones kill *Mycobacterium tuberculosis* by blocking arabinan synthesis. Science 324:801–804
284. Mikusova K, Huang H, Yagi T, Holsters M, Vereecke D, D'Haeze W, Scherman MS, Brennan PJ, McNeil MR, Crick DC (2005) Decaprenylphosphoryl arabinofuranose, the donor of the D-arabinofuranosyl residues of mycobacterial arabinan, is formed via a two-step epimerization of decaprenylphosphoryl ribose. J Bacteriol 187:8020–8025
285. Pasca MR, Degiacomi G, Ribeiro AL, Zara F, De Mori P, Heym B, Mirrione M, Brerra R, Pagani L, Pucillo L, Troupioti P, Makarov V, Cole ST, Riccardi G (2010) Clinical isolates of *Mycobacterium tuberculosis* in four European hospitals are uniformly susceptible to benzothiazinones. Antimicrob Agents Chemother 54:1616–1618
286. Nikonenko BV, Reddy VM, Protopopova M, Bogatcheva E, Einck L, Nacy CA (2009) Activity of SQ641, a capuramycin analog, in a murine model of tuberculosis. Antimicrob Agents Chemother 53:3138–3139
287. Yamaguchi H, Sato S, Yoshida S, Takada K, Itoh M, Seto H, Otake N (1986) Capuramycin, a new nucleoside antibiotic. Taxonomy, fermentation, isolation and characterization. J Antibiot (Tokyo) 39:1047–1053
288. Igarashi M, Takahashi Y, Shitara T, Nakamura H, Naganawa H, Miyake T, Akamatsu Y (2005) Caprazamycins, novel lipo-nucleoside antibiotics, from *Streptomyces* sp. II. Structure elucidation of caprazamycins. J Antibiot (Tokyo) 58:327–337
289. Hotoda H, Furukawa M, Daigo M, Murayama K, Kaneko M, Muramatsu Y, Ishii MM, Miyakoshi S, Takatsu T, Inukai M, Kakuta M, Abe T, Harasaki T, Fukuoka T, Utsui Y, Ohya S (2003) Synthesis and antimycobacterial activity of capuramycin analogues. Part 1: substitution of the azepan-2-one moiety of capuramycin. Bioorg Med Chem Lett 13:2829–2832
290. Hirano S, Ichikawa S, Matsuda A (2008) Structure-activity relationship of truncated analogs of caprazamycins as potential anti-tuberculosis agents. Bioorg Med Chem 16:5123–5133
291. Koga T, Fukuoka T, Doi N, Harasaki T, Inoue H, Hotoda H, Kakuta M, Muramatsu Y, Yamamura N, Hoshi M, Hirota T (2004) Activity of capuramycin analogues against *Mycobacterium tuberculosis*, *Mycobacterium avium* and *Mycobacterium intracellulare* in vitro and in vivo. J Antimicrob Chemother 54:755–760
292. Douthwaite S (2001) Structure-activity relationships of ketolides vs. macrolides. Clin Microbiol Infect 7(Suppl 3):11–17
293. Zhu ZJ, Krasnykh O, Pan D, Petukhova V, Yu G, Liu Y, Liu H, Hong S, Wang Y, Wan B, Liang W, Franzblau SG (2008) Structure-activity relationships of macrolides against *Mycobacterium tuberculosis*. Tuberculosis (Edinb) 88(Suppl 1):S49–S63

294. Falzari K, Zhu Z, Pan D, Liu H, Hongmanee P, Franzblau SG (2005) In vitro and in vivo activities of macrolide derivatives against *Mycobacterium tuberculosis*. Antimicrob Agents Chemother 49:1447–1454
295. Chhabria M, Jani M, Patel S (2009) New frontiers in the therapy of tuberculosis: fighting with the global menace. Mini Rev Med Chem 9:401–430
296. Buriankova K, Doucet-Populaire F, Dorson O, Gondran A, Ghnassia J-C, Weiser J, Pernodet J-L (2004) Molecular basis of intrinsic macrolide resistance in the *Mycobacterium tuberculosis* complex. Antimicrob Agents Chemother 48:143–150
297. Fleming A (1929) On the antibacterial action of cultures of a penicillium, with special reference to their use in the isolation of B. influenzae. Br J Exp Pathol 10:226–236
298. Tipper DJ, Strominger JL (1965) Mechanism of action of penicillins: a proposal based on their structural similarity to acyl-D-alanyl-D-alanine. Proc Natl Acad Sci USA 54: 1133–1141
299. Hugonnet J-E, Tremblay LW, Boshoff HI, Barry CE III, Blanchard JS (2009) Meropenem-clavulanate is effective against extensively drug-resistant *Mycobacterium tuberculosis*. Science 323:1215–1218
300. Segura C, Salvado M, Collado I, Chaves J, Coira A (1998) Contribution of beta-lactamases to beta-lactam susceptibilities of susceptible and multidrug-resistant *Mycobacterium tuberculosis* clinical isolates. Antimicrob Agents Chemother 42:1524–1526
301. Chambers HF, Moreau D, Yajko D, Miick C, Wagner C, Hackbarth C, Kocagoz S, Rosenberg E, Hadley WK, Nikaido H (1995) Can penicillins and other beta-lactam antibiotics be used to treat tuberculosis? Antimicrob Agents Chemother 39:2620–2624
302. Gupta R, Lavollay M, Mainardi J-L, Arthur M, Bishai WR, Lamichhane G (2010) The *Mycobacterium tuberculosis* protein LdtMt2 is a nonclassical transpeptidase required for virulence and resistance to amoxicillin. Nat Med 16:466–469
303. Chambers HF, Kocagoz T, Sipit T, Turner J, Hopewell PC (1998) Activity of amoxicillin/clavulanate in patients with tuberculosis. Clin Infect Dis 26:874–877
304. Holzgrabe U (2009) Meropenem-clavulanate: a new strategy for the treatment of tuberculosis? ChemMedChem 4:1051–1053
305. Maruyama T, Yamamoto Y, Kano Y, Kurazono M, Matsuhisa E, Takata H, Takata T, Atsumi K, Iwamatsu K, Shitara E (2007) CP5484, a novel quaternary carbapenem with potent anti-MRSA activity and reduced toxicity. Bioorg Med Chem 15:6379–6387
306. Barry VC, Belton JG, Conalty ML, Denneny JM, Edward DW, O'Sullivan JF, Twomey D, Winder F (1957) A new series of phenazines (rimino-compounds) with high antituberculosis activity. Nature 179:1013–1015
307. Reddy VM, Nadadhur G, Daneluzzi D, O'Sullivan JF, Gangadharam PR (1996) Antituberculosis activities of clofazimine and its new analogs B4154 and B4157. Antimicrob Agents Chemother 40:633–636
308. van Rensburg CE, Joone GK, Sirgel FA, Matlola NM, O'Sullivan JF (2000) In vitro investigation of the antimicrobial activities of novel tetramethylpiperidine-substituted phenazines against *Mycobacterium tuberculosis*. Chemotherapy 46:43–48
309. Oliva B, O'Neill AJ, Miller K, Stubbings W, Chopra I (2004) Anti-staphylococcal activity and mode of action of clofazimine. J Antimicrob Chemother 53:435–440
310. Cholo MC, Boshoff HI, Steel HC, Cockeran R, Matlola NM, Downing KJ, Mizrahi V, Anderson R (2006) Effects of clofazimine on potassium uptake by a Trk-deletion mutant of *Mycobacterium tuberculosis*. J Antimicrob Chemother 57:79–84
311. Di Santo R, Costi R, Artico M, Massa S, Lampis G, Deidda D, Pompei R (1998) Pyrrolnitrin and related pyrroles endowed with antibacterial activities against *Mycobacterium tuberculosis*. Bioorg Med Chem Lett 8:2931–2936
312. Deidda D, Lampis G, Fioravanti R, Biava M, Porretta GC, Zanetti S, Pompei R (1998) Bactericidal activities of the pyrrole derivative BM212 against multidrug-resistant and intramacrophagic *Mycobacterium tuberculosis* strains. Antimicrob Agents Chemother 42:3035–3037

313. Biava M, Cesare Porretta G, Deidda D, Pompei R, Tafi A, Manetti F (2003) Importance of the thiomorpholine introduction in new pyrrole derivatives as antimycobacterial agents analogues of BM 212. Bioorg Med Chem 11:515–520
314. Biava M, Porretta GC, Deidda D, Pompei R, Tafi A, Manetti F (2004) Antimycobacterial compounds. New pyrrole derivatives of BM212. Bioorg Med Chem 12:1453–1458
315. Biava M, Porretta GC, Poce G, Deidda D, Pompei R, Tafi A, Manetti F (2005) Antimycobacterial compounds. Optimization of the BM 212 structure, the lead compound for a new pyrrole derivative class. Bioorg Med Chem 13:1221–1230
316. Biava M, Porretta GC, Poce G, Supino S, Deidda D, Pompei R, Molicotti P, Manetti F, Botta M (2006) Antimycobacterial agents. Novel diarylpyrrole derivatives of BM212 endowed with high activity toward *Mycobacterium tuberculosis* and low cytotoxicity. J Med Chem 49:4946–4952
317. Biava M, Porretta GC, Manetti F (2007) New derivatives of BM212: a class of antimycobacterial compounds based on the pyrrole ring as a scaffold. Mini Rev Med Chem 7:65–78
318. (2008) LL-3858. Tuberculosis (Edinb) 88:126
319. Chan DCM, Anderson AC (2006) Towards species-specific antifolates. Curr Med Chem 13:377–398
320. Lange RP, Locher HH, Wyss PC, Then RL (2007) The targets of currently used antibacterial agents: lessons for drug discovery. Curr Pharm Des 13:3140–3154
321. Suling WJ, Reynolds RC, Barrow EW, Wilson LN, Piper JR, Barrow WW (1998) Susceptibilities of *Mycobacterium tuberculosis* and *Mycobacterium avium* complex to lipophilic deazapteridine derivatives, inhibitors of dihydrofolate reductase. J Antimicrob Chemother 42:811–815
322. da Cunha EFF, Ramalho TC, de Alencastro RB, Maia ER (2004) Interactions of 5-deazapteridine derivatives with *Mycobacterium tuberculosis* and with human dihydrofolate reductases. J Biomol Struct Dyn 22:119–130
323. da Cunha EFF, Ramalho TC, Reynolds RC (2008) Binding mode analysis of 2,4-diamino-5-methyl-5-deaza-6-substituted pteridines with *Mycobacterium tuberculosis* and human dihydrofolate reductases. J Biomol Struct Dyn 25:377–385
324. Suling WJ, Seitz LE, Pathak V, Westbrook L, Barrow EW, Zywno-Van-Ginkel S, Reynolds RC, Piper JR, Barrow WW (2000) Antimycobacterial activities of 2,4-diamino-5-deazapteridine derivatives and effects on mycobacterial dihydrofolate reductase. Antimicrob Agents Chemother 44:2784–2793
325. El-Hamamsy M, Smith AW, Thompson AS, Threadgill MD (2007) Structure-based design, synthesis and preliminary evaluation of selective inhibitors of dihydrofolate reductase from *Mycobacterium tuberculosis*. Bioorg Med Chem 15:4552–4576
326. Mahapatra S, Yagi T, Belisle JT, Espinosa BJ, Hill PJ, McNeil MR, Brennan PJ, Crick DC (2005) Mycobacterial lipid II is composed of a complex mixture of modified muramyl and peptide moieties linked to decaprenyl phosphate. J Bacteriol 187:2747–2757
327. Lavollay M, Arthur M, Fourgeaud M, Dubost L, Marie A, Veziris N, Blanot D, Gutmann L, Mainardi J-L (2008) The peptidoglycan of stationary-phase *Mycobacterium tuberculosis* predominantly contains cross-links generated by L,D-transpeptidation. J Bacteriol 190:4360–4366
328. Daffe M, Brennan PJ, McNeil M (1990) Predominant structural features of the cell wall arabinogalactan of *Mycobacterium tuberculosis* as revealed through characterization of oligoglycosyl alditol fragments by gas chromatography/mass spectrometry and by ^{1}H and ^{13}C NMR analyses. J Biol Chem 265:6734–6743
329. McNeil M, Daffe M, Brennan PJ (1990) Evidence for the nature of the link between the arabinogalactan and peptidoglycan of mycobacterial cell walls. J Biol Chem 265: 18200–18206
330. McNeil M, Daffe M, Brennan PJ (1991) Location of the mycolyl ester substituents in the cell walls of mycobacteria. J Biol Chem 266:13217–13223
331. Brennan PJ, Nikaido H (1995) The envelope of mycobacteria. Annu Rev Biochem 64:29–63

332. Hui J, Gordon N, Kajioka R (1977) Permeability barrier to rifampin in mycobacteria. Antimicrob Agents Chemother 11:773–779
333. Mizuguchi Y, Udou T, Yamada T (1983) Mechanism of antibiotic resistance in *Mycobacterium intracellulare*. Microbiol Immunol 27:425–431
334. David HL, Rastogi N, Clavel-Seres S, Clement F, Thorel MF (1987) Structure of the cell envelope of *Mycobacterium avium*. Zentralbl Bakteriol Mikrobiol Hyg A 264:49–66
335. Rastogi N, Goh KS, David HL (1990) Enhancement of drug susceptibility of *Mycobacterium avium* by inhibitors of cell envelope synthesis. Antimicrob Agents Chemother 34:759–764
336. Draper P (1998) The outer parts of the mycobacterial envelope as permeability barriers. Front Biosci 3:D1253–1261
337. Jarlier V, Nikaido H (1990) Permeability barrier to hydrophilic solutes in *Mycobacterium chelonei*. J Bacteriol 172:1418–1423
338. Nguyen L, Thompson CJ (2006) Foundations of antibiotic resistance in bacterial physiology: the mycobacterial paradigm. Trends Microbiol 14:304–312
339. Jarlier V, Nikaido H (1994) Mycobacterial cell wall: structure and role in natural resistance to antibiotics. FEMS Microbiol Lett 123:11–18
340. Liu J, Rosenberg EY, Nikaido H (1995) Fluidity of the lipid domain of cell wall from *Mycobacterium chelonae*. Proc Natl Acad Sci USA 92:11254–11258
341. Davis JM, Clay H, Lewis JL, Ghori N, Herbomel P, Ramakrishnan L (2002) Real-time visualization of mycobacterium-macrophage interactions leading to initiation of granuloma formation in zebrafish embryos. Immunity 17:693–702
342. Volkman HE, Pozos TC, Zheng J, Davis JM, Rawls JF, Ramakrishnan L (2010) Tuberculous granuloma induction via interaction of a bacterial secreted protein with host epithelium. Science 327:466–469
343. Makky K, Duvnjak P, Pramanik K, Ramchandran R, Mayer AN (2008) A whole-animal microplate assay for metabolic rate using zebrafish. J Biomol Screen 13:960–967
344. Flynn JL (2006) Lessons from experimental *Mycobacterium tuberculosis* infections. Microbes Infect 8:1179–1188
345. Nuermberger E (2008) Using animal models to develop new treatments for tuberculosis. Semin Respir Crit Care Med 29:542–551
346. Kramnik I (2008) Genetic dissection of host resistance to *Mycobacterium tuberculosis*: the sst1 locus and the Ipr1 gene. Curr Top Microbiol Immunol 321:123–148
347. Lenaerts AJM, Gruppo V, Brooks JV, Orme IM (2003) Rapid in vivo screening of experimental drugs for tuberculosis using gamma interferon gene-disrupted mice. Antimicrob Agents Chemother 47:783–785
348. Gaonkar S, Bharath S, Kumar N, Balasubramanian V, Shandil RK (2010) Aerosol infection model of tuberculosis in wistar rats. Int J Microbiol 2010:426035
349. McFarland CT, Ly L, Jeevan A, Yamamoto T, Weeks B, Izzo A, McMurray D (2010) BCG vaccination in the cotton rat (*Sigmodon hispidus*) infected by the pulmonary route with virulent *Mycobacterium tuberculosis*. Tuberculosis (Edinb) 90:262–267
350. McMurray DN (2001) Disease model: pulmonary tuberculosis. Trends Mol Med 7: 135–137
351. Ordway DJ, Shanley CA, Caraway ML, Orme EA, Bucy DS, Hascall-Dove L, Henao-Tamayo M, Harton MR, Shang S, Ackart D, Kraft SL, Lenaerts AJ, Basaraba RJ, Orme IM (2010) Evaluation of standard chemotherapy in the guinea pig model of tuberculosis. Antimicrob Agents Chemother 54:1820–1833
352. Abdul-Majid K-B, Ly LH, Converse PJ, Geiman DE, McMurray DN, Bishai WR (2008) Altered cellular infiltration and cytokine levels during early Mycobacterium tuberculosis sigC mutant infection are associated with late-stage disease attenuation and milder immunopathology in mice. BMC Microbiol 8:151
353. Gupta UD, Katoch VM (2009) Animal models of tuberculosis for vaccine development. Indian J Med Res 129:11–18

354. Converse PJ, Dannenberg AM Jr, Estep JE, Sugisaki K, Abe Y, Schofield BH, Pitt ML (1996) Cavitary tuberculosis produced in rabbits by aerosolized virulent tubercle bacilli. Infect Immun 64:4776–4787
355. Tsenova L, Sokol K, Freedman VH, Kaplan G (1998) A combination of thalidomide plus antibiotics protects rabbits from mycobacterial meningitis-associated death. J Infect Dis 177:1563–1572
356. Kolodny NH, Goode ST, Ryan W, Freddo TF (2002) Evaluation of therapeutic effectiveness using MR imaging in a rabbit model of anterior uveitis. Exp Eye Res 74: 483–491
357. Manabe YC, Dannenberg AM Jr, Tyagi SK, Hatem CL, Yoder M, Woolwine SC, Zook BC, Pitt MLM, Bishai WR (2003) Different strains of *Mycobacterium tuberculosis* cause various spectrums of disease in the rabbit model of tuberculosis. Infect Immun 71:6004–6011
358. Capuano SV III, Croix DA, Pawar S, Zinovik A, Myers A, Lin PL, Bissel S, Fuhrman C, Klein E, Flynn JL (2003) Experimental *Mycobacterium tuberculosis* infection of cynomolgus macaques closely resembles the various manifestations of human *M. tuberculosis* infection. Infect Immun 71:5831–5844
359. Lin PL, Rodgers M, Smith L, Bigbee M, Myers A, Bigbee C, Chiosea I, Capuano SV, Fuhrman C, Klein E, Flynn JL (2009) Quantitative comparison of active and latent tuberculosis in the cynomolgus macaque model. Infect Immun 77:4631–4642
360. Sharpe SA, Eschelbach E, Basaraba RJ, Gleeson F, Hall GA, MyIntyre A, Williams A, Kraft SL, Clark S, Gooch K, Hatch G, Orme IM, Marsh PD, Dennis MJ (2009) Determination of lesion volume by MRI and stereology in a macaque model of tuberculosis. Tuberculosis (Edinb) 89:405–416
361. Tobin DM, Vary JC Jr, Ray JP, Walsh GS, Dunstan SJ, Bang ND, Hagge DA, Khadge S, King M-C, Hawn TR, Moens CB, Ramakrishnan L (2010) The lta4h locus modulates susceptibility to mycobacterial infection in zebrafish and humans. Cell 140:717–730
362. Scanga CA, Mohan VP, Joseph H, Yu K, Chan J, Flynn JL (1999) Reactivation of latent tuberculosis: variations on the Cornell murine model. Infect Immun 67:4531–4538
363. Ilson BE, Jorkasky DK, Curnow RT, Stote RM (1989) Effect of a new synthetic hexapeptide to selectively stimulate growth hormone release in healthy human subjects. J Clin Endocrinol Metab 69:212–214
364. Pichugin AV, Yan B-S, Sloutsky A, Kobzik L, Kramnik I (2009) Dominant role of the sst1 locus in pathogenesis of necrotizing lung granulomas during chronic tuberculosis infection and reactivation in genetically resistant hosts. Am J Pathol 174:2190–2201
365. Miyazaki E, Chaisson RE, Bishai WR (1999) Analysis of rifapentine for preventive therapy in the Cornell mouse model of latent tuberculosis. Antimicrob Agents Chemother 43:2126–2130
366. Mitchison DA (2004) The search for new sterilizing anti-tuberculosis drugs. Front Biosci 9:1059–1072
367. Nuermberger EL, Yoshimatsu T, Tyagi S, Bishai WR, Grosset JH (2004) Paucibacillary tuberculosis in mice after prior aerosol immunization with *Mycobacterium bovis* BCG. Infect Immun 72:1065–1071
368. Andries K, Gevers T, Lounis N (2010) Bactericidal potencies of new regimens are not predictive for their sterilizing potencies in a murine model of tuberculosis. Antimicrob Agents Chemother 54(11):4540–4544
369. Ahmad Z, Nuermberger EL, Tasneen R, Pinn ML, Williams KN, Peloquin CA, Grosset JH, Karakousis PC (2010) Comparison of the 'Denver regimen' against acute tuberculosis in the mouse and guinea pig. J Antimicrob Chemother 65:729–734
370. Rosenthal IM, Zhang M, Almeida D, Grosset JH, Nuermberger EL (2008) Isoniazid or moxifloxacin in rifapentine-based regimens for experimental tuberculosis? Am J Respir Crit Care Med 178:989–993
371. Benator D, Bhattacharya M, Bozeman L, Burman W, Cantazaro A, Chaisson R, Gordin F, Horsburgh CR, Horton J, Khan A, Lahart C, Metchock B, Pachucki C, Stanton L, Vernon A,

Villarino ME, Wang YC, Weiner M, Weis S, Tuberculosis Trials C (2002) Rifapentine and isoniazid once a week versus rifampicin and isoniazid twice a week for treatment of drug-susceptible pulmonary tuberculosis in HIV-negative patients: a randomised clinical trial. Lancet 360:528–534

372. Mitchison DA, Chang KC (2009) Experimental models of tuberculosis: can we trust the mouse? Am J Respir Crit Care Med 180:201–202
373. Emori M, Saito H, Sato K, Tomioka H, Setogawa T, Hidaka T (1993) Therapeutic efficacy of the benzoxazinorifamycin KRM-1648 against experimental *Mycobacterium avium* infection induced in rabbits. Antimicrob Agents Chemother 37:722–728
374. Dorman SE, Hatem CL, Tyagi S, Aird K, Lopez-Molina J, Pitt MLM, Zook BC, Dannenberg AM Jr, Bishai WR, Manabe YC (2004) Susceptibility to tuberculosis: clues from studies with inbred and outbred New Zealand White rabbits. Infect Immun 72:1700–1705
375. Tsenova L, Mangaliso B, Muller G, Chen Y, Freedman VH, Stirling D, Kaplan G (2002) Use of IMiD3, a thalidomide analog, as an adjunct to therapy for experimental tuberculous meningitis. Antimicrob Agents Chemother 46:1887–1895
376. Kailasam S, Daneluzzi D, Gangadharam PR (1994) Maintenance of therapeutically active levels of isoniazid for prolonged periods in rabbits after a single implant of biodegradable polymer. Tuber Lung Dis 75:361–365
377. Manabe YC, Kesavan AK, Lopez-Molina J, Hatem CL, Brooks M, Fujiwara R, Hochstein K, Pitt MLM, Tufariello J, Chan J, McMurray DN, Bishai WR, Dannenberg AM Jr, Mendez S (2008) The aerosol rabbit model of TB latency, reactivation and immune reconstitution inflammatory syndrome. Tuberculosis (Edinb) 88:187–196
378. Flynn JL, Capuano SV, Croix D, Pawar S, Myers A, Zinovik A, Klein E (2003) Non-human primates: a model for tuberculosis research. Tuberculosis (Edinb) 83:116–118
379. Pecherstorfer M, Thiebaud D (1992) Treatment of resistant tumor-induced hypercalcemia with escalating doses of pamidronate (APD). Ann Oncol 3:661–663
380. Langermans JAM, Doherty TM, Vervenne RAW, van der Laan T, Lyashchenko K, Greenwald R, Agger EM, Aagaard C, Weiler H, van Soolingen D, Dalemans W, Thomas AW, Andersen P (2005) Protection of macaques against *Mycobacterium tuberculosis* infection by a subunit vaccine based on a fusion protein of antigen 85B and ESAT-6. Vaccine 23:2740–2750
381. Zaleskis R (2006) Adverse effects of anti-tuberculosis chemotherapy. Eur Resp Dis:47–49
382. Craig WA (2003) Basic pharmacodynamics of antibacterials with clinical applications to the use of beta-lactams, glycopeptides, and linezolid. Infect Dis Clin North Am 17:479–501
383. Andes D, Craig WA (2002) Animal model pharmacokinetics and pharmacodynamics: a critical review. Int J Antimicrob Agents 19:261–268
384. Barroso EC, Pinheiro VGF, Facanha MC, Carvalho MRD, Moura ME, Campelo CL, Peloquin CA, Guerrant RL, Lima AAM (2009) Serum concentrations of rifampin, isoniazid, and intestinal absorption, permeability in patients with multidrug resistant tuberculosis. Am J Trop Med Hyg 81:322–329
385. Davies GR, Nuermberger EL (2008) Pharmacokinetics and pharmacodynamics in the development of anti-tuberculosis drugs. Tuberculosis (Edinb) 88(Suppl 1):S65–S74
386. Jayaram R, Shandil RK, Gaonkar S, Kaur P, Suresh BL, Mahesh BN, Jayashree R, Nandi V, Bharath S, Kantharaj E, Balasubramanian V (2004) Isoniazid pharmacokinetics-pharmacodynamics in an aerosol infection model of tuberculosis. Antimicrob Agents Chemother 48:2951–2957
387. Jayaram R, Gaonkar S, Kaur P, Suresh BL, Mahesh BN, Jayashree R, Nandi V, Bharat S, Shandil RK, Kantharaj E, Balasubramanian V (2003) Pharmacokinetics-pharmacodynamics of rifampin in an aerosol infection model of tuberculosis. Antimicrob Agents Chemother 47:2118–2124
388. Gumbo T, Louie A, Deziel MR, Parsons LM, Salfinger M, Drusano GL (2004) Selection of a moxifloxacin dose that suppresses drug resistance in *Mycobacterium tuberculosis*, by use of an in vitro pharmacodynamic infection model and mathematical modeling. J Infect Dis 190:1642–1651

389. Conde MB, Efron A, Loredo C, De Souza GRM, Graca NP, Cezar MC, Ram M, Chaudhary MA, Bishai WR, Kritski AL, Chaisson RE (2009) Moxifloxacin versus ethambutol in the initial treatment of tuberculosis: a double-blind, randomised, controlled phase II trial. Lancet 373:1183–1189
390. Dorman SE, Johnson JL, Goldberg S, Muzanye G, Padayatchi N, Bozeman L, Heilig CM, Bernardo J, Choudhri S, Grosset JH, Guy E, Guyadeen P, Leus MC, Maltas G, Menzies D, Nuermberger EL, Villarino M, Vernon A, Chaisson RE, Tuberculosis Trials C (2009) Substitution of moxifloxacin for isoniazid during intensive phase treatment of pulmonary tuberculosis. Am J Respir Crit Care Med 180:273–280
391. Burman WJ, Goldberg S, Johnson JL, Muzanye G, Engle M, Mosher AW, Choudhri S, Daley CL, Munsiff SS, Zhao Z, Vernon A, Chaisson RE (2006) Moxifloxacin versus ethambutol in the first 2 months of treatment for pulmonary tuberculosis. Am J Respir Crit Care Med 174:331–338
392. Johnson JL, Hadad DJ, Dietze R, Maciel ELN, Sewali B, Gitta P, Okwera A, Mugerwa RD, Alcaneses MR, Quelapio MI, Tupasi TE, Horter L, Debanne SM, Eisenach KD, Boom WH (2009) Shortening treatment in adults with noncavitary tuberculosis and 2-month culture conversion. Am J Respir Crit Care Med 180:558–563
393. Li L, Mahan CS, Palaci M, Horter L, Loeffelholz L, Johnson JL, Dietze R, Debanne SM, Joloba ML, Okwera A, Boom WH, Eisenach KD (2010) Sputum *Mycobacterium tuberculosis* mRNA as a marker of bacteriologic clearance in response to antituberculosis therapy. J Clin Microbiol 48:46–51
394. Wallis RS, Pai M, Menzies D, Doherty TM, Walzl G, Perkins MD, Zumla A (2010) Biomarkers and diagnostics for tuberculosis: progress, needs, and translation into practice. Lancet 375:1920–1937
395. Cannas A, Goletti D, Girardi E, Chiacchio T, Calvo L, Cuzzi G, Piacentini M, Melkonyan H, Umansky SR, Lauria FN, Ippolito G, Tomei LD (2008) *Mycobacterium tuberculosis* DNA detection in soluble fraction of urine from pulmonary tuberculosis patients. Int J Tuberc Lung Dis 12:146–151
396. Reither K, Saathoff E, Jung J, Minja LT, Kroidl I, Saad E, Huggett JF, Ntinginya EN, Maganga L, Maboko L, Hoelscher M (2009) Low sensitivity of a urine LAM-ELISA in the diagnosis of pulmonary tuberculosis. BMC Infect Dis 9:141
397. Tessema TA, Bjune G, Assefa G, Svenson S, Hamasur B, Bjorvatn B (2002) Clinical and radiological features in relation to urinary excretion of lipoarabinomannan in Ethiopian tuberculosis patients. Scand J Infect Dis 34:167–171
398. Phillips M, Cataneo RN, Condos R, Ring Erickson GA, Greenberg J, La Bombardi V, Munawar MI, Tietje O (2007) Volatile biomarkers of pulmonary tuberculosis in the breath. Tuberculosis (Edinb) 87:44–52
399. Carrara S, Vincenti D, Petrosillo N, Amicosante M, Girardi E, Goletti D (2004) Use of a T cell-based assay for monitoring efficacy of antituberculosis therapy. Clin Infect Dis 38:754–756
400. Wassie L, Demissie A, Aseffa A, Abebe M, Yamuah L, Tilahun H, Petros B, Rook G, Zumla A, Andersen P, Doherty TM, Group VS (2008) Ex vivo cytokine mRNA levels correlate with changing clinical status of ethiopian TB patients and their contacts over time. PLoS ONE 3:e1522
401. Turgut T, Akbulut H, Deveci F, Kacar C, Muz MH (2006) Serum interleukin-2 and neopterin levels as useful markers for treatment of active pulmonary tuberculosis. Tohoku J Exp Med 209:321–328
402. Baylan O, Balkan A, Inal A, Kisa O, Albay A, Doganci L, Acikel CH (2006) The predictive value of serum procalcitonin levels in adult patients with active pulmonary tuberculosis. Jpn J Infect Dis 59:164–167
403. Hosp M, Elliott AM, Raynes JG, Mwinga AG, Luo N, Zangerle R, Pobee JO, Wachter H, Dierich MP, McAdam KP, Fuchs D (1997) Neopterin, beta 2-microglobulin, and acute phase proteins in HIV-1-seropositive and -seronegative Zambian patients with tuberculosis. Lung 175:265–275

404. Lee H-M, Shin JW, Kim JY, Park IW, Choi BW, Choi JC, Seo JS, Kim CW (2010) HRCT and whole-blood interferon-gamma assay for the rapid diagnosis of smear-negative pulmonary tuberculosis. Respiration 79:454–460
405. Lee SW, Jang YS, Park CM, Kang HY, Koh W-J, Yim J-J, Jeon K (2010) The role of chest CT scanning in TB outbreak investigation. Chest 137:1057–1064
406. Wang YH, Lin AS, Lai YF, Chao TY, Liu JW, Ko SF (2003) The high value of high-resolution computed tomography in predicting the activity of pulmonary tuberculosis. Int J Tuberc Lung Dis 7:563–568
407. Demura Y, Tsuchida T, Uesaka D, Umeda Y, Morikawa M, Ameshima S, Ishizaki T, Fujibayashi Y, Okazawa H (2009) Usefulness of ^{18}F-fluorodeoxyglucose positron emission tomography for diagnosing disease activity and monitoring therapeutic response in patients with pulmonary mycobacteriosis. Eur J Nucl Med Mol Imaging 36:632–639
408. Sirgel FA, Fourie PB, Donald PR, Padayatchi N, Rustomjee R, Levin J, Roscigno G, Norman J, McIlleron H, Mitchison DA (2005) The early bactericidal activities of rifampin and rifapentine in pulmonary tuberculosis. Am J Respir Crit Care Med 172:128–135
409. Jindani A, Dore CJ, Mitchison DA (2003) Bactericidal and sterilizing activities of antituberculosis drugs during the first 14 days. Am J Respir Crit Care Med 167:1348–1354
410. Ehrt S, Schnappinger D (2006) Controlling gene expression in mycobacteria. Future Microbiol 1:177–184

Top Med Chem 7: 125–180
DOI: 10.1007/7355_2011_14

Published online: 31 May 2011

Discovering New Medicines to Control and Eradicate Malaria

Jeremy N. Burrows and David Waterson

Abstract Malaria is one of the most devastating diseases the world has ever known affecting almost 250 million people a year and resulting in over 860,000 deaths, mostly of children under the age of 5 years. With increased funding and the advent of public–private partnerships such as Medicines for Malaria Venture, the search for new medicines to combat malaria has, over the last decade, undergone a renaissance. This is both as a result of increased funding and the recent development of new tools, such as high-throughput screening technology. In addition, public and political attention has been caught by calls from the Gates Foundation and World Health Organization for a strategy to eradicate malaria once and for all. This will mean having therapeutics to combat not only *Plasmodium falciparum* malaria – predominantly found in Africa – but also *Plasmodium vivax* malaria, which is endemic in many parts of Asia and South America. However, as with all infectious disease, there is a constant threat of drug resistance, and therefore the continual need for new medicines. In this review, we summarise the challenges posed by malaria drug discovery and development. We present the pipeline of existing treatments and those in clinical development. New modes of action are under investigation in preclinical discovery, and we review the opportunities as well as the risks to these early stage projects.

Keywords Antimalarial, Eradication, Malaria, *Plasmodium falciparum*, *Plasmodium vivax*, Transmission blocking, Whole cell screening

Contents

J.N. Burrows (✉) and D. Waterson
Medicines for Malaria Venture, Route de Pré-Bois 20, PO Box 1826, 1215 Geneva 15, Switzerland
e-mail: burrowsj@mmv.org

1 Introduction

Malaria is a devastating disease that predominantly affects the developing world. There are approximately 243 million cases of malaria each year and more than 860,000 deaths [1]. Tragically those most at risk are expectant mothers, babies and children under the age of 5 years. Malaria is caused by infections of the parasite *Plasmodium*, and there are five species that are known to infect humans: *Plasmodium falciparum*, *Plasmodium vivax*, *Plasmodium ovalae*, *Plasmodium malariae* and *Plasmodium knowelsi* [2]. Mortality is most often associated with falciparum malaria, and consequently the vast majority of available medicines are focused on curing patients of falciparum blood-stage parasites. This has also been aided and driven by the discovery that falciparum-infected erythrocytes can be cultured in vitro, thus providing a disease-relevant assay in which to test potential new antimalarials [3].

1.1 Plasmodium falciparum Malaria Life Cycle

The life cycle of the falciparum malaria parasite is shown in Fig. 1. Just before taking a blood meal, an infected mosquito vector typically injects <100 sporozoites into the blood of its human host. A proportion of these sporozoites can infect hepatocytes within 30 min following injection and begin to develop into liver

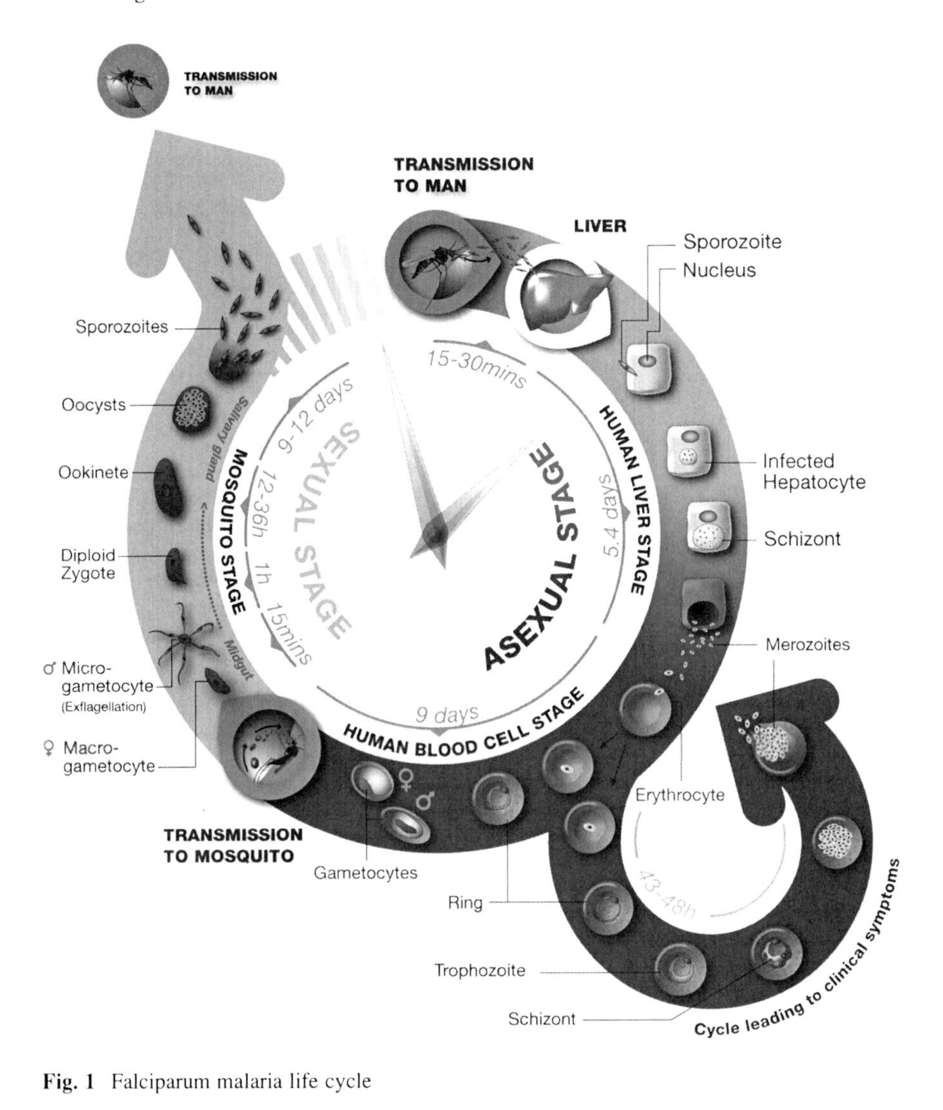

Fig. 1 Falciparum malaria life cycle

schizonts after 5–7 days. The mature liver schizont then ruptures releasing between 10,000 and 30,000 merozoites into the blood [4].

These merozoites are small and short-lived, with the ability to successfully invade an erythrocyte within 30 s. Within each infected erythrocyte, the parasite undergoes a transformation first into a ring stage, and then a trophozoite and ultimately a blood-stage schizont – generating 32 new merozoites over a 48-h period. The erythrocyte then ruptures and releases merozoites, which start a new cycle by invading new erythrocytes. This asexual cycle leads to a rapid increase in parasite numbers in the circulation. The patients start to develop symptoms once the parasitaemia reaches ten parasites per microlitre, and the fever spikes are correlated with the release of the merozoites into circulation. These numbers can rise to 200,000 per microlitre in severe disease, making a total parasite burden in excess of 10^{12} [5]. Around 1% of parasites differentiate into gametocytes, the sexual forms of the parasite – this happens late in the infection for *P. falciparum*, but in *P. vivax* infection gametocytes can often be found as soon as the patient is symptomatic.

These gametocytes take 9 days to mature, and it is only the mature gametocytes that are able to perpetuate the parasite life cycle. The gametocytes remain in the patient's blood for at least 3 weeks though in certain cases persist in the body for up to 90 days [6]. During the infective period, a female mosquito vector will ingest gametocytes within the blood meal. The resulting drop in temperature, plus local chemical stimuli such as xanthurenic acid in the insect midgut, promotes the fertilisation of the female macro-gametocyte by the micro-gametocyte to form first a diploid zygote and then an ookinete [7]. The ookinete then burrows through the gut wall to form an oocyst (vector schizont) over a period of 9–12 days. This oocyst houses many sporozoites, which migrate to the salivary gland, making the mosquito infective to the human host.

1.2 *Plasmodium vivax*

Plasmodium vivax, the dominant malaria species in India, Southeast Asia and South America, bears many similarities to *P. falciparum*. The medicines that are active against the blood stages of *P. falciparum* infection are generally found to be active in the same stage of *P. vivax* malaria. However, the key difference between the species is the ability of a sub-population of the vivax parasites to remain dormant within the liver – a stage termed the hypnozoite [8]. This dormant form can reactivate and lead to a relapse without a second mosquito bite, weeks, months or even years after the original symptoms have been cured. Sterilisation of a patient infected by *P. vivax* requires a liver acting anti-hypnozoiticidal drug. This complete elimination of all parasites is known as a radical cure.

Only one drug, primaquine (Fig. 2), is registered for this indication. In practice compliance is very hard to achieve, since primaquine requires a 14-day treatment regimen, with seemingly no symptomatic benefit. Additionally, and perhaps more importantly, it is contraindicated in patients with decreased activity of glucose-6-phosphate dehydrogenase (G6PD) [9]. G6PD deficiency is highly prevalent in

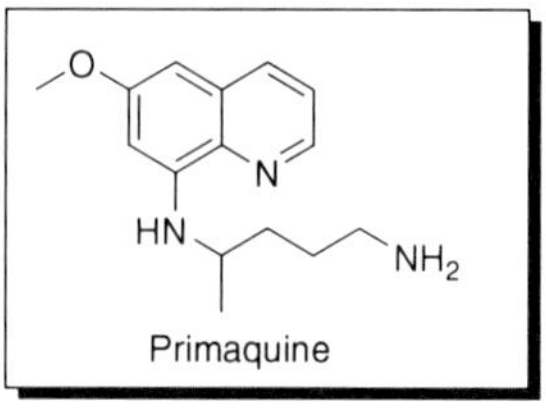

O
N
HN
NH_2
Primaquine

Fig. 2 Primaquine

malaria-endemic regions – 10–15% of the population in Africa and Southeast Asia – since it increases oxidative stress providing some degree of protection against infection. Safe anti-hypnozoiticidal agents with a shorter duration of treatment are urgently sought after, but their discovery is currently hampered because of the absence of in vitro liver stage assays or an in vivo assay in mice. Drug discovery projects targeting the radical cure of vivax malaria are an extremely important part of the eradication campaign, but require the development of additional assays before widespread progress can be made [10].

1.3 Target Product Profiles for the Eradication Agenda

A target product profile (TPP) provides a statement of the overall intent of a drug development programme and critically clarifies the objectives for the discovery phase. For malaria, given the threat of resistance, all the TPPs are combination products. Current drug combinations focused on the asexual blood stage of the parasite are extremely effective at curing patients of infected erythrocytes and saving lives. However, without a drug combination to also kill the gametocytes or vector stages, transmission will continue, to the detriment of humanity. Consequently, drug combinations focused on TPPs targeting both treatment and transmission blocking are the focus of attention. Up-to-date TPPs for antimalarial combinations to support an eradication strategy can be found at Medicines for Malaria Venture's (MMV's) website (www.mmv.org).

2 Antimalarial Medicines

2.1 Antimalarials Approved for Use in Man

The current recommended combination therapies are detailed in the World Health Organization (WHO) Treatment Guidelines 2010 [11]. For falciparum malaria, the approved combinations comprise just eight structural classes:

- Amino-alcohols
- 4-aminoquinolines
- Endoperoxides of the artemisinin family
- Aniline-sulphonamides/sulphones
- Diaminopyrimidines/diamino-dihydrotriazines
- Hydroxynaphthoquinones
- Lincosamides
- Tetracyclines

Historically, the dominant medicines were 4-amino quinolines, such as chloroquine, and anti-folate medicines such as sulphadoxine–pyramethamine. Both of these are now unusable in many parts of Africa owing to the emergence and spread of resistant strains. The current recommended first-line therapy is artemisinin combination therapy (ACT). An ACT is a fixed-dose combination of an artemisinin derivative in combination with an amino alcohol or 4-aminoquinoline, formulated to cure up to 98% of patients within 3 days of treatment. Cure, as defined by WHO, is an adequate clinical and parasitic response 28 days after administration; drugs must be able to eliminate all the blood-stage parasites from the patient to prevent recrudescence.

There are two major ACTs currently in use. The first is Coartem® (artemether–lumefantrine) and, its child-friendly formulation, Coartem® Dispersible, which have been used to treat over 300 million patients. The paediatric formulation is currently treating three million patients per month – just over 1 year post-launch. The other ACT is Coarsucam/ASAQ Winthrop (amodiaquine–artesunate), which treated 25 million patients in 2009. These treatments are provided at a cost per patient as low as $0.37 (adults) and $0.25 (infants), and this is often paid for by United Nations agencies or governments – making them affordable in disease-endemic countries. A table of the key marketed antimalarials, by class, is captured in Fig. 3 [12].

2.2 *Antimalarial Therapies in Clinical Development*

New chemical entities entering clinical development as antimalarials are first profiled as single agents in healthy volunteers (Phase I) and malaria patients (Phase IIa). Subsequent studies are then performed in combination with at least one other partner drug. The current global malaria portfolio can be viewed online at www.mmv.org, and structures of molecules under development are shown in Figs. 4 and 5[1–6] [13], (−)-NPC1161B is reported in [14].

Currently there are two other fixed-dose ACTs in late-stage clinical development (Pyramax®: pyronaridine–artesunate and Eurartesim™: piperaquine–dihydroartemisinin). Two medicines are prescribed, but do not currently conform

[1] All structures are referenced within other cited articles except the following: CDRI 97/98 is exemplifed in: Singh C, Puri SK (2001) US 6316493.

[2] CEM-101 is exemplified in: Liang C-H, Duffield J, Romero A, Chiu Y-H, Rabuka D, Yao S, Sucheck S, Marby K, Shue Y-K, Ichikawa Y, Hwang C-K (2004) WO 2004080391.

[3] P218 is exemplified in: Tarnchompoo B, Yuthavong Y, Vilaivan T, Chitnumsub P, Thongpanchang C, Kamchonwongpaisan S, Matthews D, Vivas L, Yuvaniyama J, Charman S, Charman W, Katiyar SB (2009) WO 2009048957.

[4] OZ439 is exemplified in: Vennerstrom JL, Dong Y, Charman SA, Wittlin S, Chollet J, Creek DJ, Wang X, Sriraghavan K, Zhou L, Matile H, Charman WN (2009) WO 2009058859.

[5] NITD609 is exemplified in: Ang SH, Krastel P, Leong, SY, Tan LJ, Wong WLJ, Yeung BK, Zou B (2009) WO 2009132921.

[6] BCX4208 is exemplified in: Evans GB, Furneaux RH, Lenz DH, Schramm, VL, Tyler PC, Zubkova OV (2004) WO 2004018496.

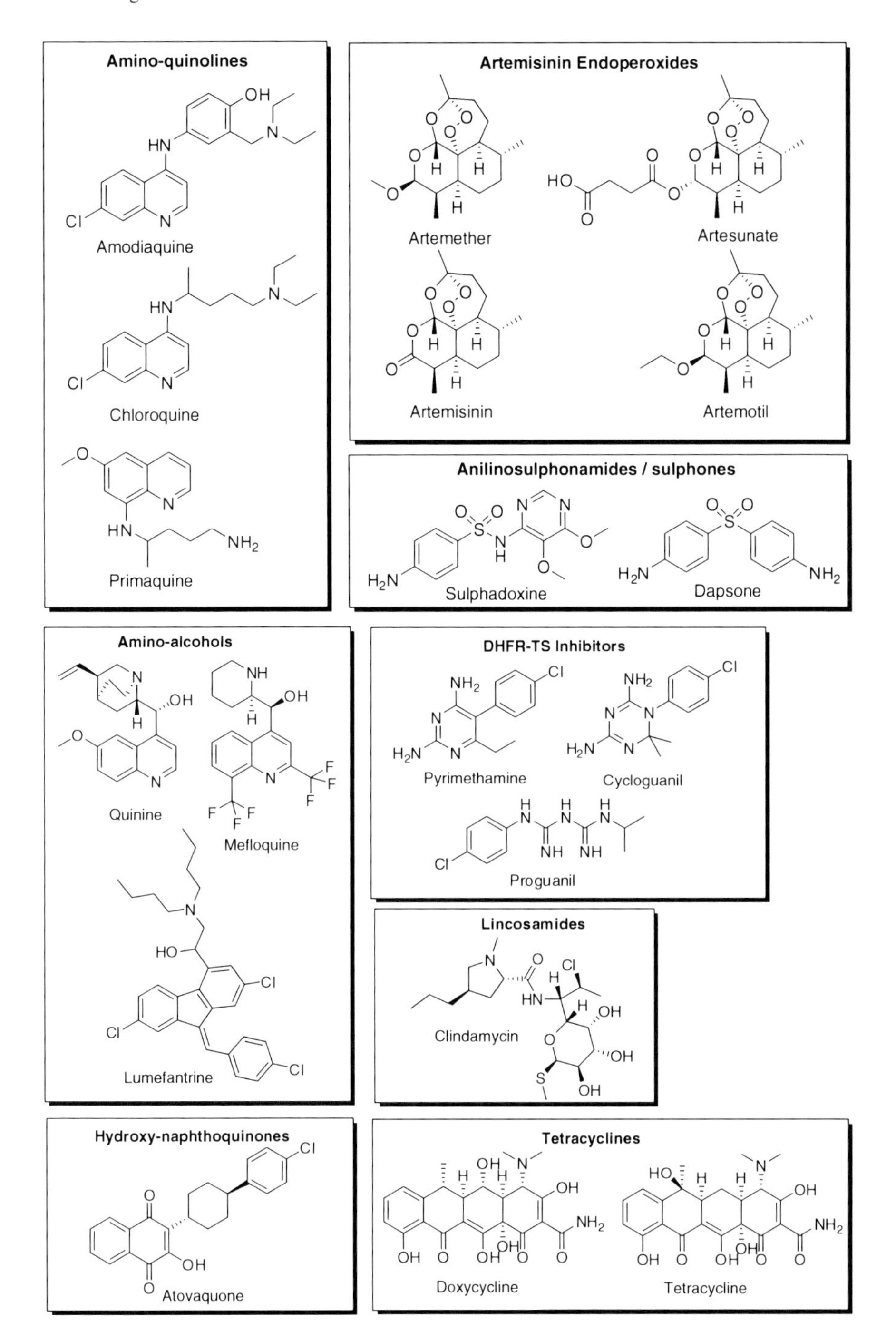

Fig. 3 Summary of marketed antimalarials, grouped by compound class

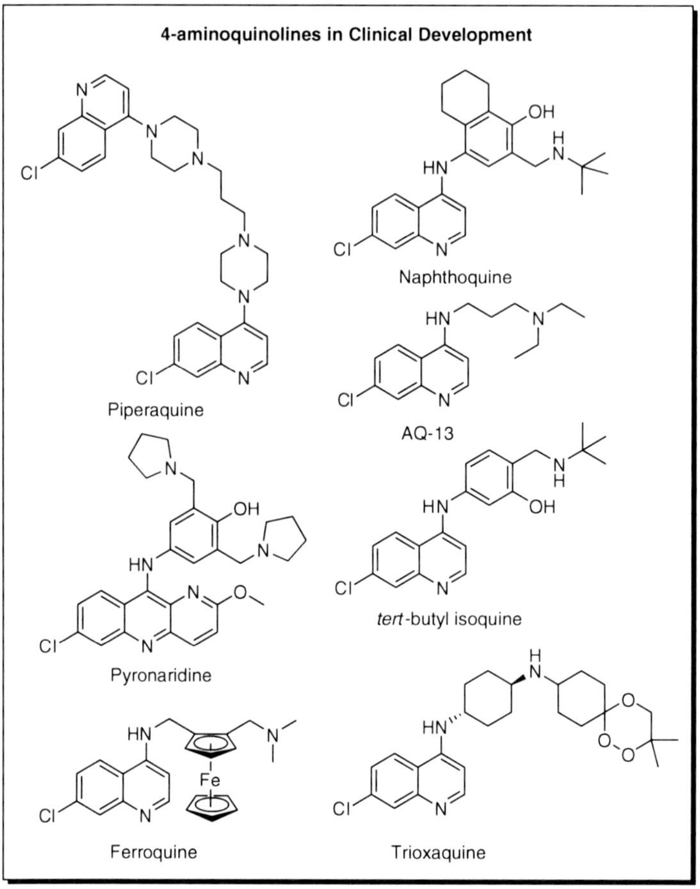

Fig. 4 4-aminoquinolines in preclinical and clinical development

to WHO standards (Arco: naphthoquine–artemisinin, for which no ICH approved clinical trial data are available, and ASMQ: artesunate–mefloquine, which is manufactured and prescribed mainly in Brazil). Generic versions of these medicines are either developed or under development by several companies. Artesunate through i.v. or i.m. administration is used extensively in cases of severe malaria, although at this stage there is no GMP supplier of medicine [15]. A rectal administration of artesunate has been developed for pre-referral treatment in remote areas, and has demonstrated benefit for patients who are more than 6 h away from a health centre [16]. In Phase IIb/III, Co-trimoxazole is a follow-on from sulphadoxine–pyrimethamine (SP) combining two anti-folates. Another antibiotic combination is azithromycin–chloroquine, being developed to reduce infant mortality due to plasmodial and bacterial infections. Azithromycin appears to work clinically as a chloroquine-resistance reversal agent.

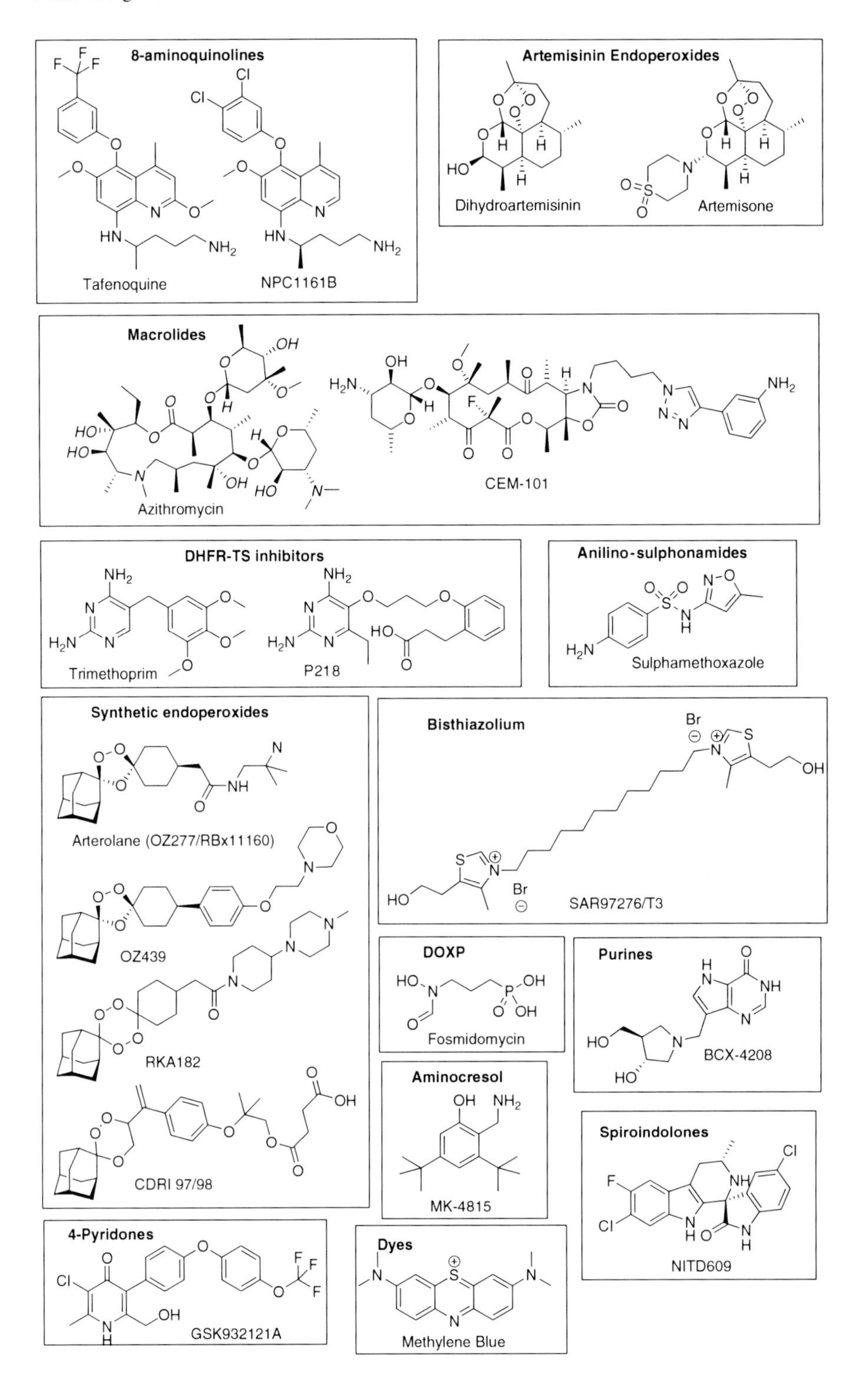

Fig. 5 Antimalarial drug classes under development

Arterolane, an ozonide, is the first in the family of fully synthetic endoperoxides to reach Phase III trials and is being developed in combination with piperaquine. It offers two potential advantages. First, being fully synthetic it should not be subject to the large variations in cost of goods that has plagued artemisinin supply. Second, it is structurally very different from artemisinin and offers the hope of activity in areas when and where artemisinin resistance emerges.

Compounds in Phase IIa studies include a novel artemisinin derivative, a next-generation ozonide, a 4-aminoquinoline, a highly novel parenteral dicationic derivative and two non-artemisinin combination therapies (NACT). Phase I and preclinical phases feature two 4-aminoquinolines, a low anticoagulant heparin derivative for severe malaria, a pyridone, three endoperoxides, a follow-up compound to pyrimethamine, several 8-aminoquinolines for vivax malaria and three additional novel chemotypes.

Therefore, of the 26 combinations or single agents in Phase III or earlier, only seven molecules present a totally novel mechanism or scaffold to currently available treatments. Furthermore, given the threat of resistance and the heavy reliance, for instance, on peroxides and certain partner drugs, it is critical that we know the extent of cross resistance between these few life-saving options. One key eagerly awaited question is whether the potentially artemisinin-resistant parasites emerging in Southeast Asia [17] are sensitive to other artemisinin derivatives and synthetic endoperoxides or not. The future will be clearly less bleak should the latter prove to be the case.

3 Recent Advances in Small Molecule Discovery

3.1 Scope

This review focuses on developments in small molecule antimalarial discovery published in the literature between 2006 and 2010. Previous generations of molecules are summarised in several comprehensive prior review articles [18–20]. We have focused on around 350 publications from a selection of key medicinal chemistry and drug discovery journals. The articles have been analysed and those chemical series and strategies of most interest, from a drug discovery perspective, are highlighted. It should be stressed that, to give a comprehensive review, there are compounds described that have issues from a medicinal chemistry perspective as a result of either their chemical functionality or their physical properties. The antimalarial patent literature has been excluded from the analysis.

Given the plethora of interesting molecules that are either starting points for chemistry, or series that have undergone extensive optimisation, this review is focused mainly on examples where activity in vitro (EC_{50}) against a *P. falciparum* culture <1 μM has been robustly achieved within a series.

Finally, except where explicitly stated, this review focuses on compound series affecting the asexual blood stage of the parasite. Over 95% of the literature analysed was based on the asexual blood stage. This illustrates the attraction of the relatively simple biology of this stage, while also underlining the dearth of active medicinal chemistry focused on liver stage falciparum or vivax malaria, gametocytes and vector stage parasites. These are precisely the targets where new medicines are urgently needed for the eradication of malaria.

3.2 *Hit Identification Strategies*

The starting point of a medicinal chemistry programme is the identification of an interesting chemical series or "hit". The definition of a hit series varies from group to group and will be determined by the TPP. Clearly the series must be potent – activity against the parasite with an EC_{50} of less than 1 μM is a useful, if arbitrary cutoff. It must be selective, in which it does not kill mammalian cell lines at the highest concentration achievable. Often, two additional criteria are applied: first, novelty (a clear intellectual property added value) and second, tractability (the compound must be favourable from a medicinal chemistry perspective and be made within the given cost constraints). There are four major antimalarial "hit identification" strategies which will be briefly described.

3.2.1 Optimisation of a Known Antimalarial Chemotype

Working in the same mechanistic or chemical space as a registered antimalarial medicine obviously brings confidence that an optimised compound will work – since the approach has previously been clinically validated in patients. The key aspect, from a medicinal chemistry drug discovery perspective, is that such a follow-up project needs to specifically address the deficiencies of its progenitor. Consequently, the frontrunner molecule must be fully profiled and have its strengths and weaknesses identified to define the specific project goals. For malaria drug development, the objective will likely be to develop a new molecule active against resistant strains produced by clinical exposure to the progenitor. However, given the changing regulatory climate, a next-generation molecule will probably also need to be safer than the progenitor. Consequently, goals such as decreasing the total dose, and decreasing risks of adverse cardiovascular, haematological, gastrointestinal and CNS effects are challenges for "next-generation" molecules. Unfortunately, many of these cannot be accurately assessed in preclinical models, which significantly increases the risk of such projects.

3.2.2 Target-Based Screening

Screening provides the opportunity to find novel chemical matter – an important feature for malaria given the need to ultimately have diverse options for managing resistance. Biochemical screening of an isolated protein target has been adopted extensively within the pharmaceutical industry and has been the source of the majority of drug discovery projects in recent years [21]. Concern with the paradigm has been raised, however, since target validation in vitro and in animal models is a very imprecise art, even in anti-infective research [22, 23].

Within the malaria field, a few projects exist that have yielded starting points for chemistry via large or small library biochemical screening – targeting enzymes such as *Pf* DHODH (dihydroorotate dehydrogenase, involved in the synthesis of nucleosides) and *Pf* falcipains (cysteine proteases involved in the degradation of haemoglobin) [24–26]. In both cases, molecules were identified that are able to inhibit the targets and were active in vivo – although neither has been shown to be active in a clinical proof-of-concept study to date. Outside of these and a few other projects, there is a dearth of advanced series focused on discrete protein targets within the global malaria portfolio.

The genome of *P. falciparum* [27] and *P. vivax* [28] has been sequenced, and there are estimated to be hundreds of druggable targets that could impact the parasite and be viable points for molecular intervention [29]. However, despite these genomic breakthroughs, the number of well-validated, druggable targets is still extremely low.

While structure-based design methods are, of course, used in antimalarial drug discovery (e.g. [30, 31]), given that there are relatively few targets of interest and, of those, relatively few structures that are available or feasible, the overall impact has been limited. Furthermore, the utility of such methods has been focused on optimising selectivity or enhancing potency of an advanced lead, rather than as a "hit generation" strategy through de novo design per se. Fragment-based methods [32] are becoming increasingly discussed, though to the authors' knowledge no examples have been pursued and published.

3.2.3 Whole Cell Phenotypic Screening

The erythrocytic stages of *P. falciparum* can be easily maintained in culture. In recent years, this has meant that the cost and speed of testing compounds against the whole parasite in human erythrocytes have improved by almost two orders of magnitude – resulting in a huge shift from 96-well to 1,536-well formats. As a result, screening large compound collections against whole parasites is an extremely attractive option with hit rates (percentage of compounds having confirmed activity typically sub 1 μM) of 0.5% typically achieved. To be active, a compound has to cross several biological membranes: erythrocytic cell membrane, parasite plasma membrane, parasitophorous vacuolar membrane and then, possibly, other parasitic sub-cellular membranes [33]. However, given the expression of transporters and the development of channels to provide nutrients to the parasite [34], it is clear that even compounds with very poor intrinsic passive permeability can enter the erythrocyte, elicit excellent antiparasitic activity and access intracellular organelles – a warning to those who assume all "whole cell actives" have the potential to be orally bioavailable.

Whole cell screening against the parasite gives no information about the mechanism of action per se. Information on blood-stage specificity (rings, trophozoites, schizonts) and rates of killing can, however, be used to prioritise or confirm similar pharmacology within a series.[7] The preferred blood-stage activity is rapid onset of action – killing all blood stages (particularly rings) such that the parasite reduction ratio, or fold reduction in parasitaemia over one life cycle, is high (1,000-fold or more). The first priority once such a hit has been identified is to understand the mechanism by which resistance could emerge: first, by showing that it is active against clinically relevant strains, and then, once considerable optimisation has been achieved, by attempting to artificially generate resistance in vitro. By virtue of the screening method, compounds identified from phenotypic screens do not necessarily have a unique target, or indeed a protein target, and so theoretically should be less susceptible to resistance generation. The molecular identity of the target may come from studies of in vitro-generated resistant clones (since the whole genomes can be rapidly sequenced). Otherwise, to identify potential targets affinity methodologies involving immobilised ligands, whole parasite extracts and mass spectrometric methods may be needed [35].

An additional approach to target identification involves screening compounds with whole cell activity against a range of available biological targets to deduce which, if any, contribute to parasite killing. Theoretically, this is an attractive method to find the target for a compound already having demonstrable activity in cells. Studies to date, however, have not been particularly fruitful [36]. This may be for a number of reasons: the chosen target may not be essential, the compound may be killing through multiple mechanisms and hence be difficult to "spot" as a true active, the molecule may undergo parasite metabolism or the compound may have a low affinity for the target, but accumulate in the parasite to a therapeutically relevant concentration, such that target-cell correlations become difficult to interpret.

Optimisation of a compound's potency and safety will clearly be aided by an understanding of its molecular mechanism. Knowing and testing against a target allows an understanding of intrinsic SAR and potentially, depending on the target, access to structural information. Furthermore, the safety risk of modulating potential-related human targets can be defined giving confidence in Phase I. However, definitive identification of the molecular target is not an absolute prerequisite in clinical development.

Phenotypic screening against other stages of the parasite life cycle (liver, gametocyte and vector stages) will become increasingly important as the focus on eradication intensifies. Before this can happen, the biology of such cellular systems must be made more reproducible to allow for screening with higher throughput.

[7]http://www.dpd.cdc.gov/dpdx/html/PDF_Files/Pfalciparum_benchaidV2.pdf

Table 1 Summary of falciparum 3D7 HTS

Organisation	Library size	3D7 screening concentration	No. of confirmed hits	Hit rate (%)
GSK [36]	1,986,056	2 μM	13,533	0.68
Novartis [38]	1,700,000	1.25 μM	17,000/5,973[a]	1/0.34
Broad Institute [41]	79,294	30 μM/6 μM	1,000/134	1.36/0.17
St. Jude [37]	309,474	7 μM	1,134	0.37

[a]17,000 primary actives were found, of which 5,973 were available and had EC50 < 1.25 μM

Published Whole Cell Screening Data

In May 2010, a paradigm-shift occurred, which changed the antimalarial drug discovery landscape forever. In a bold move, displaying exceptional leadership GlaxoSmithKline (GSK) – Tres Cantos, St. Jude Children's Research Hospital – Memphis and Novartis – NITD/GNF made available their confirmed compound hits from a combined malaria whole cell screening effort of around 4.5 million compounds. These were mirrored by publications in *Nature* of the GSK [37] and St. Jude work [38], while the Novartis disclosure supported an earlier publication in 2008 [39]. As a result of this work, around 20,000 compounds with antimalarial activity around 1 μM or less in falciparum-infected erythrocytes have been published and made available to the community via Chembl, PubChem and CDD.[8]

While the nature of the output and the decision-making process for inclusion of compounds has been questioned in some quarters [40], this represents a sea change in the availability of chemotypes as starting points for antimalarial drug discovery projects and as biological tools for mechanism and pathway exploration (Table 1).

The group from GSK led by Garcia-Bustos reported the whole cell screening of approximately two million compounds from the GSK library against sensitive (3D7) parasites at a concentration of 2 μM. The primary hits inhibiting parasite growth >80% were re-sourced and retested in duplicate to give 13,533 confirmed actives. Cytotoxicity was investigated using HepG2 and this highlighted potential issues with around 2,000 of the compounds at a concentration of 10 μM. The set was subsequently tested against multidrug resistant Dd2 at 2 μM resulting in 8,000 compounds with inhibition activity of >50%. From a physicochemical perspective, the antimalarial actives identified from the screen appear on average to be larger and more lipophilic, than the average of the entire GSK screening library. The structural characterisation was performed using clustering methodology and Murko frameworks; the former highlighted 857 clusters and 1,978 "singletons", and the latter, 416 molecular frameworks. Finally, GSK investigated several modes of action hypotheses based on the human or microbiological target data available. Around 70% of the 4,205 compounds analysed suggested G protein-coupled

[8]http://www.ebi.ac.uk/chemblntd; http://www.collaborativedrug.com/; http://pubchem.ncbi.nlm.nih.gov/

receptors (GPCR) or Kinase target bias, and the authors note that this is, at least in part, due to the relative abundance within the library of such target class inhibitors.

The team from St. Jude Children's Research Hospital led by Guy performed a similar screen of 309,474 commercial compounds against 3D7 at 7 μM concentration. The primary actives with inhibition >80% were retested against 3D7 and also profiled against K1 leading to 1,134 validated hits, of which 561 had EC_{50} values ≤2 μM. Certain series were then investigated in detail via isobologram combination studies with antimalarial drugs and reverse genetics. This was conducted via screening against 66 enzyme inhibition and thermal melt shift assays. First, combination studies highlighted two pairs with strong synergy (diaminonaphthoquinone/artemisinin and dihydropyridine/mefloquine). Second, profiling a set of 172 representatives confirmed actives against the panel of 66 assays led to 19 new inhibitors of three validated drug targets (*Pf* DHODH, Hemozoin formation and *Pf* falcipains) and 15 novel binders of novel malaria proteins (including *Pf* choline kinase, dUTPase, GSK3 and thioredoxin). One compound was even selected for in vivo work and at a dose of 100 mg/kg bid i.p. for 3 days resulted in 90% suppression in parasitemia in a *Py*-infected mouse model.

The Novartis and Broad Institute HTS workstreams have been reviewed elsewhere [19].

3.2.4 Repositioning of Clinical Candidates Developed for Other Indications

The pharmaceutical industry has a wide catalogue of compounds that were taken into the clinic for other indications, but in which development was stopped due to lack of activity. There are estimated to be several thousand such compounds. One approach to find antimalarials is to screen these advanced compounds against the whole parasite as has been done with a selection of marketed drugs [42]. Since the achievable free plasma concentrations in man are known, it is relatively easy to triage such molecules for further clinical development.

4 Moving from Hits to Leads to Candidates: Target-Independent Optimisation Using Whole Parasites

Historically, the most successful antimalarial drug discovery projects have arisen from optimising the blood-stage antiparasiticidal activity of a compound series, usually based around an existing validated chemotype (such as mefloquine from quinine or amodiaquine from chloroquine). The following section explores those series that have been identified and optimised using whole parasites, in the absence of a primary biological target.

4.1 Endoperoxides: Natural and Synthetic Peroxides

Artemisinin, a Chinese natural product, from the leaves of *Artemisia annua*, is the basis of the majority of clinical treatments prescribed for malaria [43]. The presence of the weak endoperoxide bond confers rapid, pan-erythrocytic stage activity that, so far, has not yielded to stable resistance, suggesting, potentially, that multiple mechanisms of action are at play [44, 45]. Conversely, the weak –O–O– bond in conjunction with a highly unstable hemiacetal (present in dihydroartemisinin the principle metabolite of all artemisinin species) means that the compound is rapidly broken down by the body. As such it has two major potential draw backs: first, as a natural product it is subject to fluctuations in its cost and availability, and second, it has an extremely short half-life in vivo. Medicinal chemistry strategies have therefore focused on simplifying the molecule and increasing the half-life while retaining its potency and mechanism of action.

Several strategies to achieve this have been attempted based on keeping the artemisinin scaffold. These are shown in Fig. 6 and include:

- Increasing potency by increasing the lipophilicity of the acetal side chain, such as compound 12α [46]
- Forming dimers to increase the peroxide load such as compound 5 from Posner [47]
- Transforming the acetal into an ether to improve chemical stability as is illustrated by Artefanilide [48]
- Increasing stability by transforming the lactone to a lactam, as exemplified by compound 16a [49]

Some clinical studies have suggested auditory toxicology for artemisinin in humans, and derivatives with decreased lipophilicity have been made to decrease CNS penetration and reduce the risk of any neurotoxicity, exemplified by Haynes' work culminating in Artemisone [50].

In addition, several groups have attempted to retain the biological activity, while simplifying the scaffold. The most advanced fully synthetic endoperoxide, arterolane (also known as RBx11160 or OZ277), is in Phase III [51] and a follow-up compound OZ439 is shortly to enter Phase II. These two molecules are trioxolanes (ozonides) that couple excellent efficacy with extended half-lives and good safety profiles. OZ439, in particular, has an outstanding preclinical profile and has a much longer half-life in man, compared to other endoperoxides, suggesting the possibility of a single-dose cure.

The multitude of synthetic peroxides is exemplified in Table 2 and includes:

- Dioxolanes such as compound 17A, from the group of Clardy [52]
- Trioxolanes illustrated by Arterolane [53] and OZ439 from the work of Vennerstrom et al.
- Dioxanes such as examples from Fattorusso et al. [54]
- Trioxanes such as compound 16 from the work of O'Neill et al. [55]
- Tetraoxanes exemplified by RKA182 [56]
- Trioxepanes such as compound 7A from Puri et al. [57]

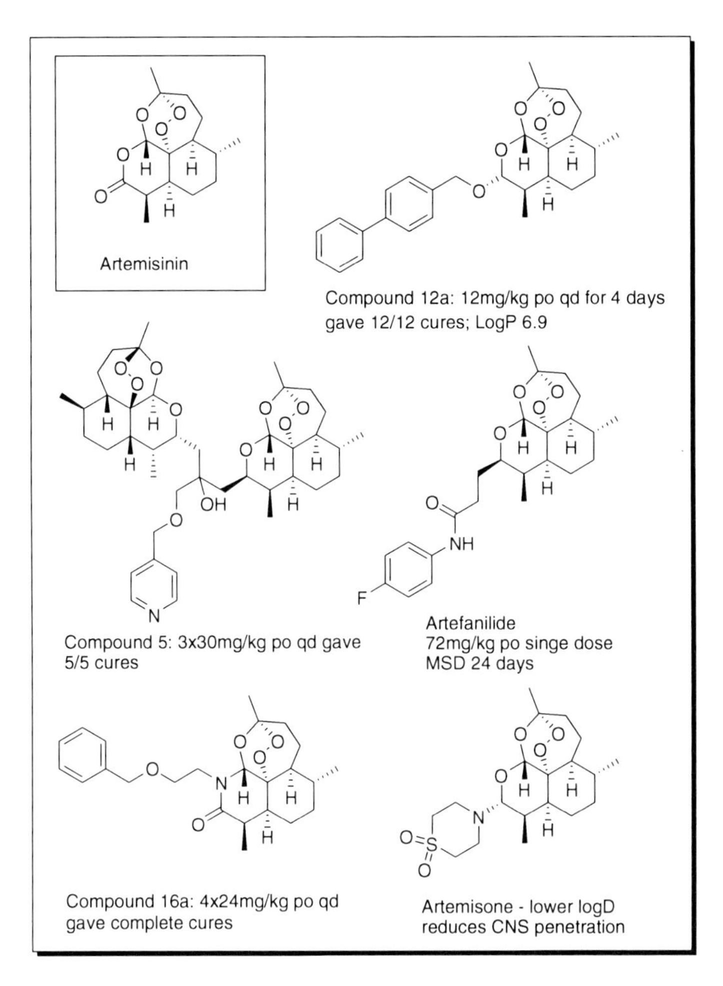

Fig. 6 Representatives of compounds and strategies to optimise artemisinin

The pharmacologically relevant functionality is believed to be the endoperoxide moiety, and the rest of the molecule has been the focus of modification to provide sufficient chemical and metabolic stability, solubility and bioavailability, while at the same time maintaining high efficacy. Interestingly, not all endoperoxides have equal intrinsic potency with the order being trioxolane > trioxane > trioxepane for similarly decorated cores [58].

A further strategy for improving on artemisinin has focused on introducing another validated activity into the molecule through linking to another antimalarial chemotype, such as a 4-aminoquinoline, exemplified by Trioxaquine [59]. The challenge in this case is to show that both parts of the molecule are active in vivo. In parallel with the artemisinin dimers, bis- or tris-synthetic endoperoxides have also been made to increase peroxide load, as illustrated by compound 12A from the work of Singh et al. [60]. From a stereochemistry complexity perspective,

Table 2 Nomenclature for different cyclic endoperoxides

Nomenclature	Structure	Number of oxygens	Ring size
1,2-Dioxolane		2	5
1,2,4-Trioxolane (or Ozonide)	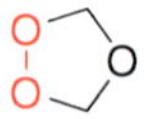	3	5
1,2-Dioxane		2	6
1,2,4-Trioxane	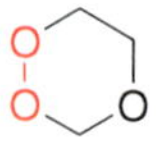	3	6
1,2,4,5-Tetraoxane	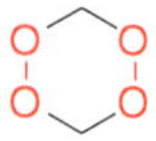	4	6
1,2-Dioxepane	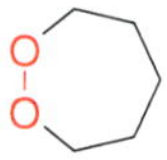	2	7
1,2,6-Trioxepane	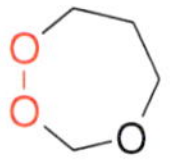	3	7

tetraoxanes are achiral, and this may lead to cheaper cost of goods later in development. Representatives of these synthetic approaches are shown in Fig. 7.

The fact that endoperoxides are clinically validated has attracted many groups to produce their own versions. However, from a medicinal chemistry perspective, it is always important to focus on demonstrating significant advantages over the synthetic frontrunners: OZ439 and OZ277 (arterolane). Such benefits could include demonstration of activity in artemisinin resistance, increases in half-life in man greater than 30 h to support a single-dose cure, or dramatically reducing the total dose in man. Improvements in safety will be difficult to achieve, given the current favourable safety profile of endoperoxides.

4.2 *Amino-Alcohols*

In the context of malaria treatment, the amino-alcohol class is exemplified by quinine and related alkaloids identified from the *cinchona* bark almost 200 years ago. Synthetic versions, such as mefloquine and lumefantrine, have since emerged as new improved versions, with half-lives of several days, compared with quinine which must be given three times per day.

Such compounds are lipophilic bases and their liabilities include low absorption (lumefantrine), nausea and vomiting (mefloquine), cardiovascular issues related to the prolongation of the QTc interval (halofantrine) and psychological side effects (mefloquine). Consequently, new drug discovery projects have focused on modifying the main cores of the structures to reduce compound-related toxicology and CNS

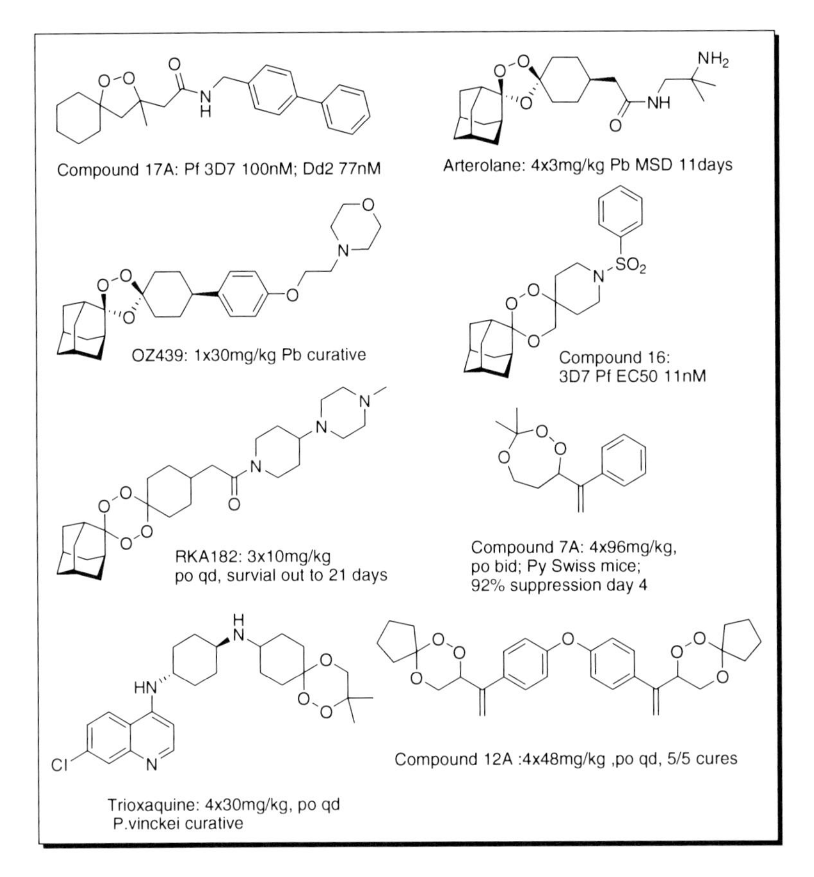

Fig. 7 Examples of different synthetic peroxides

exposure. Diamino-alcohol variants of mefloquine have also been developed with a view to reducing CNS exposure [61, 62]. Furthermore, a series of benzothiophene amino-alcohols explored by Pérez-Silanes et al. have demonstrated in vivo activity [63] as shown in Fig. 8. Interestingly, propranolol, a marketed non-selective beta-blocker and also an amino-alcohol (though with clear differences to other amino-alcohol antimalarials), has also been shown to have modest activity against 3D7 [64].

Successful delivery of any amino-alcohol clinical candidate by improving on the existing medicines still represents a significant challenge. First, achieving high potency is important: current medications have a total human dose in excess of 1 g. Second, the ideal next-generation molecule to treat blood-stage malaria should be safe enough to be given as a single human dose, but with a half-life, consistent with current therapies, of between 14 and 35 days. Balancing this need for a long half-life with measurable risk factors such as QTc prolongation and phospholipidosis is a major challenge. Third, many of the side effects, such as the CNS issues with mefloquine, are difficult to predict from preclinical models, which means that a human safety study is required for the definitive answer.

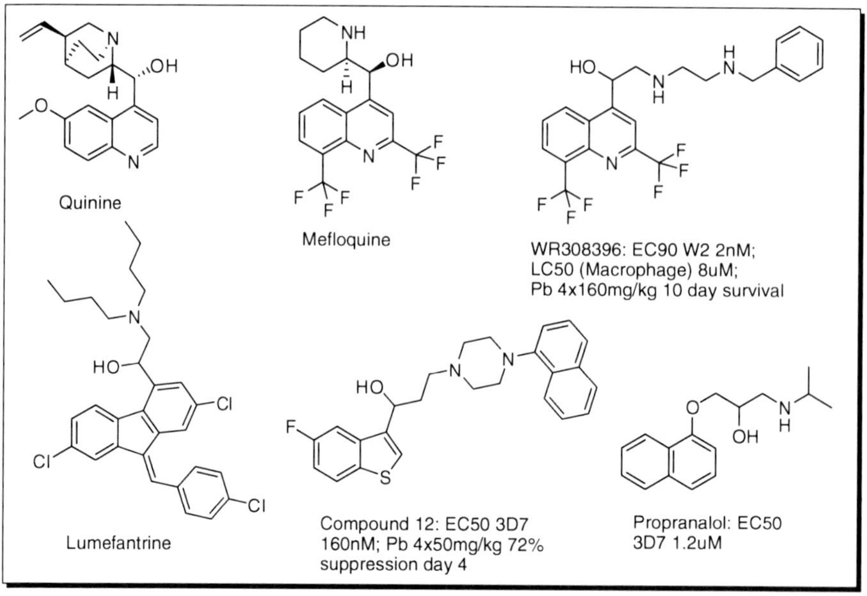

Quinine
Mefloquine
WR308396: EC90 W2 2nM; LC50 (Macrophage) 8uM; Pb 4x160mg/kg 10 day survival
Lumefantrine
Compound 12: EC50 3D7 160nM; Pb 4x50mg/kg 72% suppression day 4
Propranalol: EC50 3D7 1.2uM

Fig. 8 Examples of recent amino-alcohols

4.3 4-Aminoquinolines

Attempts to synthesise quinine in the early twentieth century serendipitously led to the discovery of the 4-aminoquinoline class, exemplified by chloroquine. While amodiaquine is the only other approved and marketed 4-aminoquinoline, there are a host of similar compounds in clinical development including: pyronaridine (actually a 4-aminobenzonaphthyridine), naphthoquine, ferroquine, isoquine, AQ-13 and trioxaquine (see Fig. 4).

The characteristics new 4-aminoquinolines must possess are activity in relevant clinical isolates resistant to chloroquine and amodiaquine and improved cardiovascular and CNS adverse events; a fatal cardiac event occurs with a 50 mg/kg dose of chloroquine compared to the therapeutic total dose of 1.5 g [65]. Other risk factors tend to be compound specific, such as reactive metabolite formation with amodiaquine – that was overcome with isoquine [66], formation of active metabolites (such as with chloroquine and amodiaquine) – that can be beneficial but complicate understandings of PK–PD relationships and toxicology. However, these risks are also off-set by some very favourable properties which the class possesses, namely rapid reductions in parasitemia and extremely long terminal half-lives.

The research on 4-aminoquinolines is immense and only a cursory overview can be given. Examples of compounds prepared to tackle the issues are described in Fig. 9 and the strategies include:

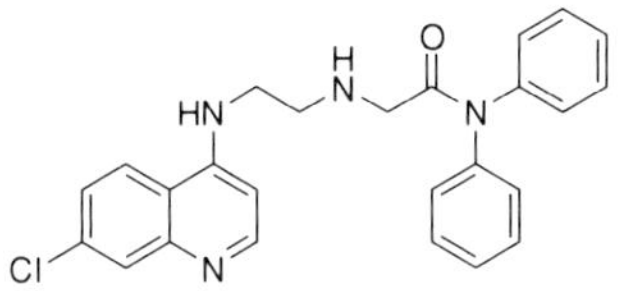

Compound 10: EC50 D6 5nM; Dd2 13nM
Cytotox (murine spleen lymphocytes) 62uM

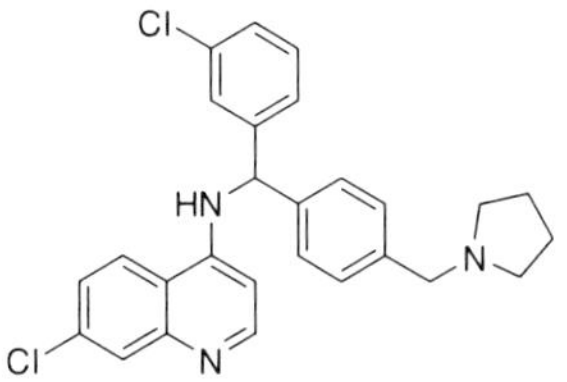

Compound 4g: EC50 D10 22nM
Pb 4x50mg/kg 98% suppression day 4;
Survival out to 18d

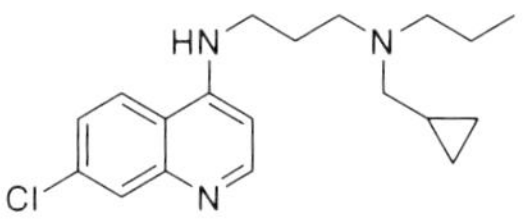

Compound 4: EC50 3D7 15nM; W2 17nM

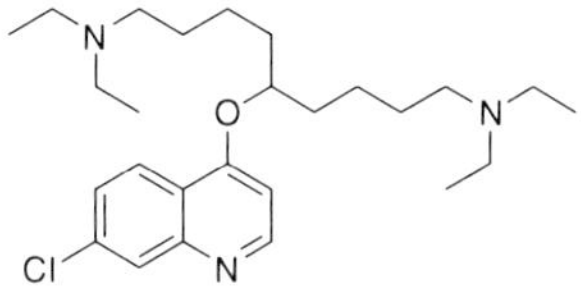

Compound 23: EC50 HB3 73nM,
Dd2 316nM

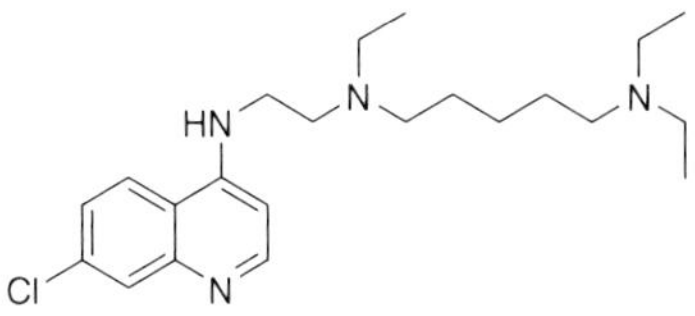

Compound 15: EC50 HB3 5nM,
Dd2 20nM

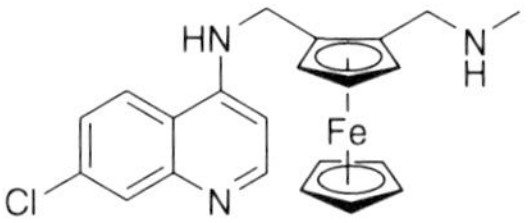

Compound 7a: EC50 Dd2 23nM

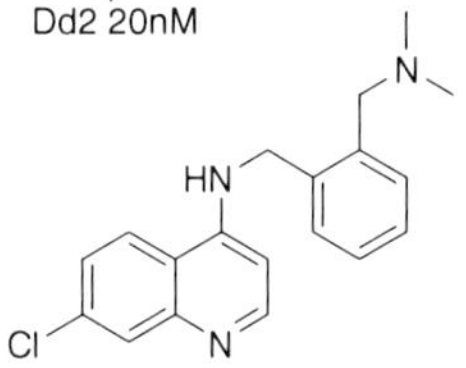

Phenylequine: EC50 D10 13nM; K1 7nM
Cytotox (murine spleen lymphocytes) 70uM
ED50 (Pb) 1.3mg/kg

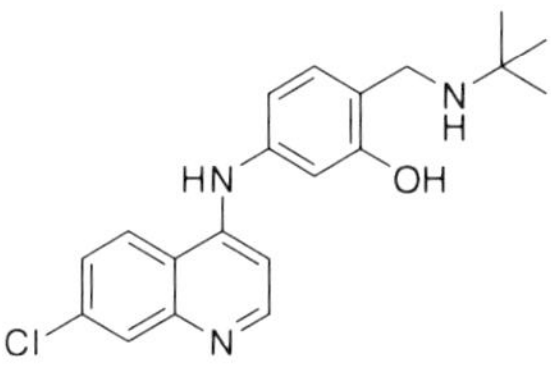

N-*tert*-butyl isoquine

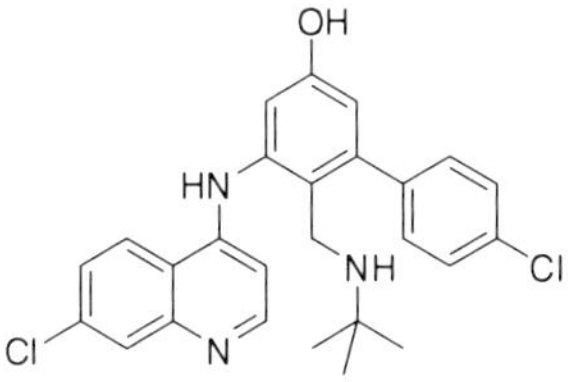

Isotebuquine Compound 1a:
EC50 D6 0.3ng/ml, W2 0.4ng/ml;
Macrophage J774 IC50 6ug/ml

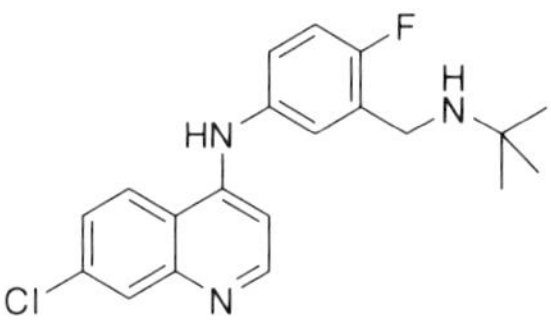

Compound 2k: EC50 HB3 15nM,
Dd2 22nM

Fig. 9 4-Aminoquinolines

- Building in side chains that inhibit the PfCRT resistance pump [67, 68] thus conferring pan-isolate activity, such as compounds 10 (Peyton et al.) and 4g (Campiani et al.)
- Extensive structure activity and property relationship investigations of the core, linker and hydrogen bonding to find compounds with improved properties and safety including compounds 4 (Guy et al.) [69], 23 and 15 (Roepe et al.) [70, 71]
- Inclusion of organometallics to impact cross resistance via potential additional pharmacology such as compound 7a [72]
- Exploiting learnings from clinical candidates, such as phenylequine, which was modelled on ferroquine [73], *N-tert*-butyl isoquine, a follow-up to amodiaquine [67], isotebuquine – which similarly learns from tebuqine [74] and compound 2k – a fluoro-replacement of the phenol hydroxyl in amodiaquine [75]
- Linking to other antimalarial motifs (such as peroxides – as discussed earlier) [76, 77]
- Considerable side chain diversity, such as cyclens [78], carbolines [79], thiazolidinediones [80], thioureas [81], triazines and oxalamides [82]

4.4 8-Aminoquinolines

Primaquine is the only approved 8-aminoquinoline. It is also the only antimalarial approved to eliminate the liver stage parasites (hypnozoites) of vivax malaria. However, as mentioned in the introduction, the drug is not without problems – principally the long duration of treatment (14 days) and the risk of haemolytic anaemia in G6PD-deficient patients.

Finding an improved backup to primaquine is complicated. First, there is no continuous culture system for vivax, adding the complication of variability of primary isolates. In addition, there is no validated in vitro vivax hypnozoite assay, and the best assays require primary liver cells. In vivo there is no murine model, and in vivo anti-relapse efficacy can only be studied using the related *Plasmodium cynomolgi* infection in rhesus monkeys. A further twist here is that this model is strictly speaking only validated for 8-aminoquinolines. Second, improving primaquine's safety profile has been problematic owing to a lack of validated in vitro and in vivo G6PD-deficiency haemolysis assays. Primaquine needs metabolic activation for both activity and the haemolytic side effects. Although several groups are now working on these model systems, this again highlights the issue faced when trying to eliminate a side effect which is only seen in humans, and thus is extremely challenging.

Tafenoquine has emerged over the decades as a potential replacement for primaquine, and was discovered by medicinal chemistry modifications and profiling. Compounds were tested in *P. cynomolgi*-infected rhesus monkeys in vivo, before proceeding to efficacy and safety studies in humans. Tafenoquine has been shown in humans to have anti-relapse efficacy and also a considerably

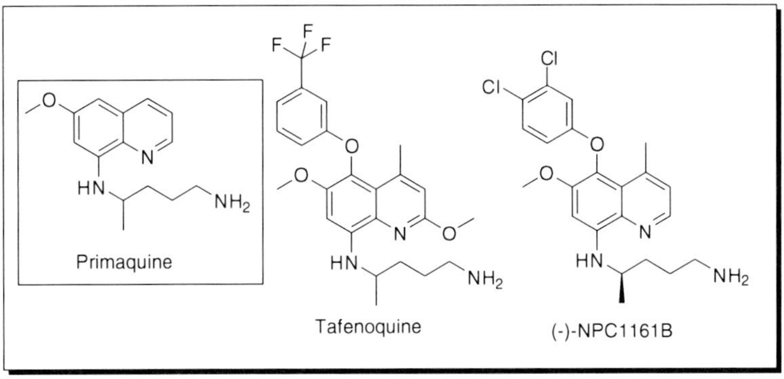

Primaquine Tafenoquine (-)-NPC1161B

Fig. 10 Advanced 8-aminoquinolines

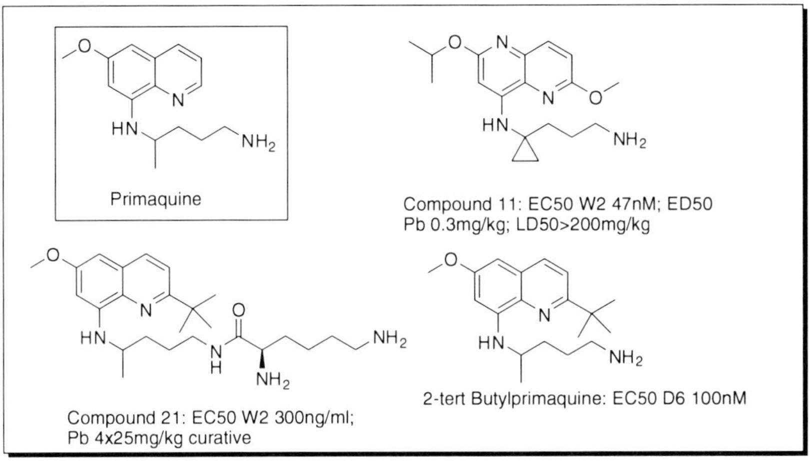

Primaquine

Compound 11: EC50 W2 47nM; ED50 Pb 0.3mg/kg; LD50>200mg/kg

Compound 21: EC50 W2 300ng/ml; Pb 4x25mg/kg curative

2-tert Butylprimaquine: EC50 D6 100nM

Fig. 11 Recent 8-aminoquinolines

longer half-life than primaquine, which could support a single or 3-day treatment regimen, which should result in an improved oral regimen. Studies are currently underway to fully define the safety profile in G6PD-deficient volunteers, before a full-dose range finding Phase II/III study. An additional analogue, (−)-NPC1161B, has also emerged having good preclinical efficacy and potential benefits over the racemate NPC1161C, primaquine and tafenoquine (Fig. 10).

Interestingly, the lack of a viable test cascade has not daunted members of the community in their attempts to deliver improved compounds. Representative examples of approaches in the last few years are shown in Fig. 11, including modifying the core to improve blood-stage activity and reduce metabolism exemplified by compound 11 from Zhu et al. [83], modulation of the side-chain pendant amine as a prodrug, such as compound 21 from Moreira et al. [84] and substitutions to reduce methaemoglobinaemia as a result of blocking formation of certain reactive metabolites, illustrated by 2-*tert*-butyl primaquine [85]. Such approaches have often illustrated progress on the blood stage or safety in preclinical assays and species, but show no evidence of improvement of liver stage activity.

4.5 Natural Products

Historically, natural products have formed the foundation for molecules used in therapy for malaria and many other infectious diseases. Consequently, there has been renewed effort to search for the "next artemisinin" from among marine and plant organism extracts: yeast, bacteria and fungi. An array of approaches exist, which broadly include either the screening of discrete natural product compounds that have been isolated and characterised (including the Queensland engine from Quinn, incorporating compounds filtered for "drug-like" characteristics [86]) or the screening of extracts, followed by fine fractionation, confirmation (parasite and cytotoxicity), purification and structure elucidation [87].

To date, the results from natural product screening have been disappointing in comparison with screening pharmaceutical diversity. Currently, there are no series in lead optimisation or clinical development within the global malaria portfolio that consist of a recently discovered, novel natural product or semi-synthetic variant. There have been hits from screening, but these have to be weighed against the available small molecule screening data when considering priorities for medicinal chemistry follow-up. Furthermore, there are a considerable number of natural product hits that exhibit unattractive chemical features (aldehydes, enones, etc.), considerable chemical complexity, moderate-to-weak potency and/or considerable cytotoxicity [88]. To date, the "new artemisinin" has not been found and whilst this would be of high value and impact, it must be seen as a rare event.

Some recent chemistry has focused on febrifugine analogues to improve the safety profile following certain hypotheses: heterocyclic replacements of the ring phenyl and contraction of the cyclic amine, such as compound 8 [89] and compound 15 [90] (Fig. 12). However, real improvements in tolerability will be difficult to demonstrate without considerable investment in preclinical toxicology, since optimisation against unknown toxicology mechanisms is always a serious challenge.

There is considerable interest and knowledge, particularly in Africa and Asia, with regard to the clinical use of natural product extracts. Such ethnopharmacological data, when verified, can be used to prioritise extracts for deconvolution to determine the key active ingredients. Indeed, this was the context in which quinine and artemisinin were found – the bark or leaf were being used for treatments and were shown to have efficacy. This highlights, perhaps, that a focus on such verified ethnopharmacological data is more likely to bear fruit than the screening of random selections of natural products with no proven link to disease [91].

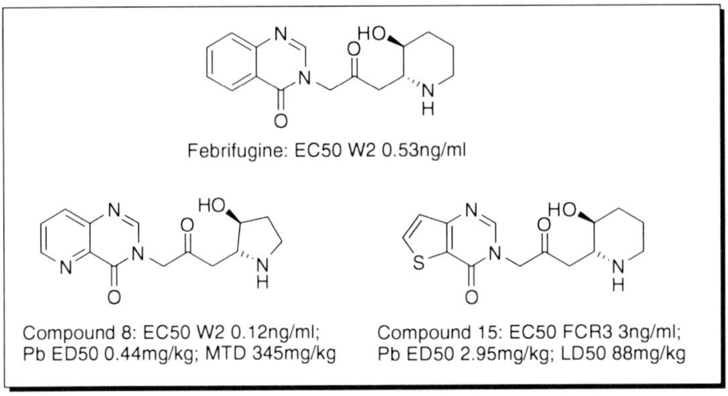

Febrifugine: EC50 W2 0.53ng/ml

Compound 8: EC50 W2 0.12ng/ml; Pb ED50 0.44mg/kg; MTD 345mg/kg

Compound 15: EC50 FCR3 3ng/ml; Pb ED50 2.95mg/kg; LD50 88mg/kg

Fig. 12 Analogues of febrifugine

4.6 Novel and Advanced Chemotypes

In the last few years, a variety of specific novel chemotypes have been advanced into hit-to-lead and lead optimisation campaigns; some even with the potential for progression into the clinic. This section will briefly review recent examples.

4.6.1 Spiroindolone

Researchers at Novartis, led by Diagana, chose a spiroindolone hit from the multitude of actives following a high-throughput screen (HTS) of the corporate collection [92]. Interestingly, the hit – compound 1 (Fig. 13) – was a singleton within the natural products library, though was confirmed to be a fully synthetic analogue that had been so designated due to it having "natural product-like" characteristics. This was rapidly profiled in vivo to demonstrate considerable parasite suppression after a single dose and consequently optimised for potency and metabolic stability. First, ring contraction from the azepine to piperidine ring and second, replacement of the lipophilic Br- with Cl- led to a considerable rise in potency and reduction in lipophilicity. Stereochemical influence on potency, given two stereogenic centres, was confirmed through the synthesis and isolation of both pairs of diastereomers, which showed one isomer to have dominant activity. Finally, substitution in the carboline ring reduced the metabolic turnover resulting in compounds with high bioavailability and extended half-lives. NITD609, one of the fruits of the optimisation process, was profiled in vivo with impressive curative effects in a mouse model infected with *Plasmodium berghei* surpassing that of marketed antimalarials.

NITD609 possesses a unique 3D structure that is novel as an antimalarial and an impressive rapid-killing phenotype [93]. NITD609 entered preclinical development at the start of 2010 and is likely to be dosed to healthy volunteers in early 2011.

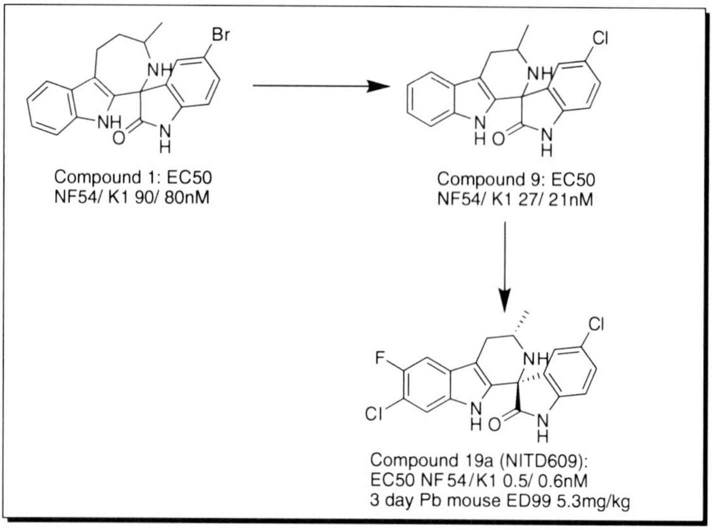

Compound 1: EC50
NF54/ K1 90/ 80nM

Compound 9: EC50
NF54/ K1 27/ 21nM

Compound 19a (NITD609):
EC50 NF54/K1 0.5/ 0.6nM
3 day Pb mouse ED99 5.3mg/kg

Fig. 13 Initial hit to candidate drug for spiroindolones

4.6.2 Aminoindole

Following the screening of the compound library at the Broad Institute, Mazitschek et al. initiated hit-to-lead activities on a novel 2-amino-3-hydroxy-indole series, illustrated by compound 1a (Fig. 14) [94]. The series showed moderate activity across strains with attractive physical properties and reasonable pharmacokinetics resulting in demonstrable in vivo efficacy. Optimisation of the molecule by researchers at Genzyme has focused on the phenyl substituent, and Genz-668764 was selected as a candidate with improved in vitro and in vivo efficacy (E Sybertz and R Barker, personal communication). The attractive features of the series are novelty, activity against resistant strains and an inability, so far, to generate resistant clones (suggesting multipharmacology).

4.6.3 Oxaborole

The team at Anacor, led by Plattner, has identified potential antimalarials from screening a selection of their proprietary oxaborole collection. Such oxaboroles have intriguing structures. Despite scepticism in the medicinal chemistry community due to the electrophilic boron, they have demonstrated efficacy and acceptable safety in preclinical and clinical settings. One such hit, the beautifully simple AN3661 (Fig. 14), was shown to have high, specific potency, with certain close analogues being devoid of activity. AN3661 has been profiled further and shown to have activity against resistant strains, apparent effects on all blood stages and impressive in vivo activity despite a short half-life, with ED_{90} (4-day Peters test

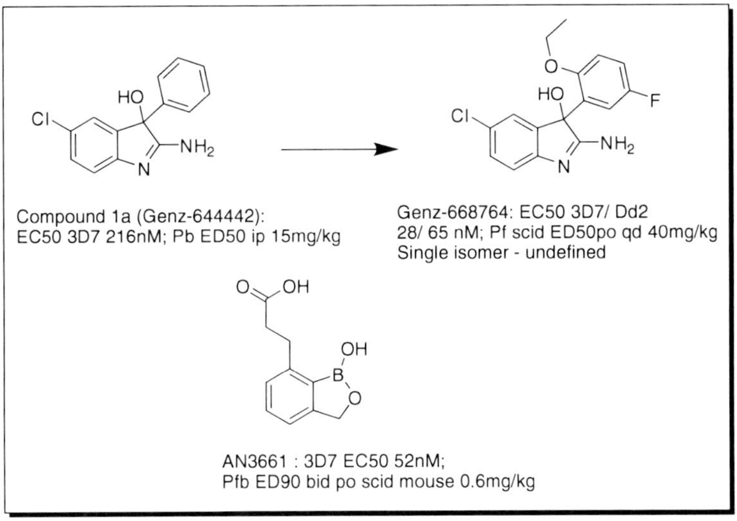

Compound 1a (Genz-644442): EC50 3D7 216nM; Pb ED50 ip 15mg/kg

Genz-668764: EC50 3D7/ Dd2 28/ 65 nM; Pf scid ED50po qd 40mg/kg Single isomer - undefined

AN3661 : 3D7 EC50 52nM; Pfb ED90 bid po scid mouse 0.6mg/kg

Fig. 14 Novel 2-amino-3-hydroxy-indoles and oxaborole AN3661

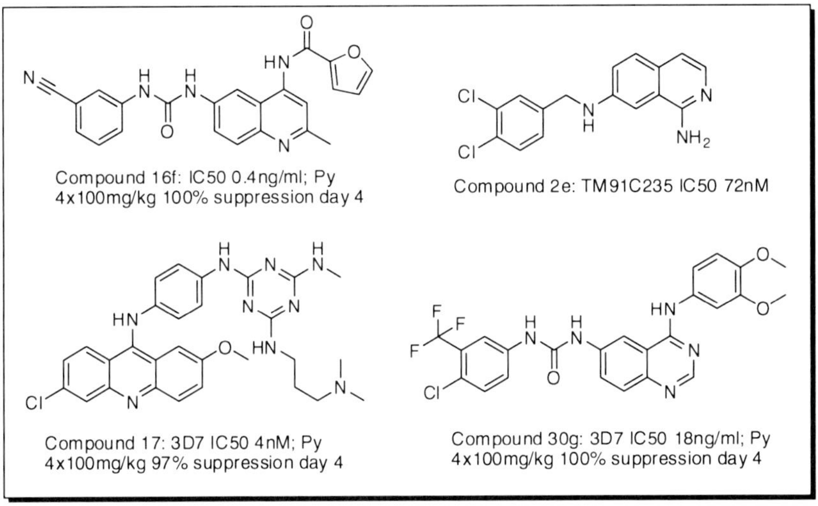

Compound 16f: IC50 0.4ng/ml; Py 4x100mg/kg 100% suppression day 4

Compound 2e: TM91C235 IC50 72nM

Compound 17: 3D7 IC50 4nM; Py 4x100mg/kg 97% suppression day 4

Compound 30g: 3D7 IC50 18ng/ml; Py 4x100mg/kg 100% suppression day 4

Fig. 15 Diverse chemotypes from phenotypic optimisation projects

in NOD SCID mice injected with *P. falciparum*-infected erythrocytes) of 0.6 mg/kg (Plattner J, 2010, ICOPA, oral presentation).

4.6.4 Diverse Chemotypes

A wide variety of chemotypes have been disclosed in the last 5 years, where the series have been optimised in the absence of the mechanism of action. They represent a range of lipophilicities and functionality and, as such, do not necessarily possess features that are consistent for lead optimisation and drug development.

Those covered include 4-amino, 6-ureidoquinolines that are distinct from traditional 4-aminoquinolines [95], amino-isoquinolines [96], large acridines [97] and anilino-quinazolines that resemble the quinoline ureas [98] – as shown in Fig. 15.

Other examples include several diaminopyrimidines such as compounds 1ap [99] and 40 [100], acridones [101], benzophenones [102], bicyclo-amine derivatives [103] and ferrocenic quinoxalines [104] – as shown in Fig. 16.

Five additional series include the in vivo active pyrazolines [105], imidazole dioxolanes [106], potent dyes such as the phenoxazinium salts [107], the intriguing indolone N-oxides with weak in vivo efficacy [108] and benzamides from researchers at GNF [109] (Fig. 17).

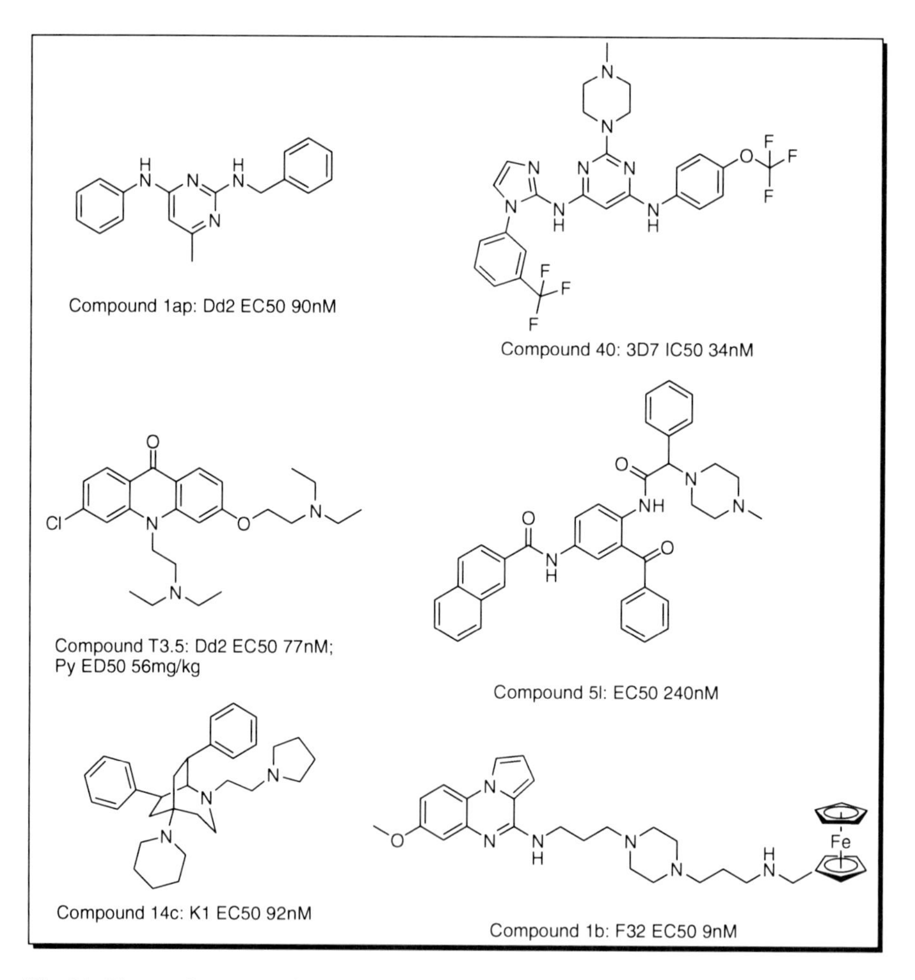

Fig. 16 Diverse chemotypes from phenotypic optimisation projects

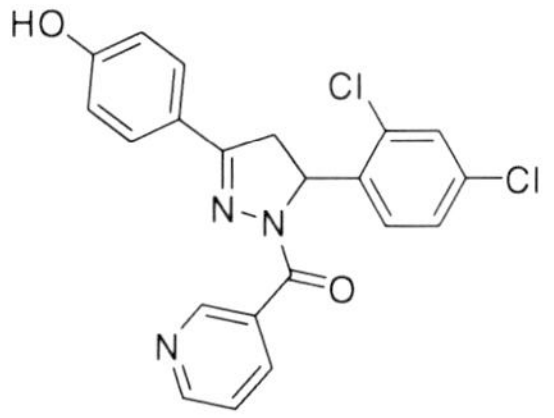

Compound 1a: MRC-02 EC50 37nM;
Pb 4x50mg/kg 69% suppression day 4

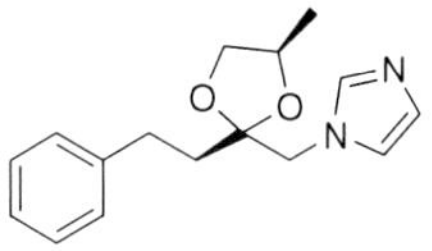

Compound QC-35: EC50 340nM

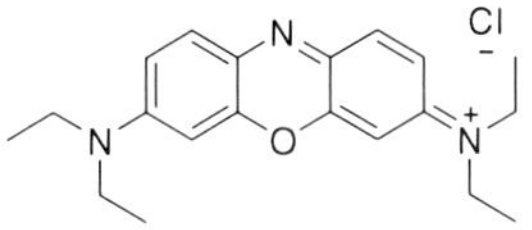

Compound 1a Basic Blue 3:K1 EC50 4nM;
Pb 4x90mg/kg po curative

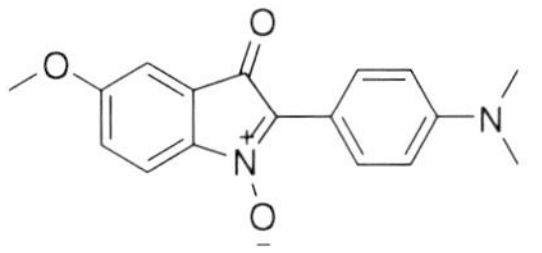

Compound 26: 3D7 EC50 1.7nM;
Pb4x30mg/kg21% suppression day4

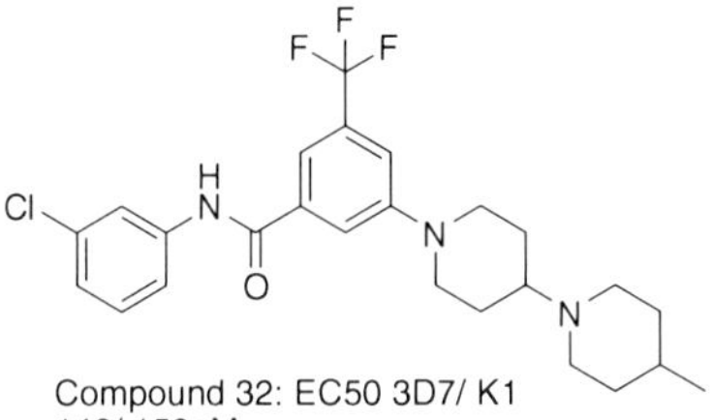

Compound 32: EC50 3D7/ K1
140/ 150nM

Fig. 17 Diverse chemotypes from phenotypic optimisation projects

4.7 Liver Stage Acting Antimalarials

Aside from the 8-aminoquinoline described earlier, all examples discussed have focused on erythrocyte stage activity. Few articles have appeared with novel chemotypes focused on liver stage parasites, and, in the main, these have involved identifying actives rather than optimisation – see Fig. 18. Mota et al. identified LY411575, an inhibitor of human γ-secretase and signal peptide peptidase as a potent inhibitor of *P. berghei* hepatic parasites [110]. Mazier et al. have reported liver stage activity (*Plasmodium yoelii* and *P. falciparum*) for a variety of well-characterised molecules, including several ionophores and HIV protease inhibitors (e.g. Saquinavir) [111]. The plant-derived Morphinan, tazopsine was also identified as killing *Py* and *Pf* liver stages [112]. Interestingly Hobbs et al. have independently reported liver stage effects of HIV protease inhibitors [113], though with different inhibitory values reflecting likely differences in the assay and absence of a standard control for comparison. Finally, bisphosphonates have been shown to have weak liver stage activity [114], and triclosan has also demonstrated activity in a *P. berghei* liver

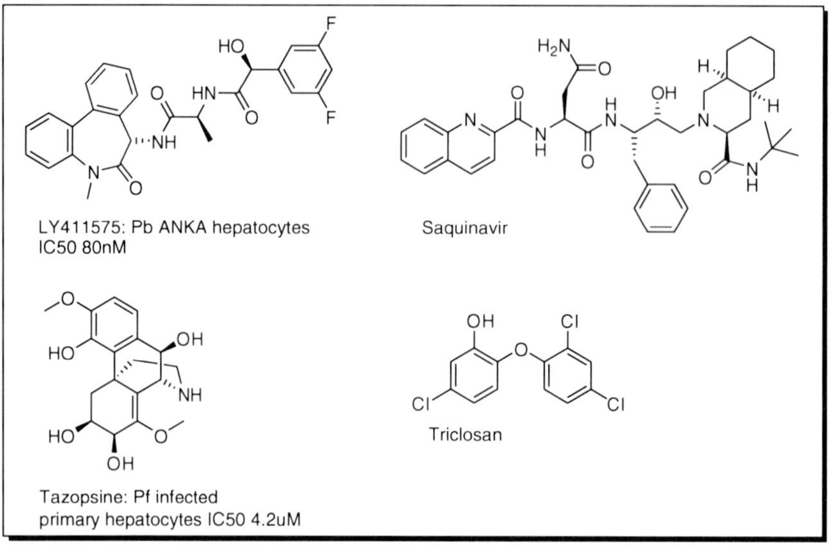

LY411575: Pb ANKA hepatocytes IC50 80nM

Saquinavir

Tazopsine: Pf infected primary hepatocytes IC50 4.2uM

Triclosan

Fig. 18 Compounds found to have rodent or falciparum malaria liver stage activity

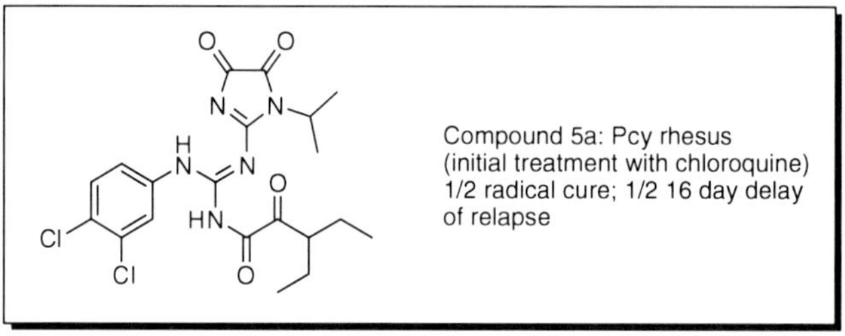

Compound 5a: Pcy rhesus (initial treatment with chloroquine) 1/2 radical cure; 1/2 16 day delay of relapse

Fig. 19 Imidazolidinedione

stage assay [115]. These reports have focused on the killing of liver stage parasites from falciparum or rodent malarias. Consequently, they give no insight on the likely impact on hypnozoites in vivax liver stage.

The only series to focus on blocking liver stage relapse are the imidazolidinediones reported by Lin et al. [116]. This work has had to rely on primary in vivo models with *P. cynomolgi* rhesus as the defining experiment, therefore using a vivax surrogate (Fig. 19). Given the difficulty working with and interpreting data from a primary in vivo model, it remains unclear what constitutes the active species and consequently optimisation has been extremely difficult, particularly since several observed rearranged/metabolised products have been observed.

4.8 *Compounds Affecting Transmission*

Transmission-blocking properties of a compound, assessed by the number of oocysts formed in the mosquito, can arise from an impact on gametocytes as well as on vector stage parasites. Given improvements in both in vitro gametocyte and vector stage technologies, compounds are starting to emerge with transmission-blocking potential.

White et al. has reported falciparum transmission-blocking potential of quinine, primaquine and artesunate based on oocyst counts [117], and Trenholme et al. have tested the gametocytogenesis of eight antimalarials [118]. Trioxaquines have demonstrated effects on falciparum Stage II–V gametocytes similar to artesunate [119], the novel DNA-binding Centamycin has also shown impacts on transmission blocking as measured by oocyst numbers [120] and gametocytocidal activity has been demonstrated with one HIV protease inhibitor [121]. Interestingly, azithromycin has shown a reduction in sporozoite production in the vector, from *P. berghei* studies, believed to be due to the impact on the gametocyte–ookinete transition [122].

For those compounds possessing dual blood stage and transmission-blocking activity, the transmission-blocking potential will depend upon the difference in activities between the erythrocytic and post-erythrocytic stages, and whether appropriate concentrations for the latter can be achieved ultimately in patients.

5 Moving from Hits to Leads to Candidates: Target-Based Optimisation

Many of the challenges associated with choosing an appropriate biological target and initiating a target-based screening approach have been described in Sect. 3.2.2.

Pursuing a drug discovery programme against a known target can be advantageous since it may allow:

- Clear understanding of the structure–activity relationship (SAR) against the target
- Rapid identification of hits via screening of compounds already known to be active against the target class
- Opportunities for structure-based design if structural data are available (X-ray/homology modelling) to rapidly improve in vitro potency or selectivity against a target
- Parallel screening against related human targets to reduce the risk of toxicity in man. The consequences of inhibiting corresponding or similar human targets should be considered in advance

During the progression of a target-based project, a broad correlation between target potency and activity in the whole parasite would be expected. Divergence from this general correlation may imply permeability/efflux issues, organelle accumulation or could be as a result of multiple mechanisms within the whole parasite.

The compounds and related targets discussed below are broadly classified into a number of different biological pathways. We have focused on publications in which activity in the whole parasite has been demonstrated below <1 μM. Whether the activity observed within the whole parasite is due to inhibition of the desired target is, however, often difficult to establish.

5.1 Nucleic Acid Synthesis Pathways

5.1.1 Pyrimidine Biosynthesis

Many antimalarial agents either directly or indirectly have an affect on pyrimidine metabolism. *Plasmodium* species are unusual in which they lack pyrimidine salvage enzymes, and so the de novo pathway provides their only source of pyrimidines for cell growth. The pyrimidine biosynthesis pathway, therefore, represents an attractive target for the identification of novel antimalarials.

Carbonic Anhydrase

The metalloenzyme carbonic anhydrase catalyses the interconversion between carbon dioxide and bicarbonate and is essential in many processes in eukaryotes and prokaryotes. Krungkrai et al. [123] describe a series of sulphonamides as PfCA inhibitors. Compound 10 (Fig. 20) shows activity in the mouse malaria model despite only moderate activity in the erythrocyte assay. The in vivo efficacy may be due to carbonic anhydrase inhibition or alternative modes of action may be important.

Cytochrome Bc1

Inhibition of parasite mitochondrial respiration through the cytochrome bc1 complex has been demonstrated as a validated approach with the introduction of atovaquone in

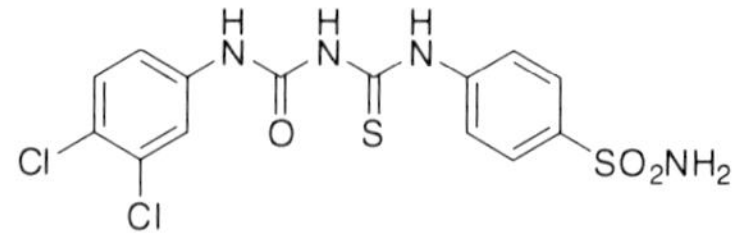

Compound 10

PfCA Ki 0.18uM
Pf IC50 1uM

Fig. 20 Carbonic anhydrase

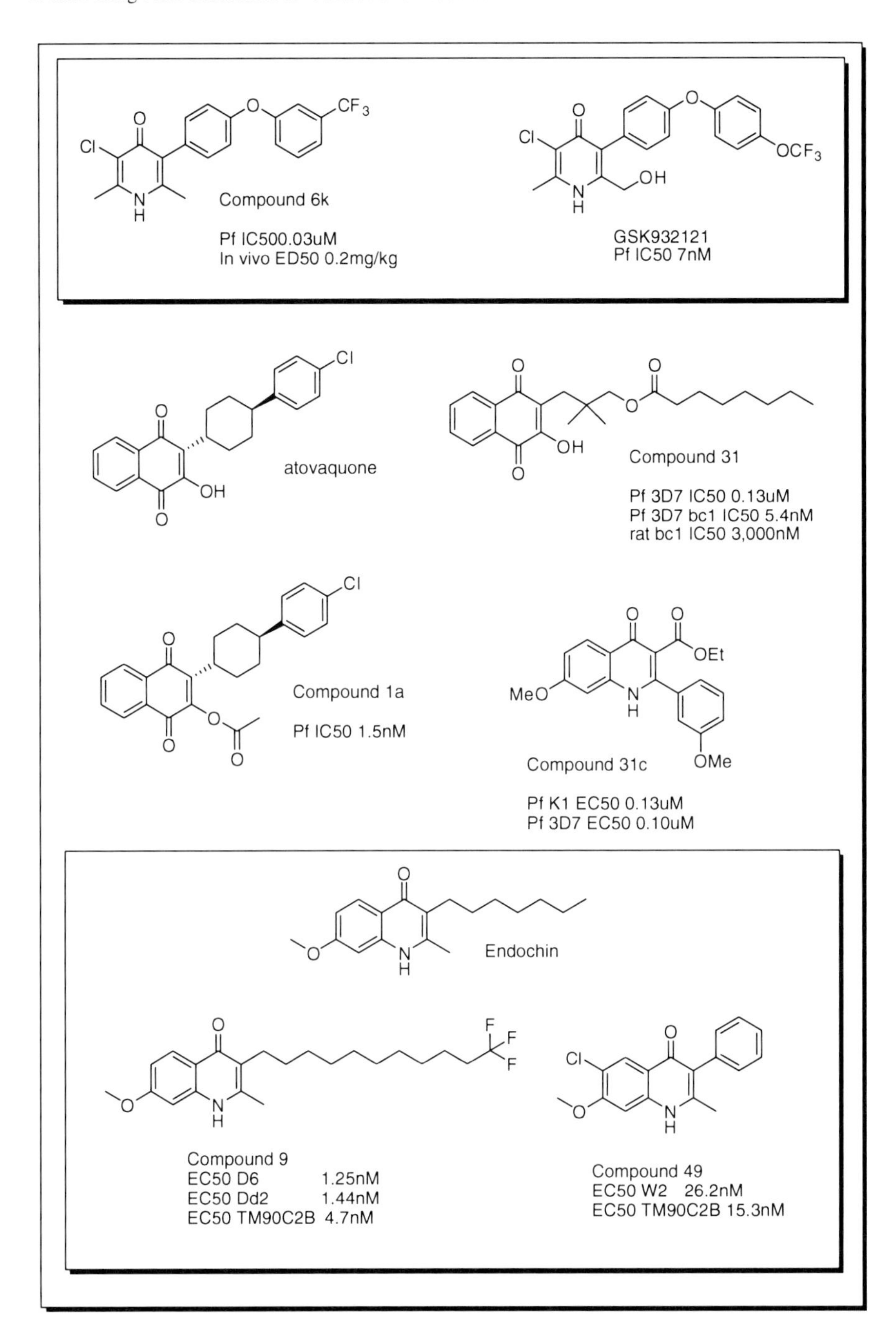

Fig. 21 Confirmed and suspected cytochrome bc1 inhibitors

1997. A key issue in identifying compounds that are active against this target is to ensure selectivity for *Pf* bc1 over human mitochondrial respiration targets.

Scientists at Wellcome and GSK have identified a series of diaryl ether-substituted 4-pyridones (e.g. compound 6k in Fig. 21). This series shows potent activity against *P. falciparum* in vitro as well as good efficacy in murine models of malaria [124]. GSK932121 is currently undergoing Phase I clinical trials in healthy volunteers [12, 125].

A number of groups describe close analogues of atovaquone (compound 31 [126], compound 1a [127]), but at this stage it is not clear whether the compounds described have any advantage over atovaquone.

Zhang et al. describe the SAR of a series of 4-oxo-3-carbonyl quinolones [128] thought to target the cytochrome bc1 complex. Compound 31c was identified as the most potent analogue and in addition showed high aqueous solubility (>100 μM). The groups of Riscoe [129] and Manetsch [130] describe modifications to the endochin scaffold that result in potent compounds showing activity in the atovaquone-resistant strain TM90C2B.

Dihydroorotate Dehydrogenase

Dihydroorotate dehydrogenase (DHODH) is the fourth enzyme in the de novo pyrimidine pathway. Inhibition of *Pf* DHODH as an antimalarial strategy has recently been reviewed by Phillips et al. [131]. HTS approaches have identified a number of chemical series that inhibit *Pf* DHODH with IC_{50}s below 100 nM and show no significant inhibition of human DHODH at 30 μM [25, 131]. Further development of these hits has led to compounds that demonstrate potent activity in the *P. falciparum* erythrocyte assay (see Fig. 22). The series show a clear correlation between enzyme activity and whole parasite activity. In vivo efficacy in the

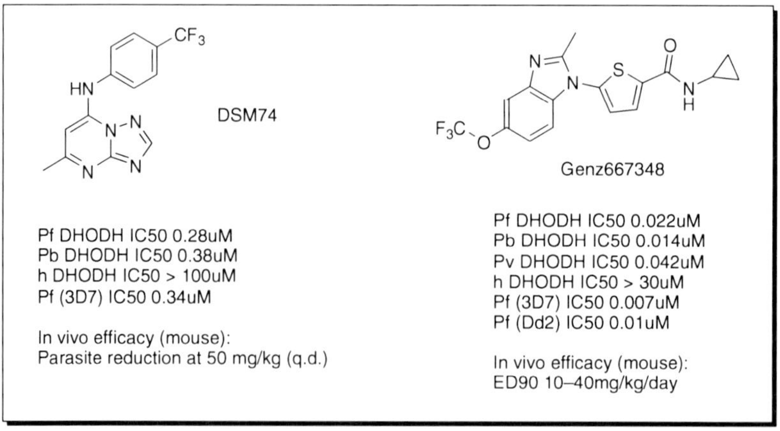

Fig. 22 DHODH inhibitors

P. berghei mouse model of malaria has also been demonstrated – exemplified by compounds DSM74 [132] and Genz667348 [133].

Orotodine 5′-Monophosphate Decarboxylase

Orotodine 5′-monophosphate decarboxylase (OMDC) plays an important role in the de novo synthesis of uridine-5′-monophosphate (UMP), which then serves as the substrate for the synthesis of other pyrimidine nucleotides. OMDC catalyses the decarboxylation of orotidine monophosphate to UMP. Bello et al. [134] report the design and synthesis of several C6 derivatives of uridine and uridine 5′-monophosphate, but as an in vitro antimalarial they are weak (IC_{50}s > 1 μM).

5.1.2 Folate Biosynthesis

Tetrahydrofolic acid is a key cofactor in the synthesis of pyrimidines such as thymidine, methionine and in 1-C metabolism in general. Inhibitors of two key enzymes in the folate biosynthesis pathway, dihydropteroate synthase (DHPS) and dihydrofolate reductase (DHFR) play an important role in the treatment of bacterial and protozoal infections.

Dihydrofolate Reductase–Thymidylate Synthase

Anti-folates such as pyrimethamine and cycloguanil are active-site inhibitors of the malarial DHFR. Resistance to pyrimethamine has arisen in parts of the world due to point mutations in DHFR. New *Plasmodium* DHFR inhibitors will need to demonstrate activity against pyrimethamine-resistant strains while maintaining good selectivity over human DHFR. Additional profiling of the known anti-folate QN254 [135] has shown that it inhibits both the wild-type and the quadruple-mutant forms of the DHFR enzyme. QN254 has also shown potent activity against clinical isolates, accompanied by oral efficacy in the mouse model; however, further development of this compound has been stopped owing to its inadequate therapeutic index. P218, a DHFR inhibitor that is also potent against resistant strains, is in clinical development in the MMV portfolio (see footnote 3), see Fig. 23. Additional details on the profile of P218 are yet to be released.

5.1.3 Deoxyuridine 5′-Triphosphate Nucleotidohydrolase

Deoxyuridine 5′-triphosphate nucleotidohydrolase (dUTPase) is a key enzyme involved in nucleotide metabolism, catalysing the hydrolysis of deoxyuridine triphosphate (dUTP) to deoxyuridine monophosphate (dUMP). Nguyen et al. [136] have reported a series of acyclic nucleoside analogues as inhibitors of *Pf* dUTPase. Compound 2a (Fig. 24) demonstrates moderate potency in both the *Pf*

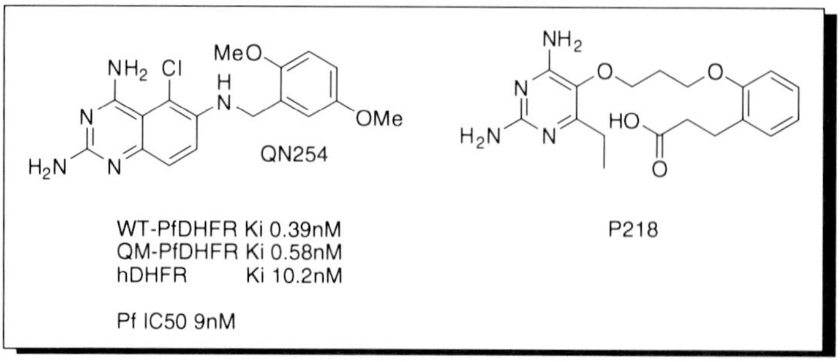

Fig. 23 DHFR inhibitors

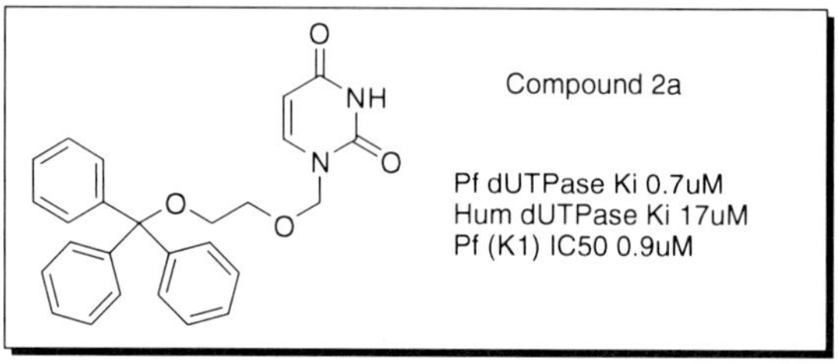

Fig. 24 dUTPase inhibitor

dUTPase assay and in the *P. falciparum* whole cell assay; the series generally shows good selectivity over the human enzyme. Improvements in potency as well as absorption, distribution, metabolism and excretion properties are likely to be required to demonstrate in vivo efficacy for the series.

5.1.4 Purine Biosynthesis

Plasmodia lack the de novo pathway for the synthesis of purine nucleotides which in turn are required for DNA and RNA synthesis. The purine bases required by the parasite have to be transported from the host. Biological processes in the parasite purine salvage pathway therefore provide attractive targets for new anti-plasmodial agents.

Too et al. [137] have reported the screening of purine analogues as potential new antimalarial compounds. Analogue 4b (Fig. 25) shows modest anti-plasmodial activity in vitro. The mode of action of the compound, however, has not been determined.

Purine Nucleoside Phosphorylase

Purine nucleoside phosphorylase (PNP) in the purine salvage pathway is an essential enzyme for the parasite. Phosphorolysis of inosine by PNP yields hypoxanthine – a major precursor for purine salvage. Hypoxanthine is the common precursor for all purine nucleotides in the parasite. The previously described Immucilin-H (forodesine)

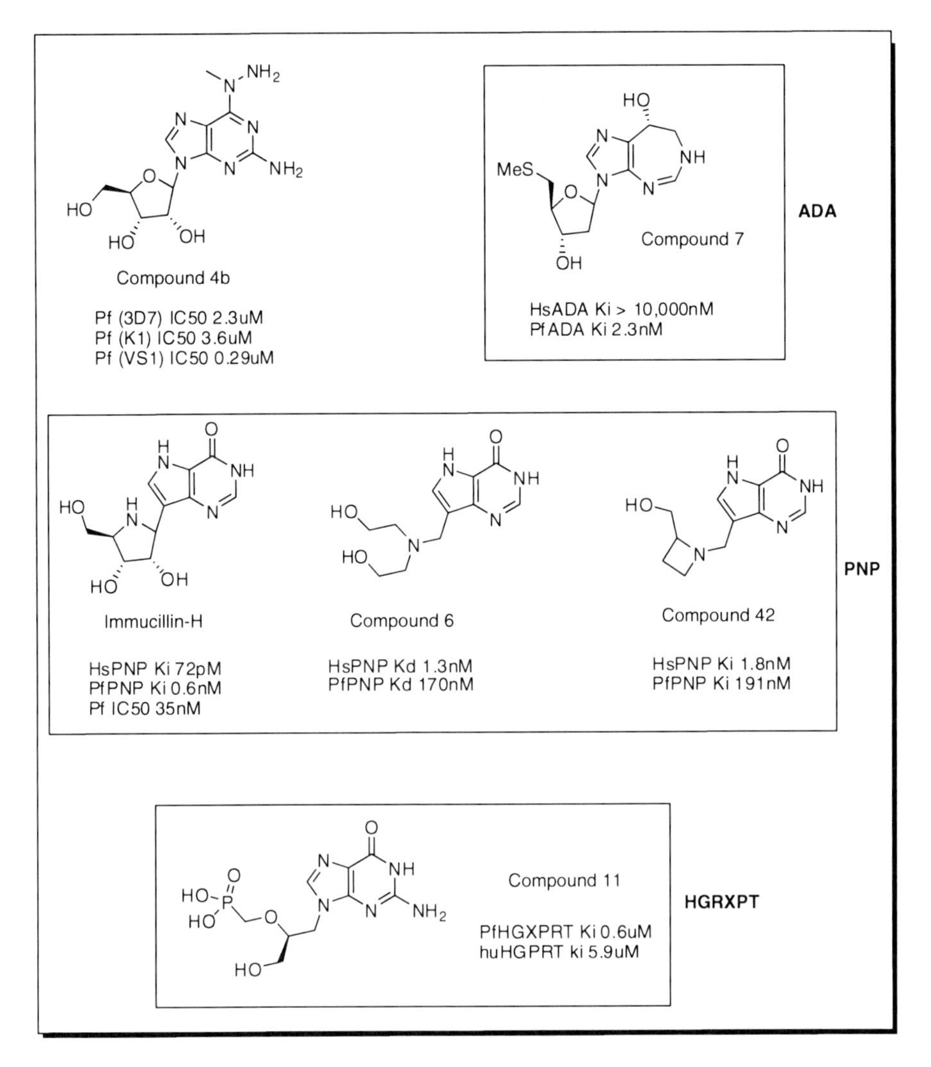

Fig. 25 Purine biosynthesis inhibitors

[138] is a potent inhibitor of human and *P. falciparum* PNP; it also shows potent activity in the *P. falciparum* blood assay. Additional analogues have also recently been reported [139, 140], with various levels of potency against the human target. No whole blood *P. falciparum* data have been reported in the latest publications.

Adenosine Deaminase

Adenosine deaminase (ADA) deaminates adenosine and plays an essential role in purine metabolism. Tyler et al. [141] have reported a series of coformycin analogues (e.g. compound 7) that display high levels of selectivity for *Pf* ADA over human ADA. No *P. falciparum* whole cell data are described.

Hypoxanthine–Guanine–Xanthine Phosphoribosyltransferase

Hypoxanthine–guanine–xanthine phosphoribosyltransferase (HGXPRT) is another key enzyme in the purine pathway. HGXRPT synthesises 6-oxopurine nucleoside monophosphates from the corresponding base. Keough et al. [142] describe a series of acyclic nucleoside phosphonates with some selectivity for HGXRPT over the corresponding human enzyme HGPRT. Activity in the *P. falciparum* erythrocyte assay is modest (~1 μM), and further improvements in potency for this approach will be required.

5.2 Degradation/Catabolic Pathways

5.2.1 Haemoglobin Processing

One of the crucial stages of the parasite life cycle during human infection is the degradation of haemoglobin – a process essential for the parasites propagation [143]. Degradation takes place within a specialised acidic food vacuole. Several enzymes have been shown to be involved in haemoglobin digestion – aspartic proteases (plasmepsins) and the closely related histo-aspartic protease (HAP), cysteine proteases (falcipains), a metalloprotease (falcilysin) and dipeptidyl aminopeptidase 1.

Cysteinyl Proteases: Falcipains

The falcipains are *P. falciparum* cysteine proteases sharing features with papain cysteine proteases. The function of falcipain-1 (FP1) remains unclear; however, the functions of FP2 and FP3 are well characterised and include degradation of host haemoglobin. FP2 and FP3 show a high level of sequence homology but different substrate specificity. Cysteine protease inhibitors that inhibit FP2 and FP3 block haemoglobin hydrolysis and parasite development. The main classes of inhibitors for the falcipains are peptides or peptidomimetics contain the standard cysteine inhibitor pharmacophores such as vinyl sulphones, halomethyl ketones and aldehydes. Safety issues may arise with these types of structures related to their potential for irreversible binding to proteins. A recent review gives a detailed account of falcipain-2 inhibitors [144].

Coteron et al. [26] have explored a series of heteroaryl nitriles (e.g. compound 253 in Fig. 26). Potent in vitro inhibitors of the FP2 and FP3 give rise to potent activity in the *P. falciparum* erythrocyte assay. In vivo profiles and selectivity against human cysteine proteases have yet to be reported.

Chipeleme et al. [145] describe a series of compounds with moderate FP2 and *P. falciparum* activity (e.g. compound 8c). The mechanism of these compounds is

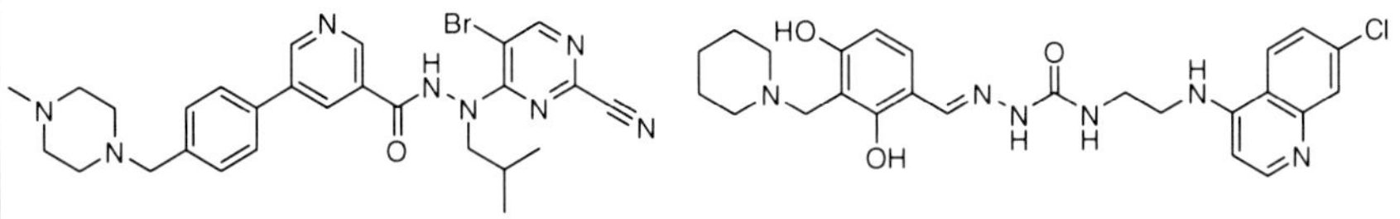

compound 253

FP2 IC50 < 0.5nM
FP3 IC50 < 3nM
Pf (W2) IC50 1nM

compound 8c

FP2 IC50 0.63uM
Pf (W2) IC50 0.27uM

Fig. 26 Falcipain inhibitors

unclear because there is no clear correlation between their FP2 activity and their anti-plasmodial activity. Their antimalarial activity is likely to be due to a combination of mechanisms.

Additional structures have been reported in the literature, but their activity is modest and they are not included in this review.

Aspartyl Proteases: Plasmepsins

At least ten genes encoding aspartyl proteases have been found in the *P. falciparum* genome; however, the precise role of each of the plasmepsins is not clear. The most extensively studied plasmepsins are those involved in the metabolism of haemoglobin within the food vacuole – Plm I, II, IV, and HAP.

Inhibitors of the plasmepsins have been described containing many of the standard aspartyl protease tetrahedral transition-state mimics such as statines and reduced amides; inhibitors of plasmepsins have been reviewed by Ersmark et al. [146].

Significant challenges in identifying novel antimalarial compounds in this field include:

- Compounds are often peptidic in nature leading to potential issues in achieving good oral bioavailability
- Potent enzyme activity frequently does not lead to good whole cell activity
- Inhibition of more than one member of the plasmepsin family may be required for good anti-plasmodial activity
- Selectivity over related human aspartyl proteases (the most similar being Cathepsin D)

In recent work, hydroxymethylcarbonyl isostere-based dipeptidomimetics (e.g. compound KNI-10527) [147, 148], shown in Fig. 27, shows good PlmII activity but modest activity in the *P. falciparum* erythrocyte assay.

A series of non-peptidic dibutylamino amides (e.g. compound 18) show activity in both a Pfm II FRET assay and a *P. falciparum* erythrocyte assay. There does not appear to be a correlation between enzyme and whole cell activity – this could

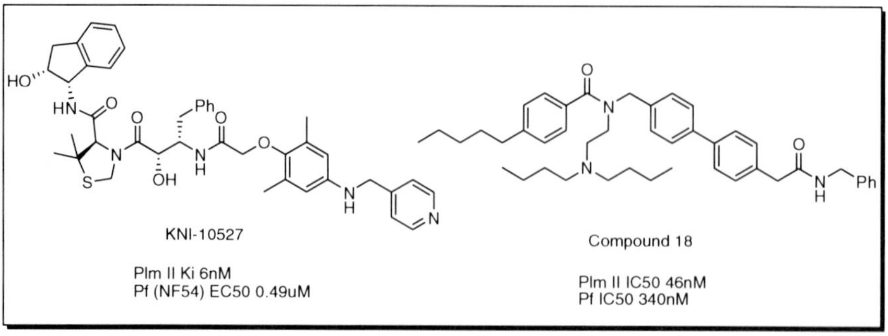

KNI-10527

Plm II Ki 6nM
Pf (NF54) EC50 0.49uM

Compound 18

Plm II IC50 46nM
Pf IC50 340nM

Fig. 27 Plasmepsin inhibitors

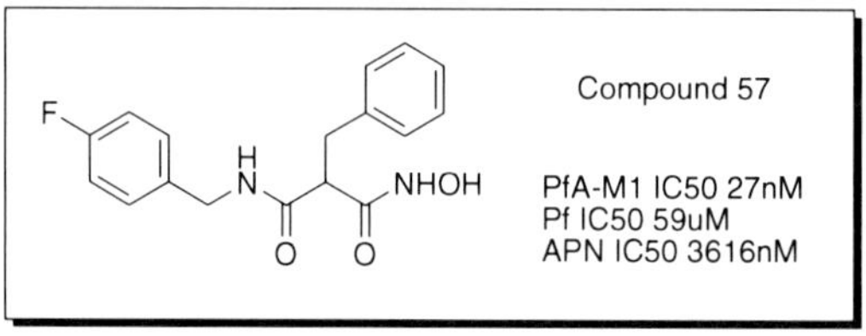

Compound 57

PfA-M1 IC50 27nM
Pf IC50 59uM
APN IC50 3616nM

Fig. 28 Aminopeptidase inhibitor

reflect the varying ability to both penetrate and accumulate in the parasite, varying plasmepsin inhibitory profiles or alternative modes of action.

Plasmodial Aminopeptidase

Plasmodial aminopeptidase (PfA-M1) is involved in haemoglobin breakdown and erythrocyte reinvasion. Flipo et al. [149] report a series of hydroxamic acids as PfA-M1 inhibitors that demonstrate selectivity over the host enzyme (APN, neutral aminopeptidase). Despite potent activity against the *P. falciparum* enzyme, the compounds (e.g. compound 57, Fig. 28) only show very weak activity in the *P. falciparum* erythrocyte assay. The reason for the discrepancy between enzyme and whole cell activity is not clear.

5.2.2 Heat Shock Proteins 70/90

Heat shock proteins (HSPs) are a class of highly conserved molecular chaperones which facilitate protein folding – malaria HSPs have recently been reviewed as drug targets [150]. Chiang et al. have shown that a series of pyrimidones inhibit Hsp70 and are active in the *P. falciparum* whole cell assay (compound MAL2-213) [151]. Shahinas et al. have used an HTS approach to identify three PfHSP90 inhibitors that are active in the *P. falciparum* whole cell assay with IC_{50}s below 100 nM and are reported to demonstrate synergistic activity in the presence of chloroquine [152]

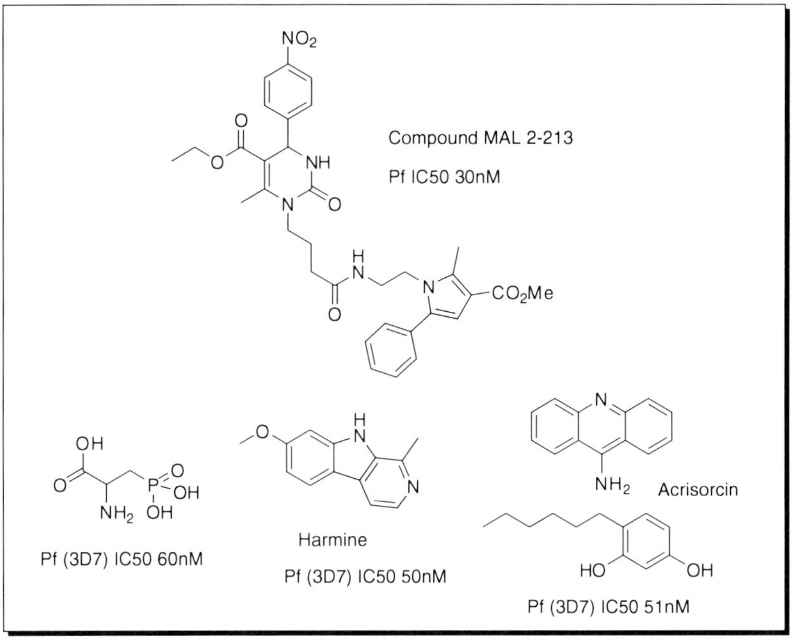

Compound MAL 2-213

Pf IC50 30nM

Pf (3D7) IC50 60nM

Harmine

Pf (3D7) IC50 50nM

Acrisorcin

Pf (3D7) IC50 51nM

Fig. 29 Heat shock protein inhibitors

(Fig. 29). Demonstration of in vivo efficacy and selectivity over related human targets will be an important milestone for this target.

5.3 Anabolism/Synthesis Pathways

5.3.1 Non-Mevalonate Pathway

The non-mevalonate pathway for isoprenoid biosynthesis is considered essential in *Plasmodium* species. The pathway starts with the condensation of pyruvate and glyceraldehyde 3-phosphate to give 1-deoxy-D-xylulose 5-phosphate (DOXP). The DOXP pathway is present in algae, higher plants, bacteria and malaria parasite. The second enzyme in the pathway, DOXP reductoisomerase (DXR), has been under significant investigation.

1-Deoxy-D-Xylulose 5-Phosphate Reductoisomerase

Fosmidomycin and its acetyl analogue (FR900098) are potent, selective inhibitors of DXR; they also inhibit *P. falciparum* in vitro and demonstrate efficacy in rodent models of malaria. Fosmidomycin also demonstrates efficacy in human trials [153]; however, it suffers from low bioavailability and a relatively short plasma half-life.

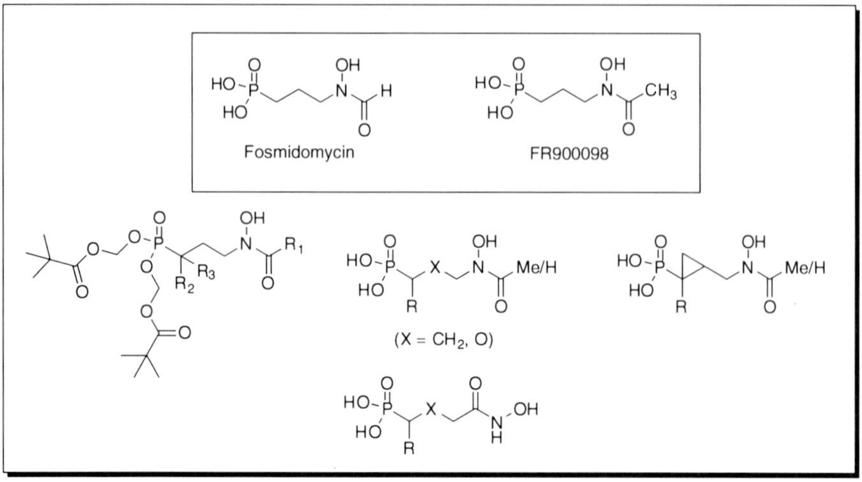

Fig. 30 DXR inhibitors

Recent work has focused on close analogues of fosmidomycin/FR900098 (substitution, ring formation and reversal of the hydroxamate group) and formation of prodrugs with the aim of improving upon these shortcomings [154–160]. The approaches are illustrated in Fig. 30. A recent review summarises developments in the area with emphasis on the structure and catalytic mechanism of DXR [161].

5.3.2 Choline Pathway

Compounds related to the bisthiazolium salt T3 [162], the bisalkylamidine M34 [163] and the bis-quaternary ammonium salt G25 [164] mimic the structure of choline and target the parasitic de novo phosphotidylcholine biosynthesis pathway. Compounds of this class are likely to accumulate within the parasite, and additional modes of action may be responsible for the death of the parasite. Not unexpectedly compounds of this type show poor oral bioavailability. A group at Montpelier describes a series of amidoximes [165] and bis-C-alkyloxadiazolones as amidine prodrugs [166]. The oral efficacy of the masked amidine 1e shows improved, but modest, efficacy when compared to the parent compound 3e. The same group also describes the synthesis and SAR of a series of bisamidines and bisguanidines – exploring the importance of the alkyl chain length (12 methylenes appear optimal) and basicity of the cationic group (pKa over 12.5 is optimal) [163] as shown in Fig. 31.

5.3.3 Fatty Acids

The type II fatty acid synthesis pathway (FAS II) in plasmodia has been assumed to be important for blood stage replication of the parasite. Inhibitors of various enzymes in the pathway, such as PfKASIII (β-ketoacyl-ACP-synthase) [167, 168]

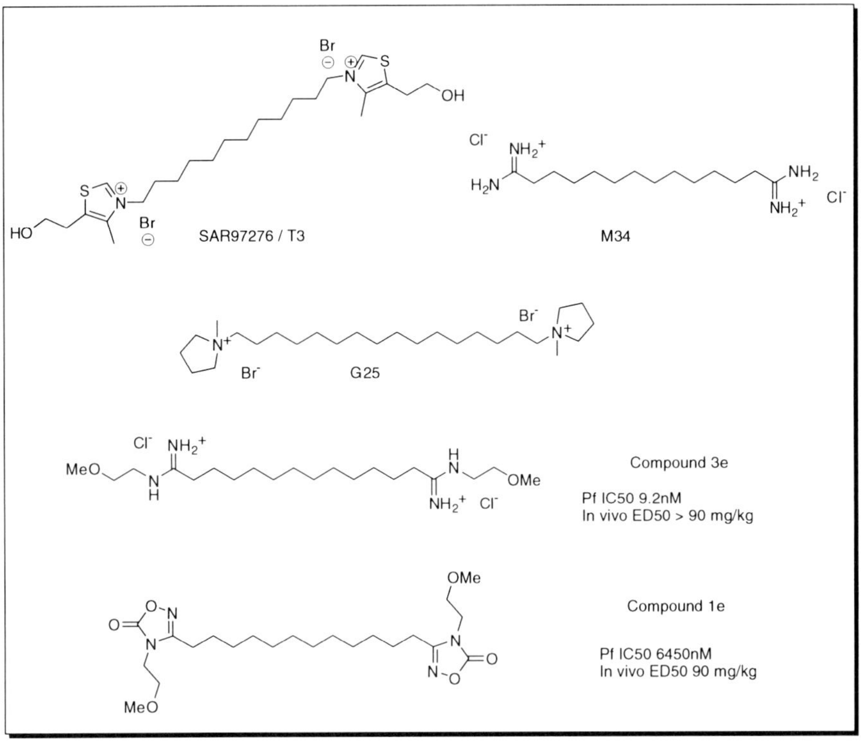

SAR97276 / T3

M34

G25

Compound 3e

Pf IC50 9.2nM
In vivo ED50 > 90 mg/kg

Compound 1e

Pf IC50 6450nM
In vivo ED50 90 mg/kg

Fig. 31 Choline mimics

and PfENR (enoyl acyl carrier protein reductase) [169], have been sought, but no potent *P. falciparum* erythrocyte activity has been found. Recent publications suggest that the pathway is not key to the parasite's asexual replication in the blood, but is important in the liver stages [170, 171].

5.4 *Signalling/Proliferation Pathways*

5.4.1 Kinases

Protein kinases play a key role in the control of proliferation and differentiation in eukaryotic cells. About 70–80 kinases can be found in the *P. falciparum* genome, and significant differences are apparent between vertebrate protein kinases and those of the parasite [172]. It is expected that many of these enzymes will be essential to parasite survival and specific inhibition of the parasite enzymes may be achievable; *Plasmodium* kinases therefore represent an attractive drug target class. Specific HTS approaches against PfPK7 and PfCDPK1 have successfully identified inhibitors that show activity in whole parasite assays.

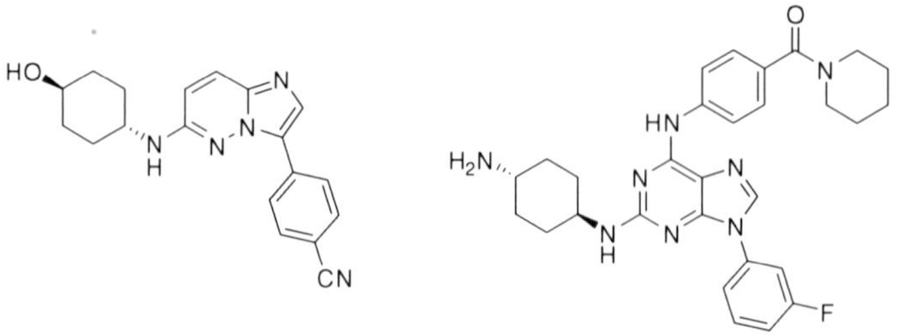

Compound 34

PfPK7 IC50 0.13uM
Pf (3D7) IC50 1uM

Compound 1

PfCDPK1 IC50 17nM
Pf EC50 (3D7) 230nM

Fig. 32 Kinase inhibitors

PfPK7 is an "orphan" plasmodial kinase, distantly related to the MAPKK family of kinases and is expressed during both the asexual and sexual stages of the parasite life cycle. Bouloc et al. describe a series of imidazopyridazines (e.g. compound 34) that inhibit PfPK7 and are active in the *P. falciparum* blood assay [173–175].

PfCDPK1 is expressed during schizogony in the erythrocytic stage as well as in the sporozoite stage. Kato et al. describe the screening of 20,000 compounds against PfCDPK1 (calcium-dependent protein kinase 1) [176]. They identified a series of 2,6,9-trisubstituted purines that inhibit PfCDPK1 and are active in the *P. falciparum* blood-stage assay (Fig. 32).

Despite the increasing interest in targeting *Plasmodium* kinases, the published compounds at this stage only show modest activity in the *P. falciparum* erythrocyte assay. As with most target-based approaches, confirmation that compounds act through the expected mechanism is not straightforward – the compounds may also target other *Plasmodium* or human kinases.

5.4.2 Histone Deacetylase

Histone acetylation plays a key role in regulating gene transcription in both eukaryotes and prokaryotes. *Plasmodium falciparum* contains at least five histone deacetylase (HDAC) homologues/orthologues – one class I HDAC, two class II HDACs and two class III sirtuins. Targeting HDAC as potential antimalarial therapy has been reviewed recently by Andrews et al. [177]. Most published inhibitors contain a zinc-binding hydroxamate that may lead to metabolic liabilities. The identification of compounds that are selective for *Plasmodium* HDACs over mammalian HDACs may be a key issue to minimise potential toxicity. The majority of recent publications relate to hydroxamate-containing inhibitors and are summarised in Fig. 33 [178–182].

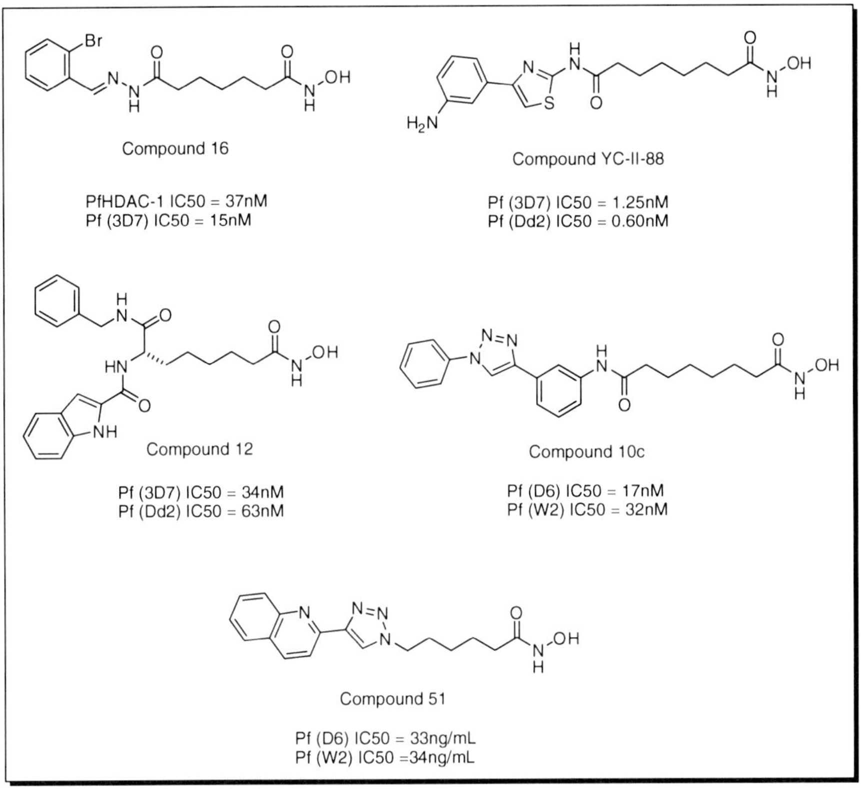

Compound 16

PfHDAC-1 IC50 = 37nM
Pf (3D7) IC50 = 15nM

Compound YC-II-88

Pf (3D7) IC50 = 1.25nM
Pf (Dd2) IC50 = 0.60nM

Compound 12

Pf (3D7) IC50 = 34nM
Pf (Dd2) IC50 = 63nM

Compound 10c

Pf (D6) IC50 = 17nM
Pf (W2) IC50 = 32nM

Compound 51

Pf (D6) IC50 = 33ng/mL
Pf (W2) IC50 =34ng/mL

Fig. 33 HDAC inhibitors

5.4.3 DNA-Binding Bisamidines

The antiparasitic mechanism of pentamidine (Fig. 34) and other diamidines is not completely clear, but is thought to involve binding to the minor groove of DNA at AT-rich regions. Not unexpectedly compounds of this type show poor oral bioavailability due to their highly charged nature. Compounds of this class are likely to accumulate within the parasite, and additional modes of action may contribute to their activity.

Farahat et al. describe a series of DAPI (4′,6-diamidino-2-phenylindole) analogues with improved *P. falciparum* erythrocyte activity (compound 21a) [183]. Oral efficacy is not reported, but bioavailability is likely to be very low in this series. In a series of papers, Bakunova et al. describe a number of further compounds in which the nature of the basic group and the linker is varied [184–186] (e.g. compound 16) [187]. Similarly, Rodriguez et al. report a series of bisguanidine and bis(2-aminoimidazoline) DNA binders with potent anti-plasmodial activity [188] – compound 4b is active in a mouse model of malaria following i.p. dosing.

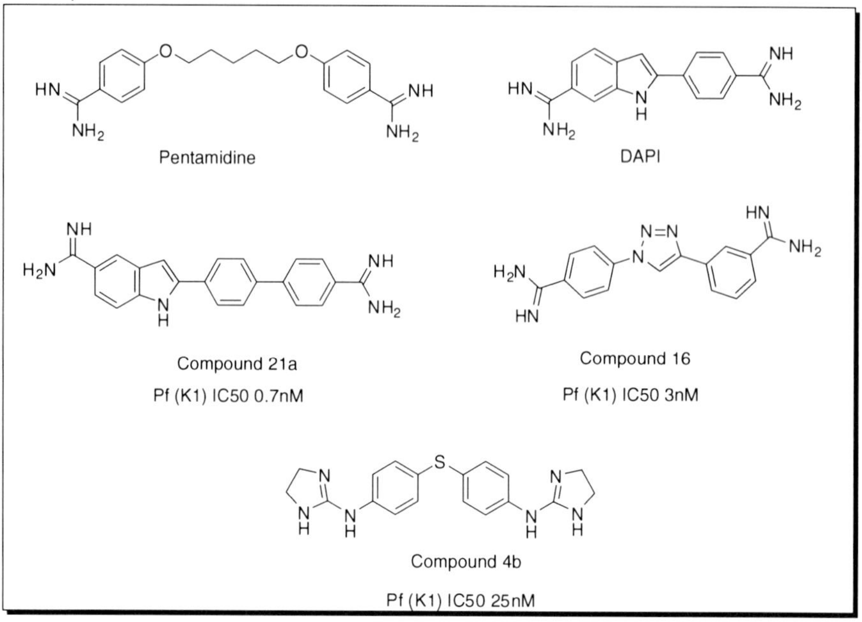

Pentamidine

DAPI

Compound 21a

Pf (K1) IC50 0.7nM

Compound 16

Pf (K1) IC50 3nM

Compound 4b

Pf (K1) IC50 25nM

Fig. 34 DNA-binding bisamidines

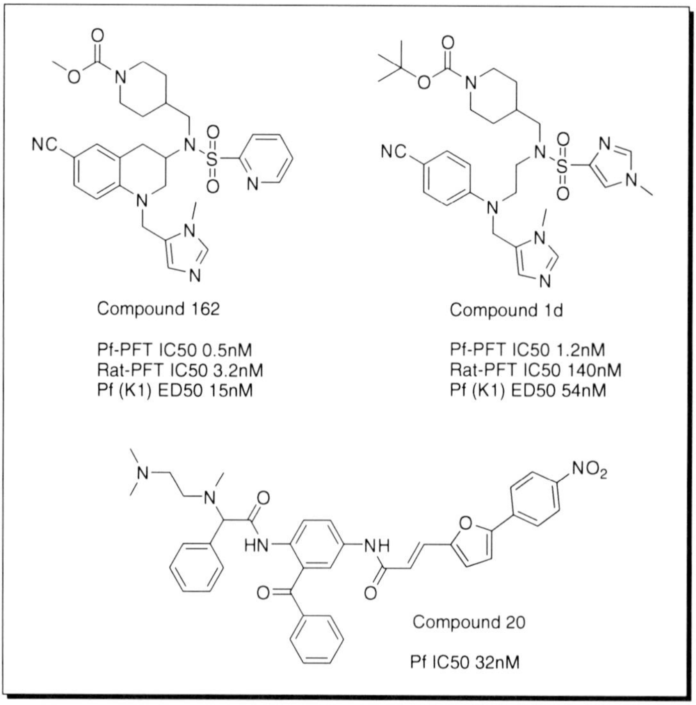

Compound 162

Pf-PFT IC50 0.5nM
Rat-PFT IC50 3.2nM
Pf (K1) ED50 15nM

Compound 1d

Pf-PFT IC50 1.2nM
Rat-PFT IC50 140nM
Pf (K1) ED50 54nM

Compound 20

Pf IC50 32nM

Fig. 35 Protein farnesyltransferase inhibitors

5.4.4 Protein Farnesyltransferase

Protein farnesyltransferase (PFT) is necessary for the post-translational modification of proteins involved in the signal transduction pathways and regulation of DNA replication and cell cycling [189]. Pf-PFT inhibitors have been shown to be cytotoxic to the parasite [190]. Based on the significant interest in the inhibition of human-PFT, Nallan et al. have demonstrated that some known human inhibitors of PFT also inhibit Pf-PFT and inhibit *P. falciparum* growth [190]. Bendale et al. have extended the SAR in a series of tetrahydroquinoline-based inhibitors such as compound 162 [191], which demonstrates potent in vitro activity. Compound 162 (Fig. 35) also shows efficacy following oral dosing in the rat *P. berghei* model [192]. Fletcher et al. [193] describe a series of ring-opened ethylenediamine-analogues (e.g. compound 1d) with good in vitro potency and improved selectivity over mammalian PFT. The group of Schlitzer has established SAR in a series benzophenones-based inhibitors demonstrating good in vitro potency and in vivo efficacy [194].

5.5 *Protein Synthesis Pathways*

Several antibiotics have antimalarial properties and recent data suggest that their target is the apicoplast within the parasite. Inhibitors of translation (such as tetracyclines, clindamycin and macrolides) and have antimalarial effects that are particularly pronounced in the progeny of treated parasites – referred to as "delayed death" [195, 196].

6 Conclusion

In the last 10 years, we have witnessed a renaissance in antimalarial drug discovery and development, with the recent approval of two new treatments – Coartem Dispersible and ASAQ (artesunate and amodiaquine). Nonetheless, the goal of eradication, set in 2007, has presented a number of new challenges to the field. Not least of these is the need for novel chemistry to tackle drug resistance as well as new molecules to halt the transmission of the parasite from person to person. But this is a really exciting time to be involved in malaria drug discovery. We have a number of strategies to identify potential drug candidates at our disposal, new technologies, a plethora of interesting targets to explore and much of the work being conducted in the spirit of partnership between the public health sector, universities, and pharmaceutical and biotechnology companies. What's more, with the recent release of 20,000 active compound hits from screening work conducted by GSK, St. Jude Children's Research Hospital and Novartis providing a catalyst to further research and development, the momentum is building. Drug discovery and development is a long process, but provided this momentum can be maintained, we will be in a strong position to provide the new medicines needed for malaria eradication over the next decade.

Acknowledgements The authors would like to thank the Central Drug Research Institute, India (CDRI), for allowing the disclosure of the structure of CDRI 97/98. Also special thanks to MMV staff, in particular Dr. T. Wells, Dr. D. Leroy and E. Poll for helpful discussions, suggestions and proof-reading.

References

1. World Malaria Report (2009) World Health Organization, Geneva, Switzerland. http://www.who.int/malaria/world_malaria_report_2009/en/index.html
2. World Health Organization, Geneva Switzerland. http://www.who.int/mediacentre/factsheets/fs094/en/
3. Trager W, Jensen JB (1976) Science 193:673–675
4. Mazier D, Renia L, Snounou G (2009) Nat Rev Drug Discov 8:854–864
5. Dondorp AM, Desakorn V, Pongtavornpinyo W, Sahassananda D, Silamut K (2005) PLoS Med 2(8):e204
6. Bousema T, Okell L, Shekalaghe S, Griffin JT, Omar S, Sawa P, Sutherland C, Sauerwein R, Ghani AC, Drakeley C (2010) Malar J 9:136
7. Sinden RE (2002) Cell Microbiol 4(11):713–724
8. Price RN, Tjiltra E, Guerra CA, Yeung S, White NJ, Anstey NM (2007) Am J Trop Med Hyg 77(Suppl 6):79–87
9. Vale N, Moreira R, Gomes P (2009) Eur J Med Chem 44:937–953
10. Wells TNC, Burrows JN, Baird JK (2010) Trends Parasitol 26(3):145–151
11. Treatment Guidelines (2010) World Health Organisation, Switzerland. http://www.who.int/malaria/publications/atoz/9789241547925/en/index.html
12. Wells TNC, Alonso PL, Gutteridge WE (2009) Nat Rev Drug Discov 8:879–891
13. Olliaro P, Wells TNC (2009) Clin Pharmacol Ther 85(6):584–595
14. Nanayakkara NPD, Ager AL, Bartlett MS, Yardley V, Croft SL, Khan IA, McChesney JD, Walker LA (2008) Antimicrob Agents Chemother 52(6):2130–2137
15. Jones KL, Donegan S, Lalloo DG (2007) Cochrane Database Syst Rev 4:CD005967
16. Gomes MF, Faiz M, Gyapong JO, Warsame M, Agbenyega T, Babiker A, Baiden F, Yunus EB, Binka F, Clerk C, Folb P, Hassan R, Hossain MA, Kimbute O, Kitua A, Krishna S, Makasi C, Mensah N, Mrango Z, Olliaro P, Peto R, Peto TJ, Rahman MR, Ribeiro I, Samad R, White NJ (2009) Lancet 373(9663):557–566
17. Dondorp AM, Nosten F, Yi P, Das D, Phyo AP, Tarning J, Lwin KM, Ariey F, Hanpithakpong W, Lee SJ, Ringwald P, Silamut K, Imwong M, Chotivanich K, Lim P, Herdman T, Sam An S, Yeung S, Singhasivanon P, Day NPJ, Lindegardh N, Socheat D, White NJ (2009) N Engl J Med 361:455–467
18. Schlitzer M (2007) ChemMedChem 2:944–986
19. Burrows JN, Chibale K, Wells TNC (2011) Curr Top Med Chem (in press)
20. Fidock DA, Rosenthal PJ, Croft SA, Brun R, Nwaka S (2004) Nat Rev Drug Discov 3:509–520
21. Bumol TF, Watanabe AM (2001) JAMA 285(5):551–555
22. Payne DJ, Gwynn MN, Holmes DJ, Pompliano DL (2007) Nat Rev Drug Discov 6:29–40
23. Hellerstein MK (2008) Metab Eng 10:1–9
24. Baldwin J, Michnoff CH, Malmquist NA, White J, Roth MG, Rathod PK, Phillips MA (2005) J Biol Chem 280(23):21847–21853

25. Patel V, Booker M, Kramer M, Ross L, Celatka CA, Kennedy LM, Dvorin JD, Duraisingh MT, Sliz P, Wirth DF, Clardy J (2008) J Biol Chem 283(50):35078–35085
26. Coteron JM, Catterick D, Castro J, Chaparro MJ, Dıaz B, Fernandez E, Ferrer S, Gamo FJ, Gordo M, Gut J, de las Heras L, Legac J, Marco M, Miguel J, Munoz V, Porras E, de la Rosa JC, Ruiz JR, Sandoval E, Ventosa P, Rosenthal PJ, Fiandor JM (2010) J Med Chem 53:6129–6152
27. Gardner MJ, Hall N, Fung E, White O, Berriman M, Hyman RW, Carlton JM, Pain A, Nelson KE, Bowman S, Paulsen IT, James K, Eisen JA, Rutherford K, Salzberg SL, Craig A, Kyes S, Chan M-S, Nene V, Shallom SJ, Suh B, Peterson J, Angiuoli S, Pertea M, Allen J, Selengut J, Haft D, Mather MW, Vaidya AB, Martin DMA, Fairlamb AH, Fraunholz MJ, Roos DS, Ralph SA, McFadden GI, Cummings LM, Subramanian GM, Mungall C, Venter JC, Carucci DJ, Hoffman SL, Newbold C, Davis RW, Fraser CM, Barrell B (2002) Nature 419(6906):498–511
28. Carlton JM, Adams JH, Silva JC, Bidwell SL, Lorenzi H, Caler E, Crabtree J, Angiuoli SV, Merino EF, Amedeo P, Cheng Q, Coulson RMR, Crabb BS, del Portillo HA, Essien K, Feldblyum TV, Fernandez-Becerra C, Gilson PR, Gueye AH, Guo X, Kang'a S, Kooij TWA, Korsinczky M, Meyer EV-S, Nene V, Paulsen I, White O, Ralph SA, Ren Q, Sargeant TJ, Salzberg SL, Stoeckert CJ, Sullivan SA, Yamamoto MM, Hoffman SL, Wortman JR, Gardner MJ, Galinski MR, Barnwell JW, Fraser-Liggett CM (2008) Nature 455(7214):757–763
29. Yeh I, Hanekamp T, Tsoka S, Karp PD, Altman RB (2004) Genome Res 14:917–924
30. Hurt DE, Widom J, Clardy J (2006) Acta Crystallogr D Biol Crystallogr D62(3):312–323
31. Deng X, Gujjar R, El Mazouni F, Kaminsky W, Malmquist NA, Goldsmith EJ, Rathod PK, Phillips MA (2010) J Biol Chem 284:26999–27009
32. Rees DC, Congreve M, Murray CW, Carr R (2004) Nat Rev Drug Discov 3:660–672
33. Kirk K (2001) Physiol Rev 81(2):495–537
34. Ginsburg H, Krugliak M, Eidelman O, Cabantchik ZI (1983) Mol Biochem Parasitol 8:177–190
35. Hart CP (2005) Drug Discov Today 10(7):513–519
36. Crowther GJ, Napuli AJ, Gilligan JH, Gagaring K, Borboa R, Francek C, Chen Z, Dagostino EF, Stockmyer JB, Wang Y, Rodenbough PP, Castaneda LJ, Leibly DJ, Bhandari J, Gelb MH, Brinker A, Engels IH, Taylor J, Chatterjee AK, Fantauzzi P, Glynne RJ, Van Voorhis WC, Kuhen KL (2011) Mol Biochem Parasitol 175:21–29
37. Gamo F-J, Sanz LM, Vidal J, de Cozar C, Alvarez E, Lavandera J-L, Vanderwall DE, Green DVS, Kumar V, Hasan S, Brown JR, Peishoff CE, Cardon LR, Garcia-Bustos J-F (2010) Nature 465:305–310
38. Guiguemde WA, Shelat AA, Bouck D, Duffy S, Crowther GJ, Davis PH, Smithson DC, Connelly M, Clark J, Zhu F, Jimenez-Dıaz MB, Martinez MS, Wilson EB, Tripathi AK, Gut J, Sharlow ER, Bathurst I, El Mazouni F, Fowble JW, Forquer I, McGinley PL, Castro S, Angulo-Barturen I, Ferrer S, Rosenthal PJ, DeRisi JL, Sullivan DJ, Lazo JS, Roos DS, Riscoe MK, Phillips MA, Rathod PK, Van Voorhis WC, Avery VM, Guy RK (2010) Nature 465:311–315
39. Plouffe D, Brinker A, McNamara C, Henson K, Kato N, Kuhen K, Nagle A, Adrian F, Matzen JT, Anderson P, Nam T-g, Gray NS, Chatterjee A, Janes J, Yan SF, Trager R, Caldwell JS, Schultz PG, Zhou Y, Winzeler EA (2008) Proc Natl Acad Sci USA 105 (26):9059–9064
40. Ekins S, Williams AJ (2010) Drug Discov Today. doi:10.1016/j.drudis.2010.08.010
41. Baniecki ML, Wirth DF, Clardy J (2007) Antimicrob Agents Chemother 51(2):716–723
42. Chong CR, Chen X, Shi L, Liu JO, Sullivan DJ (2006) Nat Chem Biol 2(8):415–416
43. O'Neill PA, Posner GH (2004) J Med Chem 47(12):2945–2964
44. Fugi MA, Wittlin S, Dong Y, Vennerstrom JL (2010) Antimicrob Agents Chemother 54:1042–1046
45. O'Neill PM, Barton VE, Ward SA (2010) Molecules 15:1705–1721
46. Singh C, Chaudhary S, Puri SK (2006) J Med Chem 49(24):7227–7233

47. Posner GH, Paik I-H, Chang W, Borstnik K, Sinishtaj S, Rosenthal AS, Shapiro TA (2007) J Med Chem 50(10):2516–2519
48. Woodard LE, Chang W, Chen X, Liu JO, Shapiro TA, Posner GH (2009) J Med Chem 52 (23):7458–7462
49. Singh AS, Verma VP, Hassam M, Krishna NN, Puri SK, Singh C (2008) Org Lett 10 (23):5461–5464
50. Haynes RK, Fugmann B, Stetter J, Rieckmann K, Heilmann H-D, Chan H-W, Cheung M-K, Lam W-L, Wong H-N, Croft SL, Vivas L, Rattray L, Stewart L, Peters W, Robinson BL, Edstein MD, Kotecka B, Kyle DE, Beckermann B, Gerisch M, Radtke M, Schmuck G, Steinke W, Wollborn U, Schmeer K, Romer A (2006) Angew Chem Int Ed 45(13):2082–2088
51. Vennerstrom JL, Arbe-Barnes S, Brun R, Charman SA, Chiu FCK, Chollet J, Dong Y, Dorn A, Hunziker D, Matile H, McIntosh K, Padmanilayam M, Santo TJ, Scheurer C, Scorneaux B, Tang Y, Urwyler H, Wittlin S, Charman WN (2004) Nature 430(7002):900–904
52. Martyn DC, Ramirez AP, Beattie MJ, Cortese JF, Patel V, Rush MA, Woerpel KA, Clardy J (2008) Bioorg Med Chem Lett 18(24):6521–6524
53. Dong Y, Wittlin S, Sriraghavan K, Chollet J, Charman SA, Charman WN, Scheurer C, Urwyler H, Santo Tomas J, Snyder C, Creek DJ, Morizzi J, Koltun M, Matile H, Wang X, Padmanilayam M, Tang Y, Dorn A, Brun R, Vennerstrom JL (2010) J Med Chem 53 (1):481–491
54. Fattorusso C, Campiani G, Catalanotti B, Persico M, Basilico N, Parapini S, Taramelli D, Campagnuolo C, Fattorusso E, Romano A, Taglialatela-Scafati O (2006) J Med Chem 49 (24):7088–7094
55. Sabbani S, Stocks PA, Ellis GL, Davies J, Hedenstrom E, Ward SA, O'Neill PM (2008) Bioorg Med Chem Lett 18(21):5804–5808
56. O'Neill PM, Amewu RK, Nixon GL, El Garah FB, Mungthin M, Chadwick J, Shone AE, Vivas L, Lander H, Barton V, Muangnoicharoen S, Bray PG, Davies J, Park BK, Wittlin S, Brun R, Preschel M, Zhang K, Ward SA (2010) Angew Chem Int Ed 49(33):5693–5697
57. Singh C, Pandey S, Sharma M, Puri SK (2008) Bioorg Med Chem 16(4):1816–1821
58. Wang X, Creek DJ, Schiaffo CE, Dong Y, Chollet J, Scheurer C, Wittlin S, Charman SA, Dussault PH, Wood JK, Vennerstrom JL (2009) Bioorg Med Chem Lett 19(16):4542–4545
59. Araujo NCP, Barton V, Jones M, Stocks PA, Ward SA, Davies J, Bray PG, Shone AE, Cristiano MLS, O'Neill PM (2009) Bioorg Med Chem Lett 19(7):2038–2043
60. Singh C, Verma VP, Naikade NK, Singh AS, Hassam M, Puri SK (2008) J Med Chem 51 (23):7581–7592
61. Milner E, McCalmont W, Bhonsle J, Caridha D, Carroll D, Gardner S, Gerena L, Gettayacamin M, Lanteri C, Luong TL, Melendez V, Moon J, Roncal N, Sousa J, Tungtaeng A, Wipf P, Dow G (2010) Bioorg Med Chem Lett 20(4):1347–1351
62. Milner E, McCalmont W, Bhonsle J, Caridha D, Cobar J, Gardner S, Gerena L, Goodine D, Lanteri C, Melendez V, Roncal N, Sousa J, Wipf P, Dow GS (2010) Malar J 9:51
63. Perez-Silanes S, Berrade L, Garcia-Sanchez RN, Mendoza A, Galiano S, Perez-Solorzano BM, Nogal-Ruiz JJ, Martinez-Fernandez AR, Aldana I, Monge A (2009) Molecules 14 (10):4120–4135
64. Murphy SC, Harrison T, Hamm HE, Lomasney JW, Mohandas N et al (2006) PLoS Med 3 (12):e528
65. Frisk-Holmberg M, Bergqvist Y, Englund U (1983) Br J Clin Pharmacol 15:502–503
66. O'Neill PM, Park BK, Shone AE, Maggs JL, Roberts P, Stocks PA, Biagini GA, Bray PG, Gibbons P, Berry N, Winstanley PA, Mukhtar A, Bonar-Law R, Hindley S, Bambal RB, Davis CB, Bates M, Hart TK, Gresham SL, Lawrence RM, Brigandi RA, Gomez-delas-Heras FM, Gargallo DV, Ward SA (2009) J Med Chem 52(5):1408–1415
67. Andrews S, Burgess SJ, Skaalrud D, Kelly JX, Peyton DH (2010) J Med Chem 53 (2):916–919
68. Gemma S, Campiani G, Butini S, Joshi BP, Kukreja G, Coccone SS, Bernetti M, Persico M, Nacci V, Fiorini I, Novellino E, Taramelli D, Basilico N, Parapini S, Yardley V, Croft S,

Keller-Maerki S, Rottmann M, Brun R, Coletta M, Marini S, Guiso G, Caccia S, Fattorusso C (2009) J Med Chem 52(2):502–513
69. Ray S, Madrid PB, Catz P, Le Valley SE, Furniss MJ, Rausch LL, Guy RK, De Risi JL, Iyer LV, Green CE, Mirsalis J (2010) J Med Chem 53(9):3685–3695
70. Natarajan JK, Alumasa JN, Yearick K, Ekoue-Kovi KA, Casabianca LB, de Dios AC, Wolf C, Roepe PD (2008) J Med Chem 51(12):3466–3479
71. Iwaniuk DP, Whetmore ED, Rosa N, Ekoue-Kovi K, Alumasa J, de Dios AC, Roepe PD, Wolf C (2009) Bioorg Med Chem Lett 17(18):6560–6566
72. Dive D, Biot C (2008) ChemMedChem 3(3):383–391
73. Blackie MAL, Yardley V, Chibale K (2010) Bioorg Med Chem Lett 20(3):1078–1080
74. Miroshnikova OV, Hudson TH, Gerena L, Kyle DE, Lin A (2007) J Med Chem 50 (4):889–896
75. O'Neill PM, Shone AE, Stanford D, Nixon G, Asadollahy E, Park BK, Maggs JL, Roberts P, Stocks PA, Biagini G, Bray PG, Davies J, Berry N, Hall C, Rimmer K, Winstanley PA, Hindley S, Bambal RB, Davis CB, Bates M, Gresham SL, Brigandi RA, Gomez-de-las-Heras FM, Gargallo DV, Parapini S, Vivas L, Lander H, Taramelli D, Ward SA (2009) J Med Chem 52(7):1828–1844
76. Araújoa NCP, Barton V, Jones M, Stocks PA, Ward SA, Davies J, Bray PG, Shone AE, Cristiano MLS, O'Neill PM (2009) Bioorg Med Chem Lett 19(7):2038–2043
77. Opsenica I, Opsenica D, Lanteri CA, Anova L, Milhous WK, Smith KS, Solaja BA (2008) J Med Chem 51(19):6216–6219
78. Khan MO, Faruk L, Mark S, Tekwani BL, Khan SI, Kimura E, Borne RF (2009) Antimicrob Agents Chemother 53(4):1320–1324
79. Gupta L, Srivastava K, Singh S, Puri SK, Chauhan PMS (2008) Bioorg Med Chem Lett 18 (11):3306–3309
80. Sunduru N, Srivastava K, Rajakumar S, Puri SK, Saxena JK, Chauhan PMS (2009) Bioorg Med Chem Lett 19(9):2570–2573
81. Mahajan A, Yeh S, Nell M, Van Rensburg CEJ, Chibale K (2007) Bioorg Med Chem Lett 17 (20):5683–5685
82. Sunduru N, Sharma M, Srivastava K, Rajakumar S, Puri SK, Saxena JK, Chauhan PMS (2009) Bioorg Med Chem 17(17):6451–6462
83. Zhu S, Zhang Q, Gudise C, Meng L, Wei L, Smith E, Kong Y (2007) Bioorg Med Chem Lett 17(22):6101–6106
84. Vale N, Nogueira F, do Rosario VE, Gomes P, Moreira R (2009) Eur J Med Chem 44 (6):2506–2516
85. Huy NT, Mizunuma K, Kaur K, Nhien NTT, Jain M, Uyen DT, Harada S, Jain R, Kamei K (2007) Antimicrob Agents Chemother 51(8):2842–2847
86. Quinn RJ, Carroll AR, Pham NB, Baron P, Palframan ME, Suraweera L, Pierens GK, Muresan S (2008) J Nat Prod 71:464–468
87. Mishra KP, Ganju L, Sairam M, Banerjee PK, Sawhney RC (2008) Biomed Pharmacother 62:94–98
88. Kaur K, Jain M, Kaur T, Jain R (2009) Bioorg Med Chem 17:3229–3256
89. Zhu S, Zhang Q, Gudise C, Wei L, Smith E, Zeng Y (2009) Bioorg Med Chem 17:4496–4502
90. Kikuchi H, Yamamoto K, Horoiwa S, Hirai S, Kasahara R, Hariguchi N, Matsumoto M, Oshima Y (2006) J Med Chem 49(15):4698–4706
91. Wells TNC (2011) Malar J 10(Suppl 1):S3
92. Yeung BKS, Zou B, Rottmann M, Lakshminarayana SB, Ang SH, Leong SY, Tan J, Wong J, Keller-Maerki S, Fischli C, Goh A, Schmitt EK, Krastel P, Francotte E, Kuhen K, Plouffe D, Henson K, Wagner T, Winzeler EA, Petersen F, Brun R, Dartois V, Diagana TT, Keller TH (2010) J Med Chem 53(14):5155–5164
93. Rottmann M, McNamara C, Yeung BKS, Lee MCS, Zou B, Russell B, Seitz P, Plouffe DM, Dharia NV, Tan J, Cohen SB, Spencer KR, Gonzalez-Paez GE, Lakshminarayana SB, Goh

A, Suwanarusk R, Jegla T, Schmitt EK, Beck H-P, Brun R, Nosten F, Renia L, Dartois V, Keller TH, Fidock DA, Winzeler EA, Diagana TT (2010) Science 329(5996):1175–1180
94. Urgaonkar S, Cortese JF, Barker RH, Cromwell M, Serrano AE, Wirth DF, Clardy J, Mazitschek R (2010) Org Lett 12(18):3998–4001
95. Madapa S, Tusi Z, Sridhar D, Kumar A, Siddiqi MI, Srivastava K, Rizvi A, Tripathi R, Puri SK, Shiva Keshava GK, Shukla PK, Batra S (2009) Bioorg Med Chem 17:203–221
96. Gutteridge CE, Hoffman MM, Bhattacharjee AK, Gerena L (2007) J Heterocycl Chem 44 (3):633–637
97. Kumar A, Srivastava K, Raja Kumar S, Puri SK, Chauhan PMS (2009) Bioorg Med Chem Lett 19(24):6996–6999
98. Madapa S, Tusi Z, Mishra A, Srivastava K, Pandey SK, Tripathi R, Puri SK, Batra S (2009) Bioorg Med Chem 17(1):222–234
99. Martyn DC, Nijjar A, Celatka CA, Mazitschek R, Cortese JF, Tyndall E, Liu H, Fitzgerald MM, O'Shea TJ, Danthi S, Clardy J (2010) Bioorg Med Chem Lett 20(1):228–231
100. Deng XD, Nagle A, Wuc T, Sakata T, Henson K, Chen Z, Kuhen K, Plouffe D, Winzeler E, Adrian F, Tuntland T, Chang J, Simerson S, Howard S, Ek J, Isbell J, Tully DC, Chatterjee AK, Gray NS (2010) Bioorg Med Chem Lett 20:4027–4031
101. Kelly JX, Smilkstein MJ, Brun R, Wittlin S, Cooper RA, Lane KD, Janowsky A, Johnson RA, Dodean RA, Winter R, Hinrichs DJ, Riscoe MK (2009) Nature 459 (7244):270–273
102. Altenkaemper M, Bechem B, Perruchon J, Heinrich S, Maedel A, Ortmann R, Dahse H-M, Freunscht E, Wang Y, Rath J, Stich A, Hitzler M, Chiba P, Lanzer M, Schlitzer M (2009) Bioorg Med Chem 17(22):7690–7697
103. Faist J, Seebacher W, Kaiser M, Brun R, Saf R, Weis R (2010) Eur J Med Chem 45:179–185
104. Guillon J, Moreau S, Mouray E, Sinou V, Forfar I, Fabre SB, Desplat V, Millet P, Parzy D, Jarry C, Grellier P (2008) Bioorg Med Chem 16(20):9133–9144
105. Acharya BN, Saraswat D, Tiwari M, Shrivastava AK, Ghorpade R, Bapna S, Kaushik MP (2010) Eur J Med Chem 45(2):430–438
106. Vlahakis JZ, Kinobe RT, Nakatsu K, Szareka WA, Crandall IE (2006) Bioorg Med Chem Lett 16:2396–2406
107. Takasu K, Shimogama T, Satoh C, Kaiser M, Brun R, Ihara M (2007) J Med Chem 50 (10):2281–2284
108. Nepveu F, Kim S, Boyer J, Chatriant O, Ibrahim H, Reybier K, Monje M-C, Chevalley S, Perio P, Lajoie BH, Bouajila J, Deharo E, Sauvain M, Tahar R, Basco L, Pantaleo A, Turini A, Arese P, Valentin A, Thompson E, Vivas L, Petit S, Nallet J-P (2010) J Med Chem 53: 699–714
109. Wu T, Nagle A, Sakata T, Henson K, Borboa R, Chen Z, Kuhen K, Plouffe D, Winzeler E, Adrian F, Tuntland T, Chang J, Simerson S, Howard S, Ek J, Isbell J, Deng X, Gray NS, Tully DC, Chatterjee AK (2009) Bioorg Med Chem Lett 19(24):6970–6974
110. Parvanova I, Epiphanio S, Fauq A, Golde TE, Prudencio M, Mota MM (2009) PLoS ONE 4 (4):e5078
111. Mahmoudi N, Garcia-Domenech R, Galvez J, Farhati K, Franetich J-F, Sauerwein R, Hannoun L, Derouin F, Danis M, Mazier D (2008) Antimicrob Agents Chemother 52 (4):1215–1220
112. Carraz M, Jossang A, Franetich J-F, Siau A, Ciceron L, Hannoun L, Sauerwein R, Frappier F, Rasoanaivo P, Snounou G, Mazier D (2006) PLoS Med 3(12):2392–2402
113. Hobbs CV, Voza T, Coppi A, Kirmse V, Marsh K, Borkowsky W, Sinnis P (2009) J Infect Dis 199:134
114. Singh AP, Zhang Y, No J-H, Docampo R, Nussenzweig V, Oldfield E (2010) Antimicrob Agents Chemother 54(7):2987–2993
115. Singh AP, Surolia N, Surolia A (2009) IUBMB Life 61(9):923–928
116. Guan J, Wang X, Smith K, Ager A, Gettayacamin M, Kyle DE, Milhous WK, Kozar MP, Magill AJ, Lin AJ (2007) J Med Chem 50:6226–6231

117. Chotivanich K, Sattabongkot J, Udomsangpetch R, Looareesuwan S, Day NPJ, Coleman RE, White NJ (2006) Antimicrob Agents Chemother 50:1927–1930
118. Peatey CL, Skinner-Adams TS, Dixon MWA, McCarthy JS, Gardiner DL, Trenholme KR (2009) J Infect Dis 200:1518–1521
119. Benoit-Vical FB, Lelievre J, Berry A, Deymier C, Dechy-Cabaret O, Cazelles J, Loup C, Robert A, Magnaval JF, Meunier B (2007) Antimicrob Agents Chemother 51:1463–1472
120. Yanow SK, Purcell LA, Pradel G, Sato A, Rodríguez A, Lee M, Spithill TW (2008) J Infect Dis 197:527–534
121. Peatey CL, Andrews KT, Eickel N, MacDonald T, Butterworth AS, Trenholme KR, Gardiner DL, McCarthy JS, Skinner-Adams TS (2010) Antimicrob Agents Chemother 54:1334–1337
122. Shimizu S, Osada Y, Kanazawa T, Tanaka Y, Arai M (2010) Malar J 9:73
123. Krungkrai J, Krungkrai SR, Supuran CT (2008) Bioorg Med Chem Lett 18:5466–5471
124. Yeates CL, Batchelor JF, Capon EC, Cheesman NJ, Fry M, Hudson AT, Pudney M, Trimming H, Woolven J, Bueno JM, Chicharro J, Fernandez E, Fiandor JM, Gargallo-Viola D, Gomez de las Heras F, Herreros E, Leon ML (2008) J Med Chem 51(9):2845–285
125. Garcia-Bustos JF (2008) Progress and challenges in building an antimalarial drug discovery Portfolio. Am Soc Trop Med & Hygiene meeting, Oral Presentation, December 2008
126. Kongkathip N, Pradidphol N, Hasitapan K, Grigg R, Kao W-C, Hunte C, Fisher N, Warman AJ, Biagini GA, Kongsaeree P, Chuawong P, Kongkathip B (2010) J Med Chem 53:1211–1221
127. El Hage S, Ane M, Stigliani J-L, Marjorie M, Vial H, Baziard-Mouysset G, Payard M (2009) Eur J Med Chem 44:4778–4782
128. Zhang Y, Guiguemde WA, Sigal M, Zhu F, Connelly MC, Nwaka S, Guy RK (2010) Bioorg Med Chem 18:2756–2766
129. Winter RW, Kelly JX, Smilkstein MJ, Dodean R, Hinrichs D, Riscoe MK (2008) Exp Parasitol 118:487–497
130. Cross RM, Monastyrskyi A, Mutka TS, Burrows JN, Kyle DE, Manetsch R (2010) J Med Chem 53:7076–7094
131. Phillips MA, Rathod PK (2010) Infect Dis Drug Targ 10:226–239
132. Gujjar R, Marwaha A, El Mazouni F, White J, White KL, Creason S, Shackleford DM, Baldwin J, Charman WN, Buckner FS, Charma S, Rathod PK, Phillips MA (2009) J Med Chem 52:1864–1872
133. Booker ML, Bastos CM, Kramer ML, Barker RH Jr, Skerlj R, Bir Sidhu A, Deng X, Celatka C, Cortese JF, Guerrero Bravo JE, Krespo Llado KN, Serrano AE, Angulo-Barturen I, Jimenez-Diaz MB, Viera S, Garuti H, Wittlin S, Papastogiannidis P, Lin JW, Janse CJ, Khan SM, Duraisingh M, Coleman B, Goldsmith EJ, Phillips MA, Munoz B, Wirth DF, Klinger JD, Wiegand R, Sybertz E (2010) J Biol Chem 285:33054–33064
134. Bello AM, Poduch E, Liu Y, Wei L, Crandall I, Wang X, Dyanand C, Kain KC, Pai EF, Kotra LP (2008) J Med Chem 51:439–448
135. Nzila A, Rottmann M, Chitnumsub P, Kiara SM, Kamchonwongpaisan S, Maneeruttanar-ungroj C, Taweechai S, Yeung BKS, Goh A, Lakshminarayana SB, Zou B, Wong J, Ma NL, Weaver M, Keller TH, Dartois V, Wittlin S, Brun R, Yuthavong Y, Diagana TT (2010) Antimicrob Agents Chemother 54:2603–2610
136. Nguyen C, Ruda GF, Schipani A, Kasinathan G, Leal I, Musso-Buendia A, Kaiser M, Brun R, Ruiz-Perez LM, Sahlberg B-L, Johansson NG, Gonzalez-Pacanowska D, Gilbert IH (2006) J Med Chem 49:4183–4195
137. Too K, Brown DM, Bongard E, Yardley V, Vivas L, Loakes D (2007) Bioorg Med Chem 15:5551–5562
138. Kicska GA, Tyler PC, Evans GB, Furneaux RH, Schramm VL, Kim K (2002) J Biol Chem 277:226–3231
139. Taylor EA, Clinch K, Kelly PM, Li L, Evans GB, Tyler PC, Schramm VL (2007) J Am Chem Soc 129:6984–6985
140. Evans GB, Furneaux RH, Greatrex B, Murkin AS, Schramm VL, Tyler PC (2008) J Med Chem 51:948–956

141. Tyler PC, Taylor EA, Froehlich RFG, Schramm VL (2007) J Am Chem Soc 129:6872–6879
142. Keough DT, Hockova D, Holy A, Naesens LMJ, Skinner-Adams TS, de Jersey J, Guddat LW (2009) J Med Chem 52:4391–4399
143. Goldberg DE (1993) Semin Cell Biol 4:355–361
144. Ettari R, Bova F, Zappala M, Grasso S, Micale N (2010) Med Res Rev 30:136–167
145. Chipeleme A, Gut J, Rosenthal PJ, Chibale K (2007) Bioorg Med Chem 15:273–282
146. Ersmark K, Samuelsson B, Hallberg A (2006) Med Res Rev 26:626–666
147. Hidaka K, Kimura T, Ruben AJ, Uemura T, Kamiya M, Kiso A, Okamoto T, Tsuchiya Y, Hayashi Y, Freire E, Kiso Y (2008) Bioorg Med Chem 16:10049–10060
148. Hidaka K, Kimura T, Tsuchiya Y, Kamiya M, Ruben AJ, Freire E, Hayashi Y, Kiso Y (2007) Bioorg Med Chem Lett 17:3048–3052
149. Flipo M, Beghyn T, Leroux V, Florent I, Deprez BP, Deprez-Poulain RF (2007) J Med Chem 50:1322–1334
150. Pesce E-R, Cockburn IL, Goble JL, Stephens LL, Blatch GL (2010) Infect Dis Drug Targ 10:147–157
151. Chiang AN, Valderramos J-C, Balachandran R, Chovatiya RJ, Mead BP, Schneider C, Bell SL, Klein MG, Huryn DM, Chen XS, Day BW, Fidock DA, Wipf P, Brodsky JL (2009) Bioorg Med Chem 17:1527–1533
152. Shahinas D, Liang M, Datti A, Pillai DR (2010) J Med Chem 53:3552–3557
153. Na-Bangchang K, Ruengweerayut R, Karbwang J, Chauemung A, Hutchinson D (2007) Malar J 6:70
154. Schlueter K, Walter RD, Bergmann B, Kurz T (2006) Eur J Med Chem 41:1385–1397
155. Kurz T, Schlueter K, Kaula U, Bergmann B, Walter RD, Geffken D (2006) Bioorg Med Chem 14:5121–5513
156. Verbrugghen T, Cos P, Maes L, Van Calenbergh S (2010) J Med Chem 53:5342–5346
157. Devreux V, Wiesner J, Goeman JL, Van der Eycken J, Jomaa H, Van Calenbergh S (2006) J Med Chem 49:2656–2660
158. Haemers T, Wiesner J, Van Poecke S, Goeman J, Henschker D, Beck E, Jomaa H, Van Calenbergh S (2006) Bioorg Med Chem Lett 16:1888–1891
159. Devreux V, Wiesner J, Jomaa H, Rozenski J, Van der Eycken J, Van Calenbergh S (2007) J Org Chem 72:3783–3789
160. Haemers T, Wiesner J, Giessmann D, Verbrugghen T, Hillaert U, Ortmann R, Jomaa H, Link A, Schlitzer M, Van Calenbergh S (2008) Bioorg Med Chem 16:3361–3371
161. Singh N, Cheve G, Avery MA, McCurdy CR (2007) Curr Pharm Des 13:1161–1177
162. Nicolas O, Margout D, Taudon N, Wein S, Calas M, Vial HJ, Bressolle FMM (2005) Antimicrob Agents Chemother 49:3631–3639
163. Calas M, Ouattara M, Piquet G, Ziora Z, Bordat Y, Ancelin ML, Escale R, Vial H (2007) J Med Chem 50:6307–6315
164. Wengelnik K, Vidal V, Ancelin ML, Cathiard A-M, Morgat JL, Kocken CH, Calas M, Herrera S, Thomas AW, Vial HJ (2002) Science 295:1311–1314
165. Ouattara M, Wein S, Calas M, Hoang YV, Vial H, Escale R (2007) Bioorg Med Chem Lett 17:593–596
166. Degardin M, Wein S, Durand T, Escale R, Vial H, Vo-Hoang Y (2009) Bioorg Med Chem Lett 19:5233–5236
167. Alhamadsheh MM, Waters NC, Sachdeva S, Lee P, Reynolds KA (2008) Bioorg Med Chem Lett 18:6402–6405
168. Lee PJ, Bhonsle JB, Gaona HW, Huddler DP, Heady TN, Kreishman-Deitrick M, Bhattacharjee A, McCalmont WF, Gerena L, Lopez-Sanchez M, Roncal NE, Hudson TH, Johnson JD, Prigge ST, Waters NC (2009) J Med Chem 52:952–963
169. Freundlich JS, Wang F, Tsai H-C, Kuo M, Shieh H-M, Anderson JW, Nkrumah LJ, Valderramos J-C, Yu M, Kumar TRS, Valderramos SG, Jacobs WR Jr, Schiehser GA, Jacobus DP, Fidock DA, Sacchettini JC (2007) J Biol Chem 282:25436–25444

170. Vaughan AM, O'Neill MT, Tarun AS, Camargo N, Phuong TM, Aly ASI, Cowman AF, Kappe SHI (2009) Cell Microbiol 11:506–520
171. Yu M, Kumar TRS, Nkrumah LJ, Coppi A, Retzlaff S, Li CD, Kelly BJ, Moura PA, Lakshmanan V, Freundlich JS, Valderramos J-C, Vilcheze C, Siedner M, Tsai JH-C, Falkard B, Sidhu AS, Purcell LA, Gratraud P, Kremer L, Waters AP, Schiehser G, Jacobus DP, Janse CJ, Ager A, Jacobs WR Jr, Sacchettini JC, Heussler V, Sinnis P, Fidock DA (2008) Cell Host Microbe 4:567–578
172. Doerig C (2004) Biochim Biophys Acta 1697:155–168
173. Bouloc N, Large JM, Smiljanic E, Whalley D, Ansell KH, Edlin CD, Bryans JS (2008) Bioorg Med Chem Lett 18:5294–5298
174. Doerig C, Meijer L (2007) Expert Opin Thera Targ 11:279–290
175. Doerig C, Billker O, Haystead T, Sharma P, Tobin AB, Waters NC (2008) Trends Parasitol 24:570–577
176. Kato N, Sakata T, Breton G, Le Roch KG, Nagle A, Andersen C, Bursulaya B, Henson K, Johnson J, Kumar KA, Marr F, Mason D, McNamara C, Plouffe D, Ramachandran V, Spooner M, Tuntland T, Zhou Y, Peters EC, Chatterjee A, Schultz PG, Ward GE, Gray N, Harper J, Winzeler EA (2008) Nat Chem Biol 4:347–356
177. Andrews KT, Tran TN, Wheatley NC, Fairlie DP (2009) Curr Top Med Chem 9:292–308
178. Patel V, Mazitschek R, Coleman B, Nguyen C, Urgaonkar S, Cortese J, Barker RH Jr, Greenberg E, Tang W, Bradner JE, Schreiber SL, Duraisingh MT, Wirth DF, Clardy J (2009) J Med Chem 52:2185–2187
179. Agbor-Enoh S, Seudieu C, Davidson E, Dritschilo A, Jung M (2009) Antimicrob Agents Chemother 53:1727–1734
180. Andrews KT, Tran TN, Lucke AJ, Kahnberg P, Le GT, Boyle GM, Gardiner DL, Skinner-Adams TS, Fairlie DP (2008) Antimicrob Agents Chemother 52:1454–1461
181. Chen Y, Lopez-Sanchez M, Savoy DN, Billadeau DD, Dow GS, Kozikowski AP (2008) J Med Chem 51:3437–3448
182. Patil V, Guerrant W, Chen PC, Gryder B, Benicewicz DB, Khan SI, Tekwani BL, Oyelere AK (2010) Bioorg Med Chem 18:415–425
183. Farahat AA, Kumar A, Say M, Barghash AE-DM, Goda FE, Eisa HM, Wenzler T, Brun R, Liu Y, Mickelson L, Wilson WD, Boykin DW (2010) Bioorg Med Chem 18:557–566
184. Bakunova SM, Bakunov SA, Patrick DA, Kumar EVKS, Ohemeng KA, Bridges AS, Wenzler T, Barszcz T, Kilgore JS, Werbovetz KA, Brun R, Tidwell RR (2009) J Med Chem 52:2016–2035
185. Bakunova SM, Bakunov SA, Wenzler T, Barszcz T, Werbovetz KA, Brun R, Tidwell RR (2009) J Med Chem 52:4657–4667
186. Bakunov SA, Bakunova SM, Bridges AS, Wenzler T, Barszcz T, Werbovetz KA, Brun R, Tidwell RR (2009) J Med Chem 52:5763–5767
187. Bakunov SA, Bakunova SM, Wenzler T, Ghebru M, Werbovetz KA, Brun R, Tidwell RR (2010) J Med Chem 53:254–272
188. Rodriguez F, Rozas I, Kaiser M, Brun R, Nguyen B, Wilson WD, Garcia RN, Dardonville C (2008) J Med Chem 51:909–923
189. Chakrabarti D, Da Silva T, Barger J, Paquette S, Patel H, Patterson S, Allen CM (2002) J Biol Chem 277:42066–42073
190. Nallan L, Bauer KD, Bendale P, Rivas K, Yokoyama K, Horney CP, Pendyala RP, Floyd D, Lombardo LJ, Williams DK, Hamilton A, Sebti S, Windsor WT, Weber PC, Buckner FS, Chakrabarti D, Gelb MH, Van Voorhis WC (2005) J Med Chem 48:3704–3713
191. Bendale P, Olepu S, Suryadevara PK, Bulbule V, Rivas K, Nallan L, Smart B, Yokoyama K, Ankala S, Pendyala PR, Floyd D, Lombardo LJ, Williams DK, Buckner FS, Chakrabarti D, Verlinde CLMJ, Van Voorhis WC, Gelb MH (2007) J Med Chem 50:4585–4605
192. Van Voorhis WC, Rivas KL, Bendale P, Nallan L, Horney C, Barrett LK, Bauer KD, Smart BP, Ankala S, Hucke O, Verlinde CLMJ, Chakrabarti D, Strickland C, Yokoyama K,

Buckner FS, Hamilton AD, Williams DK, Lombardo LJ, Floyd D, Gelb MH (2007) Antimicrob Agents Chemother 51:3659–3671
193. Fletcher S, Cummings CG, Rivas K, Katt WP, Horney C, Buckner FS, Chakrabarti D, Sebti SM, Gelb MH, Van Voorhis WC, Hamilton AD (2008) J Med Chem 51:5176–5197
194. Kohring K, Wiesner J, Altenkaemper M, Sakowski J, Silber K, Hillebrecht A, Haebel P, Dahse H-M, Ortmann R, Jomaa H, Klebe G, Schlitzer M (2008) ChemMedChem 3:1217–1231
195. Dahl EL, Rosenthal PJ (2008) Trends Parasitol 24:279–284
196. Sidhu ABS, Sun Q, Nkrumah LJ, Dunne MW, Sacchettini JC, Fidock DA (2007) J Biol Chem 282:2494–2504

Top Med Chem 7: 181–242
DOI: 10.1007/7355_2011_17

Published online: 22 July 2011

Kinetoplastid Parasites

Tomas von Geldern, Michael Oscar Harhay, Ivan Scandale, and Robert Don

Abstract Kinetoplastid parasites, most notably of the genus *Trypanosoma* and *Leishmania*, have a major impact on the quality of life in the developing world. Human African trypanosomiasis (HAT, or African sleeping sickness), Chagas disease, and Leishmaniasis typically afflict the poorest and most neglected segments of these populations, and as such have historically been of little interest to the modern pharmaceutical industry. As a consequence, the few treatments that exist for these diseases are to some extent effective, however, often toxic, and not adapted to the field conditions. More recently, however, a resurgence of interest and an increase in funding have created an environment where multi-partner consortia are applying modern drug discovery methods to kinetoplastid diseases. These efforts have already produced new potential therapies for each of the kinetoplastid diseases. Clinical evaluation of these novel drug candidates (in many cases, ongoing) will determine their potential to transform the therapeutic landscape. This review will summarize the therapeutic challenges presented by each of the kinetoplastid diseases, the "state of the art" of current treatments, and will focus on evolving opportunities, emphasizing the role of the Drugs for Neglected Diseases *initiative* (DNDi) as a preferred development partner Product Development Partnership (PDP) providing "the best science for the most neglected."

Keywords American trypanosomiasis (Chagas disease), Aminoquinolines, Cell cycle inhibitors, Dihydrofolate reductase inhibitors, Fexinidazole, Human African

T. von Geldern (✉)
Embedded Medicinal Chemistry Consultant, Richmond, IL 60071, USA
and
Drug for Neglected Diseases initiative, Geneva 1202, Switzerland
e-mail: Tom@dndi.org

M.O. Harhay
Graduate Program in Public Health Studies, University of Pennsylvania School of Medicine, Philadelphia, PA 19104, USA

I. Scandale and R. Don
Drug for Neglected Diseases initiative, Geneva 1202, Switzerland

trypanosomiasis, Leishmaniasis, Lipid biosynthesis inhibitors, Myristoylation, Nifurtimox–eflornithine combination therapy, Nitroheterocycles, Oxaboroles, Pathway of isoprenoid biosynthesis, Pediatric benznidazole, Protease inhibitors, Pteridine reductase inhibitors, Quinolines, Trypanothione reductase

Contents

1 Introduction

Parasitic diseases such as leishmaniasis and trypanosomiasis (known as Chagas' disease (CD) in Latin America and human African trypanosomiasis (HAT) in Africa) are caused by protozoan parasites of the Kinetoplastida order, and represent a chronic challenge and serious impediment to improving the population health and economy of many developing countries. Clinical manifestations of the *Leishmania* species cause a variety of diseases, from self-healing cutaneous lesions to life-threatening visceral infections (VL). In the poorest communities of India, Nepal, Bangladesh, Sudan, and Brazil, where about 90% of all VL cases worldwide are found, the disease is a leading cause of illness and economic distress to families [1–3]. CD is one of the major causes of morbidity and mortality due to cardiovascular diseases in Latin America [4]. More recently, CD has emerged as a global public health threat as migration has introduced the parasite into donated blood supplies throughout the world [5, 6]. The lethal threat of HAT hampers the agricultural cultivation of some of most fertile land in sub-Saharan Africa [7], while also disabling or killing those who live and work close to the endemic regions [8]. Known epidemiology suggests that the tsetse fly vector covers approximately 10 million square kilometers, equating to one-third of the total landmass in Africa. This translates to an almost uncultivable area in Africa, due to tsetse fly infestation, that is slightly larger than the continental USA.

While these three diseases collectively span over 70 countries in three continents, there are similarities that bind them together beyond their Kinetoplastida order. Foremost, they are diseases that invariably inflict the poorest members in their endemic regions, imposing what economists term the "poverty trap." The high endemicity of these, as well as other so-called neglected tropical diseases, in rural and in impoverished urban areas of low-income countries has a pervasive cyclical reinforcing effect. Beyond their disabling clinical features, these infections can impair childhood growth and hinder intellectual development, thus creating a negative impact on education, as well as worker productivity. They become in consequence both poverty-promoting and poverty-sustaining diseases [9, 10]. This is further exacerbated with the kinetoplastids because of the high direct costs (e.g., the costs of diagnosis and treatment) and indirect costs (e.g., loss of household income) of these infections.

Furthermore, and where they differ from the other so-called neglected tropical diseases, such as soil transmitted helminthaisis, the chemotherapeutic options for the kinetoplastidas are (1) not well understood pharmacodynamically or pharmacokinetically, (2) unaffordable either to ministries of health or those infected (many STH drugs are donated), and (3) unavailable due to a lack of sustained demand. Another hurdle for the common chemotherapeutic agents currently used for these diseases is that they are often inadequate since they require long courses of parenteral administration that can have toxic (sometimes fatal) side effects. Resistance is another major concern and has been documented in a number of monotherapy regimens for VL and HAT. Since no vaccination is available, or will be in the foreseeable future for any of these infectious diseases, treatment is solely dependent on the available chemotherapeutic drugs [11]. Therefore, new, effective, and inexpensive drugs that can be used to treat these diseases are clearly urgently required. The long neglect that these diseases and their endemic populations have endured has elevated these inflictions from a local public health problem to global humanitarian tragedy.

1.1 Leishmania

1.1.1 Epidemiology and Clinical Characteristics

Leishmaniasis is endemic in areas of the tropics, subtropics, and southern Europe, from rain forests in Latin America, East Africa, to Western Asia, and in both rural and peri-urban areas [12]. Brazil especially has seen a more recent change in the ecology of infection into urban areas [13]. It is not unforeseeable that such a change in disease pattern could occur in urbanizing settings of Asia. Several clinical syndromes are subsumed under the term leishmaniasis. The three most prominent are visceral, cutaneous, and mucosal leishmaniasis, which result from replication of the parasite in macrophages in the mononuclear phagocyte system, dermis, and naso-oropharyngeal mucosa, respectively [14]. These syndromes are caused by a total of about 21 leishmanial species, which are transmitted by about 30 species of phlebotomine sand flies [15]. There are an estimated 1.5–2.0 million new cases of leishmaniasis yearly, of which approximately 500,000 belong to the visceral form, which is fatal if untreated [16]. However, these numbers are dated and a number of local reports suggest that limited epidemiological surveillance in endemic areas leads to significant under-reporting [17, 18].

Clinical cases of VL are characterized by prolonged fever, anemia, weakness, splenomegaly, and, to a lesser extent, lymphadenopathy and malaise [19, 20]. The parasites transform into amastigotes by losing their flagella within human dendritic cells and macrophages and then multiply and survive in phagolysosomes through a complex parasite–host interaction [21]. The clinical manifestation of visceral leishmaniasis generally occurs after an incubation period that generally lasts between 2 and 6 months and entails a parasitic invasion into the blood and reticulo-endothelial system. Thereafter, the parasites continue to migrate through the lymphatic and vascular systems creating a complicated symptomology that can appear as or be conducive to other bacterial co-infections such as pneumonia, diarrhea, or tuberculosis. This further complicates the differential diagnosis and treatment. It is not uncommon for VL symptoms to persist for several weeks to months before patients either seeks medical care or die from bacterial co-infections, massive bleeding, or severe anemia [12]. At the advanced stage, wasting is prominent, but once a patient responds well to treatment, disability can be substantially dampened [22]. VL has also emerged as an important opportunistic infection among people with HIV-1 infection [23]. Diagnosis may be achieved by direct methods, such as cytology, parasite isolation, and cultivation, or molecular detection of the parasite with polymerase chain reaction (PCR); however, diagnostics can be difficult to apply in field conditions [24, 25].

1.1.2 Brief History of Control

Contemporary military motivations have served to develop much of our understanding of human infection from *Leishmania* [26, 27]. Although leishmaniasis had existed for hundreds of years, the first scientific report of parasites in skin lesions was made by British Major D.D. Cunningham in 1885. James Homer Wright has been credited with the first clinical description of an Armenian patient treated at the Massachusetts General Hospital (1903) [28]. Yet, the disease was named after British Colonel W.B. Leishman, who described the organism in 1903 while examining the spleen during autopsy of a soldier with visceral leishmaniasis.

In the same year, a second British military physician, Colonel C. Donovan, linked the clinical disease to parasites recovered from splenic puncture of a living patient. The organisms have subsequently been referred to as Leishman–Donovan bodies. Both during, and outside wartime, soldiers stationed in Latin America and more recently, the Middle East remain at risk for leishmaniasis. For example, during World Wars I and II, thousands of soldiers contracted leishmaniasis [29]. More recently, US activity and deployments to the Middle East (Iraq and Afghanistan) and North Africa resulted in numerous leishmaniasis cases among US military members and as a consequence has allowed R&D to sustain, albeit at a smaller scale than needed [30–34].

There is currently a VL elimination campaign in the Indian subcontinent; however, its success has not yet been evaluated [35]. In Africa, VL remains a serious threat capable of surging rapidly. For instance in Southern Sudan, in a context of civil war and famine, VL killed an estimated 100,000 people out of a population of 280,000 between 1984 and 1994 [36]. There are currently no antileishmanial vaccines, and as such, current control strategies focus on reservoir and vector control, the use of insecticide-impregnated bed nets, and active case detection and treatment.

1.1.3 Current Treatments for Leishmaniasis

Pentavalent Antimonials [37]

Since the 1940s, pentavalent antimonials have been the standard first-line treatment for visceral leishmaniasis. Two treatments with different concentrations of SbV are available: meglumine antimoniate and sodium stibogluconate (**1** and **2**, Scheme 1). These medicines are administrated intravenously or intramuscularly at 20 milligrams per kilograms of body weight during 28–30 days. Administration of antimonials is painful and also causes severe side effects such as cardiotoxicity and pancreatitis. Overall cure rate of sodium stibogluconate and meglumine antimoniate is usually superior to 90%. But drug resistance is a major concern in the Bihar state and in Nepal, where up to 60% of patients does not respond to treatment.

Amphotericin B [37]

In the last 10 years, Amphotericin B (**3**, Scheme 1), a polyene antibiotic, has been introduced for treatment of visceral leishmaniasis. Administration by infusion causes severe side effects such as nephrotoxicity, hypokalemia, and myocarditis. In order to reduce toxicity, several formulations, including liposomal amphotericin B, amphotericin B lipid complex, and amphotericin B colloidal dispersion, have been developed. For treatment of visceral leishmaniasis, the liposomal formulation is the most widely used. The total efficacious dose varies by regions. In India,

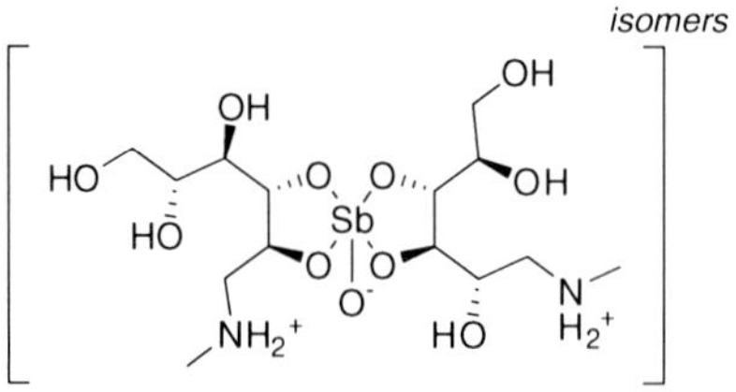

1. Meglumine antimoniate (Glucantime)

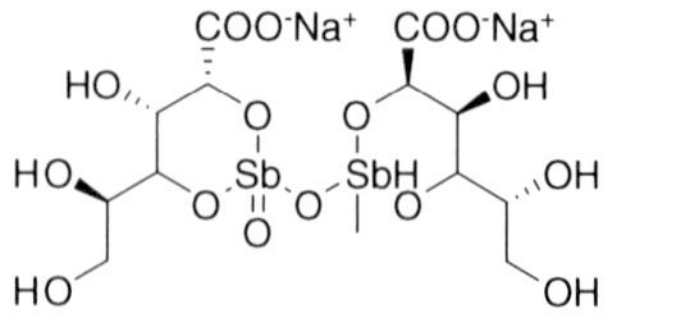

2. Sodium stibogluconate (Pentostam)

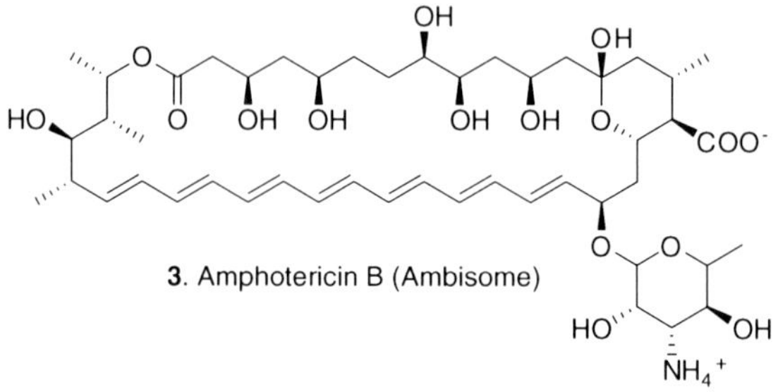

3. Amphotericin B (Ambisome)

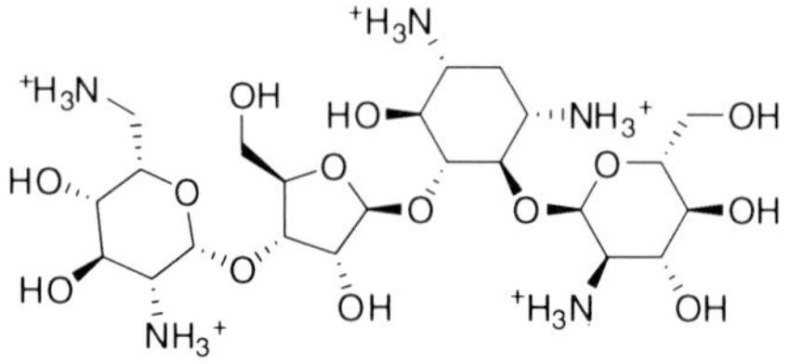

4. Paromomycin

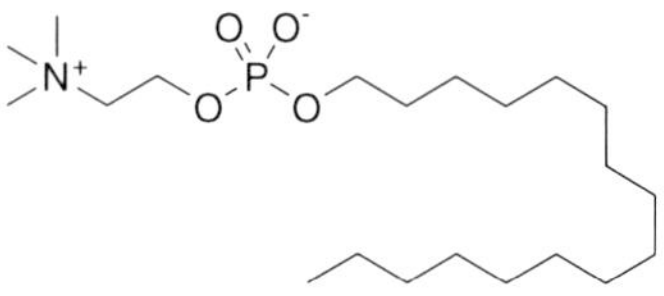

5. Miltefosine

Scheme 1 Current drugs for the treatment of leishmaniasis

a single dose of 10 mg/kg was found curative in 98% of patients. However, the use of Amphotericin B in endemic regions is limited by the high cost of treatment.

Paromomycin [37]

Paromomycin (aminosidine; compound **4**, Scheme 1) is an aminoglycoside antibiotic, usually administered intramuscularly. Side effects are relatively mild; the most common is pain at the injection site. Rarely, reversible ototoxicity, hepatotoxicity, or renal toxicity is observed.

A major concern with the use of Paromomycin is a variable efficacy between and within regions. For example, a cure rate of visceral leishmaniasis was found significantly higher in India compare to East Africa. In India, a cure rate of 93–95% was obtained at a dose of 15 mg/kg administrated per day for 21 days. In comparison, a cure rate of 85% was achieved at a dose of 20 mg/kg per day administrated for

21 days in East Africa. A topical formulation of Paromomycin is available for cutaneous leishmaniasis.

Miltefosine [37]

Miltefosine (**5**, Scheme 1) is an alkyl phospholipid, which was developed as an oral anticancer drug. Miltefosine is the first oral drug efficacious for treatment of visceral leishmaniasis. The treatment duration is relatively long: 28 days at a dose of 150 mg/day for adults. The cure rate shown in India is of 94%, in Ethiopa the cure rate is lower: 90%.

The main limitation of Miltefosine is related to potential teratogenic effects. As a result, WHO has recommended that Miltefosine shall not be used for the treatment of pregnant women. The administration of Miltefosine in child-bearing age women should be done only if adequate contraception is assured for the duration of treatment and for 3 months afterward.

1.2 American Trypanosomiasis (Chagas Disease)

1.2.1 Epidemiology and Clinical Characteristics

Trypanosoma cruzi is the etiological agent of CD which is transmitted by various insect vectors that belong to the Reduviidae family [38]. In Latin America, the prevalence rate is approximately 1.4%, and CD is estimated to kill 14,000 people every year, which is more people in the region each year than any other parasite-born disease, including malaria [39]. Eight to fourteen million people are assumed to be infected with CD [40]. Through its impact on worker productivity, and by causing premature disability and death, CD annually costs an estimated 667,000 disability-adjusted life years lost [41]. In Brazil alone, losses of over US$1.3 billion in wages and industrial productivity were estimated due to the disabilities of workers with CD [42]. There is not currently a vaccine or appropriate drugs available for large-scale public-health interventions. As a consequence, the main control strategy for CD entails the prevention of parasite transmission by eliminating the insect vectors from human dwellings. However, this is a complex zoonosis involving a heterogeneous parasite population that consists of multiple strains divided into two major phylogenetic groups [43], with numerous vectors and a wide range of wild and domestic mammals that can act as parasite reservoirs. Moreover, as *T. cruzi* often establishes a chronic infection in humans, there is always the risk of parasite transmission by contaminated blood, either from mother to baby during pregnancy or delivery, or by blood transfusion [44].

The trypomastigote forms of *T. cruzi* enter a mammal through contact of an infected vector with a skin wound or a mucosal membrane where they then invade many different cell types including macrophages and muscle cells [45].

The parasites then go through a parasite-amplification process known as the acute phase in which the parasitemia is often high and parasitological diagnosis can be made by direct microscopic examination of fresh blood [46]. This disease phase (in which 2%–8% of children die from cardiac involvement and encephalomyelitis) frequently passes undiagnosed in the absence of active screening programs, as CD manifests itself with a febrile and toxemic illness having nonspecific symptoms reminiscent of any childhood infection [47]. If untreated, the disease transitions into a clinically dormant, indeterminate chronic phase, where parasite numbers drop dramatically. However, low levels of parasites can still be detected in certain tissues throughout this chronic quiescent phase [48]. Later, 10–30 years after the initial infection, approximately 30% of infected people will experience the symptomatic, chronic stage characterized by severe organ pathologies primarily involving cardiomyopathy and digestive tract abnormalities such as mega colon and mega esophagus [49]. Broadly speaking, the pathology of CD is not entirely well understood and appears to manifest with different clinical presentation, severity, and prevalence from region to region. Such variation may be related to the impact of past control strategies and different living ecologies and concomitant adaptive mechanisms of both host and parasite [45].

1.2.2 Brief History of Control

Considerable progress has been made in large areas of Latin America in the control of natural transmission since Carlos Chagas first identified CD in 1909 [50]. However, while evidence of human infection has been found in mummies up to 9,000 years old, endemic CD became established as a zoonosis only in the last 200–300 years, as triatomines (the most common CD insect vector, known as the "kissing bug") adapted to domestic environments [51]. More recently, CD control has become in many ways a success story since the launch of the Southern Cone initiative in 1991 [52]. Yet in terms of socioeconomic impact, CD remains one of the most destructive infectious diseases in the region [53]. According to Ramsey (2003), Uruguay, Chile, most of Southern and Central Brazil, and four previously endemic provinces of Argentina have been formally certified by PAHO as free of Chagas disease transmission. Large areas of Paraguay and Southern Bolivia are also free of vector-borne transmission, with some progress also in Southern Peru (Tacna). In Central America, CD has been eliminated from El Salvador and from large parts of the previously endemic regions of Guatemala, Honduras, and Nicaragua. Mexico also, although only just beginning to plan large-scale campaigns against CD, appears already to have almost eliminated the vector from the southern states. These regional and multinational initiatives will likely continue to herald success. However, even while domestic bug populations continue to be minimized, new cases continue to occur, and in new locations. As a result, new cases of human infection will continue to remain a challenge, and in the absence of sustained surveillance CD could surge. Another ongoing, more challenging problem is the relative ease by which human-to-human transmission can occur through exposure

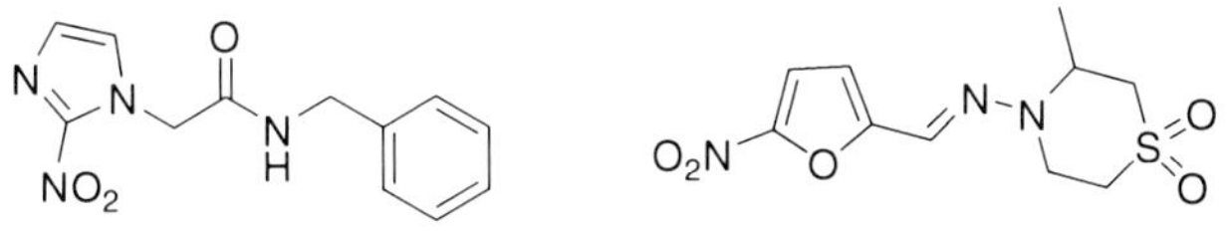

6, Benznidazole

7, Nifurtimox

Scheme 2 Current drugs for the treatment of Chagas disease

to contaminated blood during pregnancy or blood transfusion [54]. Infected blood supplies especially have become an issue of great concern in certain developed countries such as the USA and Europe, following the large number of Latin American immigrants worldwide [55].

1.2.3 Current Treatment for Chagas Disease

Patients with Chagas disease have two therapeutic options, the nitroheterocycles benznidazole and nifurtimox (compounds **6** and **7**, Scheme 2) used to treat the vast majority of cases [56]. The drugs are orally deliverable, but treatment courses are extended, typically requiring 2–3 months of dosing. Despite this regimen, disease recurrence is common; the drugs are considered to be relatively ineffective. Side effect profiles are significant and include severe gastrointestinal and dermatological adverse events. Surprisingly, the drugs are generally much better tolerated in children, who for this reason are sometimes given higher doses.

1.3 Human African Trypanosomiasis

1.3.1 Epidemiology and Clinical Characteristics

HAT, also known as sleeping sickness, once the second (CNS) stage of the infection has been reached, is a fatal parasitic disease endemic in 36 sub-Saharan countries, where more than 60 million people are at risk of becoming infected through the bite of the *Glossina* tsetse fly. Two subspecies of the kinetoplastid protozoan parasite *Trypanosoma brucei* (*T. b.*) are found to infect humans. *T. b. gambiense* causes the chronic form of the human disease which represents ~97% of reported HAT cases, whereas *T. b. rhodesiense* is primarily zoonotic (having a huge animal reservoir) and is responsible for the more virulent form of the disease. Both forms of the disease involve two stages, with the second stage marked by parasitic invasion of the central nervous system (CNS) and, ultimately, death [57–59]. *T. b. gambiense* HAT is a highly focalized disease (>85% of cases are found in the Democratic Republic of the Congo (DRC), northern Angola, and southern Sudan) but has been reported in 24 countries across sub-Saharan Africa. Typically occurring in remote, rural, and poor areas, HAT places a large burden on both communities and individual households. In 2002 WHO estimated that approximately 1.5 million disability-adjusted life years (DALYs) were lost due to HAT [60]. A research study by Lutumba et al. found that a HAT outbreak over a 2-year period in the Buma region of DRC cost each household the equivalent of 5 months' income, highlighting the profound financial impact the disease can have on poor populations [61].

After infection from blood meals taken by various species of the tsetse fly, patients experience a variety of intermittent, nonspecific, and mostly mild symptoms such as malaise, fever, arthralgia, and weight loss during a hemo-lymphatic phase (stage 1). The parasites spread into the bloodstream and lymphatic system, where they avoid the immune system through antigenic variation of the variant surface glycoprotein (VSG) coat. As a result, parasitemia occurs in waves as the immune system successfully reduces one variant only to have another variant replace it [62]. This is followed by a meningo-encephalitic phase (stage 2) initiated when parasite penetration of the blood–brain barrier occurs [59, 63–65]. The ensuing inflammatory process leads to relentless CNS deterioration: associated behavioral and neurologic changes include severe and protracted headache, apathy, confusion, sensory disturbances, poor coordination, tremor, paralysis, sleep disorders, irritability, antisocial behavior, speech impediment, coma, and, without treatment, eventually death. Opportunistic infections and disease-associated malnutrition further complicate the clinical picture [66]. If untreated, stage 1 and stage 2 each may last about 18 months, if not longer, although *T. b. rhodesiense* typically progresses more rapidly than *T. b. gambiensis* [67]. The trypanosomes disrupt the natural circadian rhythm, causing the nocturnal insomnia and diurnal somnolence that gives the disease its common name of sleeping sickness. The disease invariably progresses in severity until the patient suffers from stupor, becoming increasingly difficult to awaken until falling comatose, and all stage 2 patients die if left untreated [68]. Because few specific signs and symptoms are associated with HAT, particularly in the early stage, diagnosis of HAT is complicated and requires specific laboratory tests [69]. Mass screening of the population at risk by mobile teams is key to detect patients at an earlier stage of the illness. Screening is based on the field-designed card agglutination test for trypanosomiasis (CATT), which detects antibodies against *T. b. gambiense*. Diagnosis subsequently needs to be confirmed by microscopical visualization of the trypanosomes. If parasites are found, disease staging is then mandatory and performed by examining the cerebrospinal fluid (CSF) obtained by lumbar puncture for the presence of trypanosomes and/or elevated white blood cells count.

1.3.2 Brief History of Control

David Livingston first suggested that HAT is caused by the bite of tsetse flies in 1852, but descriptions of the disease can be found in ancient Egyptian writings [70]. African animal trypanosomiasis, important in domestic livestock such as cattle, was first shown to be caused by *T. brucei* by David Bruce in 1899 [71]. Subsequent work by Aldo Castellani enabled the identification of trypanosomes in the blood and CSF in human patients with HAT in 1903. During the first half of the twentieth century, HAT caused by *T. b. gambiense* ravaged much of central Africa. HAT was almost eliminated in the 1960s, but a cessation in major vector control activities has resulted in epidemics that ravaged the African countryside in the 1990s [72, 73]. During this period, HAT spread steadily across sub-Saharan Africa, including Angola, southern Sudan, the Central African Republic, and Uganda, with up to 30% of the communities infected in some regions [74, 75]. In 2007, the total number of reported new cases of *T. b. gambiense* HAT was 10,446[1] from a peak incidence of 37,000 per year in the late 1990s. However, because large endemic regions go without effective health surveillance, it is likely that there is considerable underreporting of the condition [76, 77].

Although it is still difficult to provide accurate estimates of disease incidence and prevalence [78], more recent WHO estimates have suggested that as a result of more efficient surveillance, a significant improvement has occurred with currently as low as 70,000 existing cases, mainly infected with *T. b. gambiense*. However, HAT historically has repeatedly demonstrated its ability to recur even after it had been virtually brought under control. Because the prevalence of the disease ebbs and flows as implementation of control measures and local politics fluctuate, drugs have been central to most control programs and chemotherapeutic intervention will remain at the forefront of anti-HAT campaigns for the foreseeable future. Ideally, sustained active surveillance programs would enable more use of the drugs able to treat early stage disease. In reality, however, it is probable that the majority of cases will continue to be treated for stage 2 of the disease when patients appear at treatment centers.

1.3.3 Current Treatments for HAT

Suramin

The first pharmaceutical agent used for the treatment of HAT, the polysulfonate suramin (**8**, Scheme 3), is still in use today. First tested in 1921, suramin remains a first-line therapy for the treatment of stage 1 infections of the *T. brucei rhodesiense* parasite [79], which produces a particularly aggressive form of the disease. It is also used on occasion to treat stage 2 disease in patients who cannot tolerate melarsoprol, though with limited success. Its mechanism of action is unclear; it has been proposed that it may block the uptake of low-density lipoprotein (LDL), or alternatively that it acts as an inhibitor of parasitic glycolysis.

[1]NTD statement on the inclusion of Nifurtimox–Eflornithine combination as a treatment for Stage 2 *T. b. gambiense* human African trypanosomiasis (sleeping sickness) in the WHO Model List of Essential Medicines. 17th Expert Committee on the Selection and Use of Essential Medicines. Submitted February 2009 by the Director of NTD. Available at: http://www.who.int/selection_medicines/committees/expert/17/application/nifurtimox/en/index.html.

Stage 1:

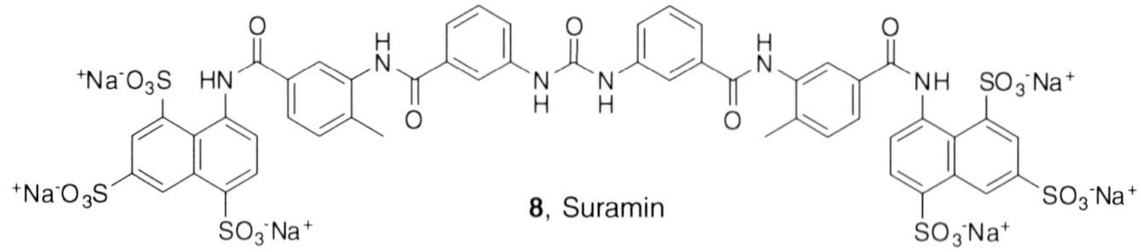

8, Suramin

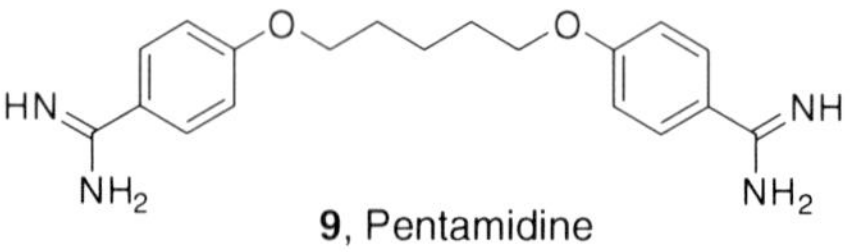

9, Pentamidine

Stage 2 :

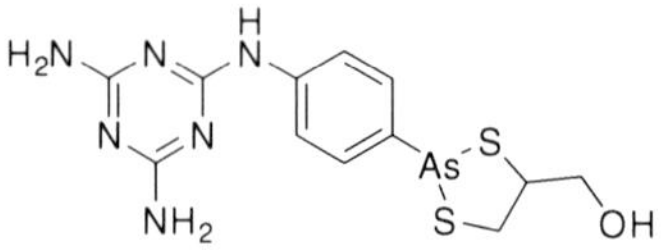

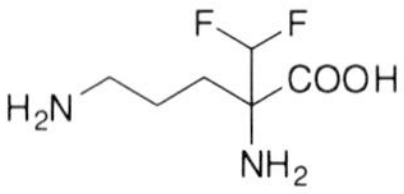

10, Melarsoprol

11, Eflornithine (DFMO)

Scheme 3 Current drugs for the treatment of human African trypansomiasis (HAT)

Suramin therapy is unpleasant and inconvenient; large doses of the drug (on the order of 1 g per dose) must be given, i.v. or i.m., at least five times over a 3-week period. The treatment is associated with a number of adverse side effects [79]; essentially all patients receiving the drug experience nausea and vomiting, abdominal distress, and a rash. Half of these suffer from adrenocortical damage; a rapid destruction of cutaneous microfilaria can also occur in some patients, triggering anaphylaxis. Suramin therapy causes neuropathic pain in some patients and produces renal damage which can lead to renal failure. Despite these concerns, the drug, which is cheap and readily available through a WHO partnership with Bayer AG, remains in common use in endemic regions [80].

Pentamidine

Pentamidine (compound **9**, Scheme 3) is the prototype of a class of drugs known as bis-amidines, originally developed as insulin mimetics [81] but subsequently discovered to have antiparasitic activity. Pentamidine is used almost exclusively in treating stage 1 infections, and is the first-line therapy for *T. brucei gambiense*. The drug is ineffective against stage 2 disease. Pentamidine is taken up and concentrated in parasites through the agency of the P2 nucleotide transporter; its mechanism of action remains unclear, though inhibition of nucleic acid metabolism (via inhibition of P2) may play a role [82]. Developed in 1941, the drug is administered through a series of intramuscular injections, once daily for 14 days. Drug treatment is associated with a number of serious adverse events including severe skin reactions (which are common), nephrotoxicity (which occurs in ~25% of patients), and hepatotoxicity. The drug is available through a private partnership between WHO and Sanofi-Aventis [80].

The related diamidine berenil, developed for the treatment of bovine trypanosomiasis, is sometimes used for the treatment of stage 1 HAT, despite the fact that it has not been approved for human use and has not been evaluated clinically [82].

Melarsoprol

For many decades, the only treatment available for stage 2 (meningo-encephalitic) disease was the organoarsenical melarsoprol (**10**, Scheme 3). Its mode of action has not been elucidated, though it is believed to act via inhibition of parasitic glycolysis [82]. Melarsoprol is delivered as an intravenous infusion, given once daily for 10 days. As might be anticipated for a therapy based on arsenic, melarsoprol treatment is associated with a number of severe adverse events, including convulsions, fever, and reactive encephalopathy. *Melarsoprol treatment is fatal in 5–10% of patients*, a situation that is only tolerable when treating a condition (like stage 2 HAT) that is 100% fatal. The drug remains as first-line therapy for stage 2 disease throughout most of the endemic region.

Eflornithine

The only "modern" treatment for HAT is eflornithine (**11**, Scheme 3), a mechanism-based inhibitor of ornithine decarboxylase (ODC) first explored clinically in the 1980s [83]

Compared with the drugs noted above, eflornithine is well tolerated, and as a result has seen increasing use for the treatment of stage 1 and stage 2 disease. In particular, the drug has demonstrated efficacy in patients who are refractory to melarsoprol treatment [83]. Unfortunately, eflornithine, provided through a private partnership between WHO and Sanofi-Aventis, is only effective against *T. b. gambiense* infections. In order to maintain continuously high drug levels in all compartments, eflornithine must be dosed via 30-min intravenous infusions, given every 6 h for 14 days. This regimen places burdens on patients, their families, and health care providers, and creates a significant access issue, as one patient-course of therapy requires a volume of 1 m^3. The recent demonstration that eflornithine can be used as part of a two-drug combination (*vide infra*) offers the opportunity to reduce this burden substantially.

2 Therapeutic Approaches to Kinetoplastid Diseases: Established Mechanisms

As noted in the previous section, the current state of the art of therapeutic intervention for the kinetoplastid diseases is generally abysmal. Currently available drugs are often ineffective and cause significant adverse reactions; despite these flaws, there have been few improvements in the chemoarsenal in recent decades. Chagas disease is still treated with the same two medications, nifurtimox and benznidazole, that were used in the 1960s. For HAT, only eflornithine has been introduced in the past 60 years. In many ways, the "modern" era of drug discovery has passed by these diseases. This failure is not due to a complete lack of effort. Much has been learned about these parasites, and a plethora of novel approaches and agents have been described in recent years, many targeting specific kinetoplastid-critical pathways. This section will not review all of these strategies, but will focus on those that have moved closest to clinical exploration, and will attempt to analyze the reasons why many of these have not (yet) led to improved therapies for patients. For organizational clarity, these approaches will be grouped based on the target class, where a specific target or group of targets has been explored across several kinetoplastid diseases, and these will be discussed together [84].

2.1 *Nucleotide Uptake*

The success of pentamidine as a first-line therapy for the treatment of HAT has resulted in an extensive effort to develop second-generation agents which act through a similar mechanism, while addressing the flaws in the profile of the "parent" drug. In particular, it would be desirable to eliminate some or all of the adverse reactions (including hypoglycemia [85], nephrotoxicity [86, 87], and cardiovascular events including hypotension [88]) linked to pentamidine treatment, while an orally deliverable and CNS penetrant second-generation agent would dramatically simplify the treatment of patients in these under-developed settings. The development of optimized bis-amidines has been one primary focus of the Consortium for Parasitic Drug Development (CPDD).

Pentamidine is believed to act via inhibition of parasitic nucleotide uptake [89]. Lead-optimization efforts, targeting the identification of a second-generation agent from this class, have used two distinct screening strategies, (1) either target-based approaches focusing directly on inhibition of the nucleotide transporter (2) or using a holistic approach that measures trypanosome survival as the primary outcome. These optimization efforts have been reviewed [90–92]; we will focus here on the preclinical and clinical profiles of the most advanced agents of this class, DB-75 and DB-289.

DB-75, or furamidine (compound **12**, Scheme 4), is a bis-amidine with a conformationally constrained linker moiety consisting of three aromatic rings.

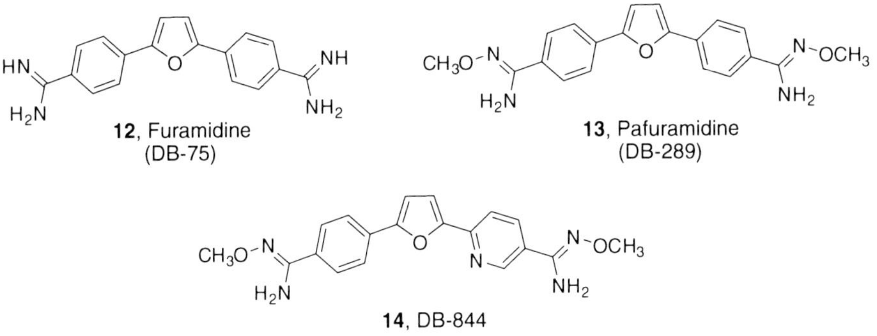

12, Furamidine (DB-75)

13, Pafuramidine (DB-289)

14, DB-844

Scheme 4 Nucleotide uptake inhibitors

These modifications produce a significant improvement in in vitro antitrypanosomal activity [93], as measured against *T. brucei rhodesiense*. Furamidine is also active in vivo, approximately fourfold more potent than pentamidine in a mouse model of HAT [94]. But, like pentamidine, furamidine lacks oral bioavailability and can only be delivered parenterally. Additionally, toxicity issues with **12** eventually led to its abandonment as a potential candidate. To address the bioavailability issue, the CPDD team borrowed a prodrug strategy originally developed by Clement and Raether (for pentamidine [95]), and applied it to DB-75. In vivo and metabolic screening of a family of "protected amidine" analogs of DB-75 [93] led to the identification of the bis-methoxyamidine **13** (DB-289, or pafuramidine) [96]. Oral bioavailability of **13** is modest in rat and monkey pharmacokinetic studies ($F = 10$–40%, with significant dose- and species-dependence) though vastly superior to other agents of this class [97]. While the prodrug is well absorbed, first-pass metabolism limits systemic exposure. The metabolic fate of **13** is complex [98], but active metabolite **12** (DB-75) is the most persistent component, and the primary species in key target tissues such as liver and brain. Production of DB-75 from DB-289 is mediated through the agency of CyP4F and b5 reductase enzymes [99, 100]. Consistent with this profile, pafuramidine is highly effective in a mouse model of stage 1 disease, curing animals with a single treatment [101]. Similarly, monkeys treated with DB-289 for 5 days are cleared of parasites and remain parasite-free for 3 months posttreatment [102]. Although the prodrug is able to cross the blood–brain barrier [103], and may be converted locally to active parent [103], DB-289 is relatively ineffective against stage 2 disease [102].

Other aromatic diamidines, most notably the azafuramidine DB-844 (**14**, Scheme 4), are superior at penetrating the CNS. DB-844 is superior to DB-289 in an acute mouse model of *T. brucei* infection [104], and notably is able to cure a chronic infection when dosed orally, a first for the class. However, toxicity issues [104] have limited the ability to dose this agent in larger-animal HAT models [104], and, at the tolerated doses, DB-844 has modest ability to cure a stage 2 infection.

The in vivo models accurately predict the clinical profile of pafuramidine, which is effective at treating stage 1 disease in humans [105]. Phase I studies confirm that the drug is taken up from an oral dose and converted to the active principle **12**. The plasma levels of A1 achieved at the clinical doses were sufficient to predict clearance of the parasite (based on the preclinical in vivo work). Phase II evaluation was useful in optimizing the treatment paradigm, with the result that Phase III trials were performed using BID dosing (100 mg/dose) for 10 days. These studies indicated that the efficacy of oral pafuramidine was not inferior to that of i.v. pentamidine.

Extensive clinical evaluation of DB-289 has revealed significant flaws in its profile. In particular, longer treatment periods are associated with more serious liver toxicity, and in particular with a severe delayed renal toxicity [105]. These results, which eventually triggered a more detailed Phase I safety evaluation that confirmed the findings, have led to the discontinuation of the pafuramidine clinical program [106].

The use of diamidines to treat other protozoal infections has also been explored. Both DB-75 and its N-aryl derivative DB-569 have been studied for the treatment of Chagas disease [107]; the compound has micromolar activity in vitro against several strains of *T. cruzi*. DB-569 is also active in an in vivo mouse model, improving survival while reducing the cardiac toxicity associated with chronic infection [108]. DB-75 and related compounds also show activity against the Leishmania parasite *Leishmania donovani*, for which pentamidine is a second-line therapy [109, 110]. Unfortunately, none of the diamidines is effective in clearing parasites from Leishmania-infected macrophages [109]. In light of the persistent observations of toxicity noted with agents of the diamidine class, these agents have not moved beyond the preclinical stage.

2.2 *Polyamine Synthesis*

2.2.1 Ornithine Decarboxylase

The most obvious validation for the critical role of polyamines in kinetoplastid survival comes from the success of the drug eflornithine. Eflornithine [difluoro-methylornithine, or DFMO, (compound **15**, Scheme 5)] is an irreversible covalent inhibitor of ODC, a key enzyme responsible for generating the putrescine core of the kinetoplastid polyamine family [83]. Eflornithine has proven to be the safest and most effective agent for the treatment of stage 2 HAT, though a complicated treatment regimen limits its impact on the clinical management of the disease. Several aspects of this mechanism, established through mechanistic studies with eflornithine, make ODC (retrospectively) a less-appealing target than other enzymes in this pathway. In particular, eflornithine acts via trypanostatic mechanism, thus requiring longer dosing regimens (typically 2 weeks or more) to be effective. In addition, ODC is rapidly regenerated in response to drug-induced depletion; so high drug levels must be maintained for extended periods of time to be effective, a particular challenge in the brain. Apart from his pharmacokinetic profile, the use

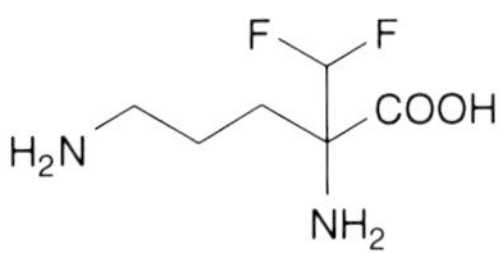

15, Eflornithine (DFMO)

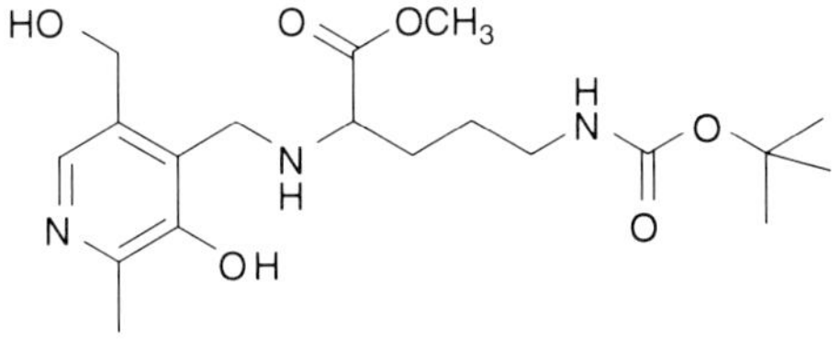

16, POB

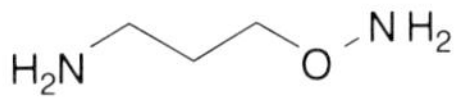

17, APA

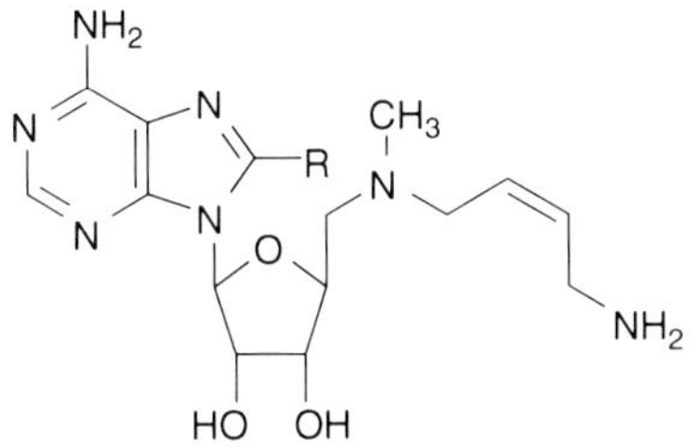

18, MDL73811 (R = H)

19, Genz-644131 (R = Me)

Scheme 5 Polyamine pathway inhibitors

of eflornithine is limited because of his low efficacy against the *T. b. rhodesiense* strain; the drug is not used to treat infections with this strain. For these reasons, only a limited effort has been made to identify second-generation ODC inhibitors. Eflornithine analogs with improved in vitro activity against ODC have been identified, but these face similar challenges with regard to pharmacokinetic properties and have not advanced [111]. In an alternative approach, multisubstrate inhibitors [112, 113] including POB (**16**, Scheme 5) block ODC activity and inhibit cell growth; however, the requirement that this prodrug survives plasma and organ metabolism while converting efficiently in cells makes it an unlikely candidate for oral delivery.

ODC inhibitors have also proven effective for the treatment of *Leishmania* infection. 3-Aminooxy-1-aminopropane (APA, compound **17**, Scheme 5) is able to inhibit the growth of *L. donovani* promastigotes and amastigotes [114] with IC50 values in the micromolar range. Unfortunately, the effectiveness of this agent is reduced when ODC is over-expressed, or when putrescine/spermidine levels are elevated, conditions that are typical of many clinical presentations. Thus, questions regarding its in vivo stability and pharmacokinetic profile have not been explored.

Other enzymes in the polyamine pathway have also been targeted for the development of antitrypanosomal agents. MDL 73811 (compound **18**, Scheme 5) is a potent product-analog inhibitor of *S*-adenosylmethionine decarboxylase (SAM-DC), an enzyme that generates a critical intermediate for polyamine elongation [115]. The compound, which is also a substrate for the trypanosomal purine uptake transporter and is highly concentrated in the parasite [116], cures *T. brucei* infections when given i.p. to mice at elevated doses. However, the pharmacokinetic profile of this compound is insufficient to make it a viable oral candidate for the treatment of HAT. More recently, the trypanocidal properties of Genz-644131 (compound **19**, Scheme 5), an optimized analog of MDL 73811, have been reported [117]. Addition of a methyl group at the 8-position on the purine ring simultaneously improves in vitro antitrypanosomal potency (by a factor of 100-fold) and stabilizes the molecule against metabolism. As a consequence, Genz-644131 is active in a murine model of stage 1 HAT, at doses as low as 2 mg/kg/day (given i.p., once daily). Unfortunately the compound is not effective in a model of stage 2 disease, suggesting that CNS penetration is poor. While further optimization would clearly be necessary to produce a clinical candidate from this class, the experiments to date have validated SAM-DC as a therapeutically relevant target.

2.3 *Trypanothione Reductase*

Kinetoplastids are unique in using a thiol-based shuttle system for reducing equivalents that originally derive from NADPH [118, 119]. A key to this system is the glutathione analog trypanothione, which is a required cofactor for a number of critical parasitic enzymes. Trypanothione redox cycling is mediated by the flavin-containing enzyme trypanothione reductase (TryR). Thus, TryR serves as a bulwark against the oxidative stress that kinetoplastids must endure throughout their life cycle; and so presents an attractive therapeutic target.

TryR is one of the more intensively studied trypanosomal drug targets [120, 121]. Target-based screening has identified numerous chemical classes that demonstrate TryR inhibitory activity, including alkaloidal and polycyclic natural products [122], as well as a variety of agents that interact with thiol groups in either a reversible or irreversible manner. None of these molecules are particularly drug-like; there is a general disconnect between their TryR inhibition and antitrypanosomal activities, indicative of potential issues with permeability, stability, and/or off-target activity. A series of spermine/spermidine analogs including MDL 27695 (**20**, Scheme 6) micromolar inhibitors of *L. donovani* in vitro were originally hypothesized to bind DNA [123], but have subsequently been shown to act as inhibitors of TryR as well [124]; their activity against kinetoplastids may have multiple mechanistic origins. Compounds like **20** have anti-leishmanial activity in vitro and in vivo [125, 126], and are even able to suppress *L. donovani* infection in mice when dosed by oral gavage, but their high molecular weight, lipophilicity, and polybasic nature make them unlikely drug candidates, and studies to compare dose potencies with different modes of administration [126] suggest that uptake from an oral dose is poor. Aminoacridines such as mepacrine (**21**) [127] and quaternized analogs of tricyclic antidepressants such as chlorpromazine (**22**) [128] are interesting lead structures with micromolar potencies against TryR, but little effort has been expended to improve their activity or (in particular) to make them

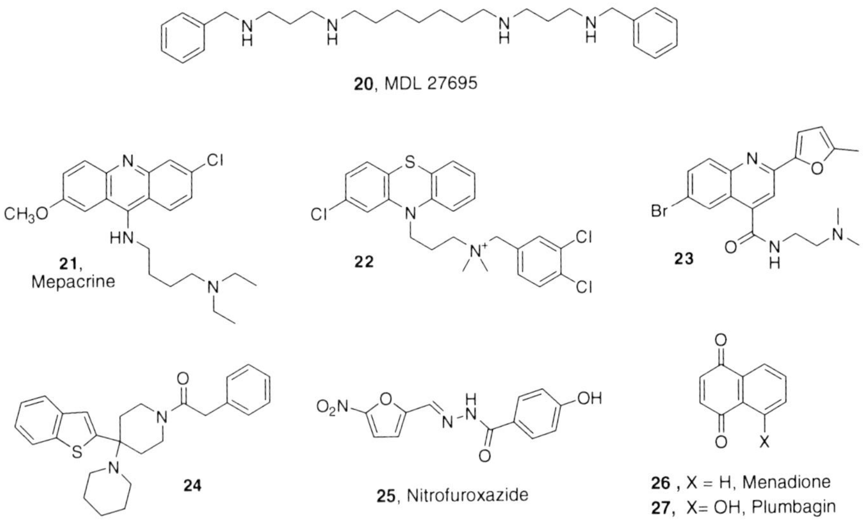

20, MDL 27695

21, Mepacrine

22

23

24

25, Nitrofuroxazide

26 , X = H, Menadione
27, X= OH, Plumbagin

Scheme 6 Trypanothione reductase (TryR) inhibitors

TryR-selective (vs. their original targets). One notable exception is a study of mepacrine-aryl sulfide conjugates [129], which have target affinity superior to either parent structure. These molecules also demonstrate in vitro antiprotozoal activity, but the high molecular weight of these conjugates, coupled with their charged nature, makes them unlikely leads for developing an oral antitrypanosomal agent.

More recently, researchers from the University of Dundee (Scotland) have highlighted the antitrypanosomal activity of several novel, drug-like inhibitors of TryR, including the quinoline-4-carboxamides (e.g., **23**) and benzothienylcyclohexylpiperidines (e.g., **24**), identified through screening and optimized through SAR studies [130, 131]. Compounds of these class have micromolar activities against *T. brucei* TryR, with selectivities of 10- to 100-fold against the related counter-target glutathione reductase (human). The activity of these analogs against the parasite is fairly low, however, with IC_{50} values typically in the tens of micromolar range, and selectivity indices versus mammalian cell cytotoxicity are typically low. As a result, these leads have not progressed further.

Another class of TryR inhibitors are referred to as "subversive substrates." Compounds of this class are capable of undergoing TryR-mediated reduction, producing intermediates that trap oxygen resulting in the production of superoxide. Thus, a reductive enzyme is "subverted" to produce oxidizing equivalents, disrupting the redox balance of the parasite. Nitroheterocycles such as nitrofuroxazide (**25**) [132] and quinones such as menadione and plumbagin (**26/27**) are examples of compounds in this category. While some of these molecules have drug-like properties, their redox activity is rather nonspecific, creating concerns regarding host-vs.-tryp selectivity and thus eliminating many of the theoretical benefits that TryR offers as a parasite-specific target.

2.4 Aminoquinolines and Quinolines for Leishmaniasis

Shortly after the introduction of the aminoquinoline antimalarials in the 1950s, broad-based antiparasitic screening efforts at the Walter Reed Army Institute uncovered the antiprotozoal activity of this class of agents. In particular, the 8-aminoquinoline primaquine (**28**, Scheme 7) was demonstrated to have in vivo activity against *L. donovani* in the gold-standard hamster model. Subsequent optimization of this activity [133] led to the selection of sitamaquine (**29**) [134] as a preclinical candidate, dramatically superior to antimonials in this assay. The anti-leishmanial activity of sitamaquine is species-dependent, suggesting that metabolites play a significant role in mediating its effect [135]. The compound is noticeably *less* active against *L. tropica*, the parasite responsible for cutaneous leishmaniasis.

GlaxoSmithKline is currently developing sitamaquine for the treatment of visceral leishmaniasis. The results of Phase II studies in India [136] and Kenya [137] were positive, with cure rates approaching 85% in both groups. The most significant adverse event associated with treatment was renal impairment, which was observed in ~2–3% of patients. Although this renal damage may simply be a consequence of the disease itself, some instances were considered to be drug related. To clarify this result, GSK has launched a parallel Phase IIb study evaluating safety and tolerability of sitamaquine as compared with amphotericin B (AmpB), the current standard of care. Preliminary results from this study suggest that sitamaquine is better tolerated than AmpB,[2] and the clinical evaluation of sitamaquine as an anti-leishmanial agent continues today.

In addition to aminoquinolines, several literature reports have described the anti-leishmanial activity of 2-substituted quinolines [138, 139]. DNDi, partnering with Advinus (India) in a Visceral Leishmaniasis Lead Optimization Consortium, selected this class for further evaluation and optimization. While the reported activity of 2-propylquinoline (compound **30**, Scheme 8) could not be confirmed in an in vitro assay for proliferation of *L. donovani*, the related analog **31** [140]

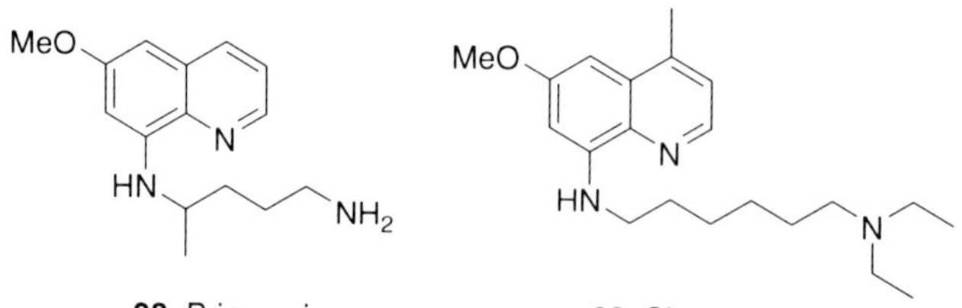

MeO HN N NH_2

28, Primaquine

MeO HN N N

29, Sitamaquine

Scheme 7 Aminoquinoline antiprotozoals

[2] IFPMA Developing World Partnerships web site, http://www.ifpma.org/index.php?id=2170, referenced 10 May 2010.

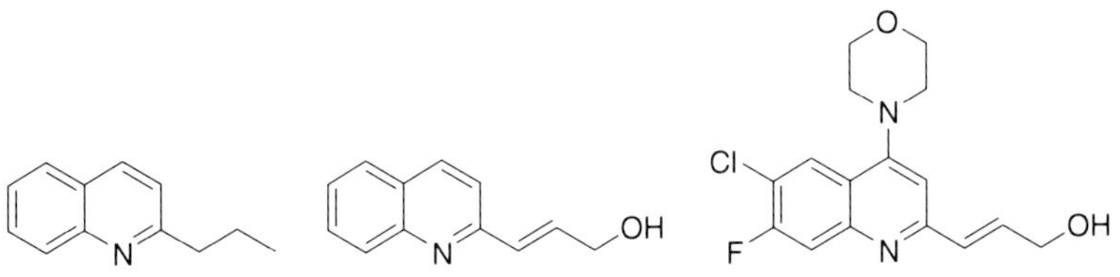

Scheme 8 Anti-leishmanial quinolines

showed modest anti-leishmanial properties, and became a lead for structure–activity studies.

Appropriate substitution on the quinoline ring provided improvements in potency, while halogen blocking of putative metabolic sites led to improvements in metabolic stability, as measured with an in vitro microsomal system (unpublished data). An optimized analog, **32**, was evaluated in vivo in a hamster model of visceral leishmaniasis. Administered orally at a dose of 50 mg/kg, BID, for 5 days postinfection, **32** is able to slow but not eliminate the rise in parasitemia that occurs over the course of this experiment. Notably, this compound is clearly inferior to the gold-standard miltefosine in this testing paradigm. These results are consistent with data generated in a subsequent pharmacokinetic analysis of **32**, which indicates that the compound is rapidly cleared from the systemic circulation, resulting in a short plasma half-life and low drug exposures. These results have led the team to de-prioritize the quinoline SAR program, in favor of other compound classes that have proven more promising *(vide infra)*.

3 Therapeutic Approaches to Kinetoplastid Diseases: Novel Mechanisms

3.1 Lipid Biosynthesis Inhibitors

Trypanosomes have a unique and absolute requirement for certain sterols to maintain proper membrane morphology and function [141, 142], suggesting that the sterol biosynthesis pathways might provide valuable targets for the development of novel antitrypanosomal agents. The Chagas parasite *T. cruzi* is particularly susceptible to disruption of its *de novo* sterol synthetic capability, and ergosterol biosynthesis inhibitors (EBIs), including the azole antifungal agents and allylamines like terbinafine (**33**, Scheme 9), have been of interest as anti-Chagas agents since the early 1980s (see [143, 144], and references therein]. The evaluation of these agents has progressed steadily. However, questions persist regarding the eventual clinical efficacy of agents from this class. First-generation EBIs like ketoconazole (**34**) have shown limited efficacy against *T. cruzi* infection in animal

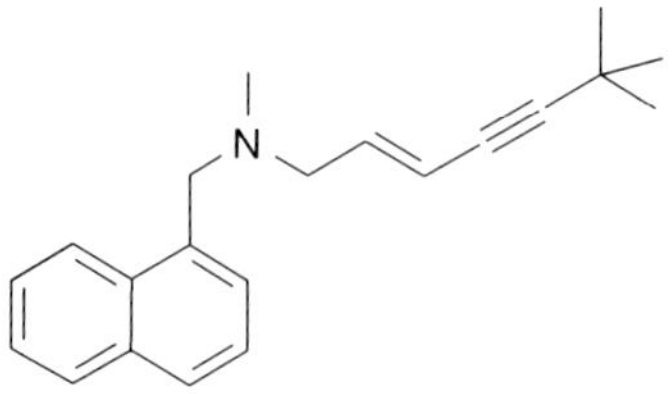

33, Terbinafine

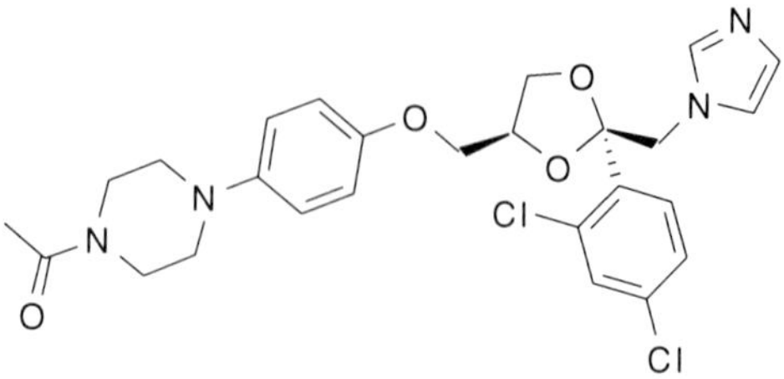

34, Ketoconazole

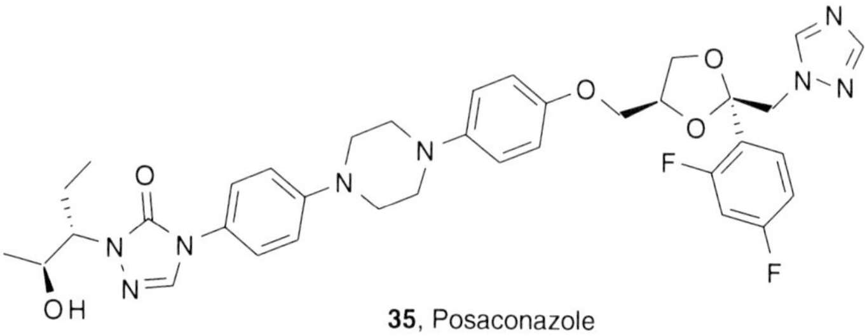

35, Posaconazole

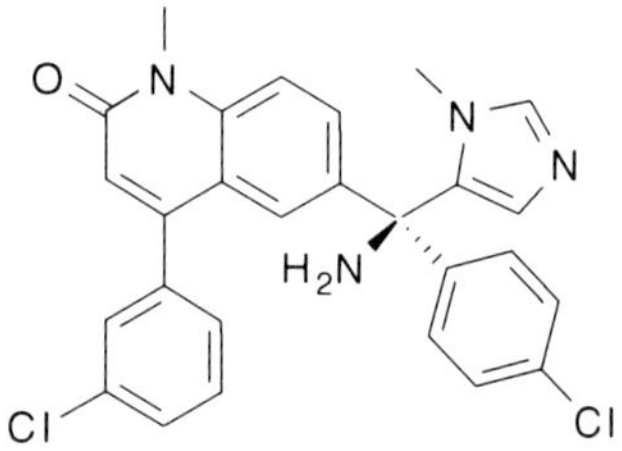

36, Tipifarnib

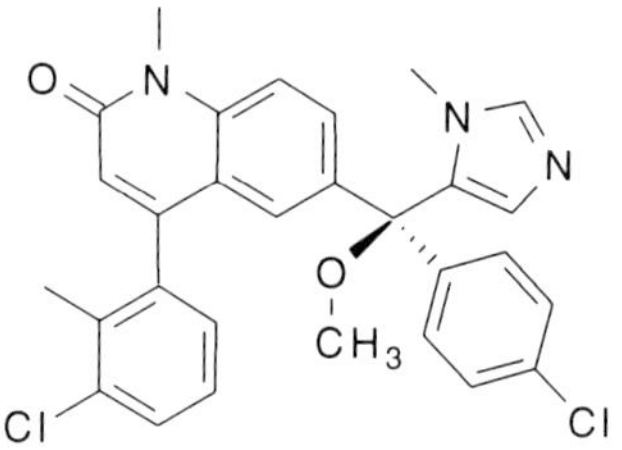

37

Scheme 9 Sterol biosynthesis inhibitors

models [145] and in human clinical trials [146]. Second-generation agents of this class have been selected with a focus on optimizing potency and pharmacokinetic profiles, but were still not developed primarily as antitrypanosomals. In the most advanced of these efforts to date, posaconazole (**35**) [147], a potent azole antifungal with superior pharmacokinetic properties, has been shown to work well in combination with existing Chagas treatments [148], and has been demonstrated to be superior to benznidazole at preventing the cardiac damage associated with chronic *T. cruzi* infection [149]. Expanding on these positive results in in vivo rodent models, Merck has recently initiated a clinical proof-of-concept study for the use of posaconazole in the treatment of Chagas disease [150].

The high level of validation for this pathway as a source of drug targets, coupled with the challenges experienced in progressing the existing agents, has led to a burst of activity in recent years. Several groups have explored the potential of

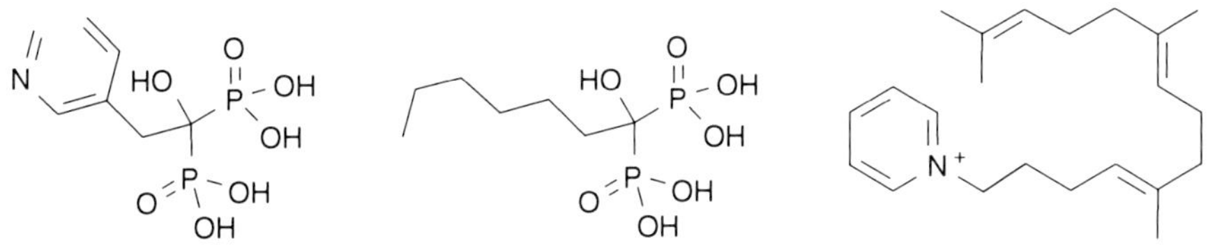

38, Risedronate **39** **40**

Scheme 10 Farnesyl pyrophosphate and oxidosqualene synthase inhibitors

bisphosphonates as antitrypanosomals, with the majority of this effort focused on the treatment of *T. cruzi* Infection. The bisphosphonate class has seen extensive clinical evaluation (with several approved drugs) for the treatment of osteoporosis and other bone disorders, suggesting an attractive safety profile. While they act via several distinct mechanisms to inhibit bone metabolism, their antitrypanosomal activity appears to be linked to the inhibition of farnesyl-pyrophosphate synthase (FPPS), and the resultant interruption of the prenylation pathway [151]. The nitrogenous bisphosphonate risedronate (**38**, Scheme 10) and non-nitrogenous bisphosphonates like **39** have been shown to inhibit *Tc*FPPS with submicromolar potencies [151, 152]. These compounds also inhibit the growth of *T. cruzi* epimastigotes in vitro, though substantially higher drug concentrations are required in these assays. Furthermore, risedronate dosed parenterally for 7–14 days decreased parasitemia levels and improved survival in an acute murine model of Chagas disease [153]. Preliminary evidence suggests that bisphosphonates might have activity against other kinetoplastids as well. Despite these hopeful observations, the class has not moved further down the discovery pathway. Several barriers loom. The non-nitrogenous bisphosphonates lack activity against the clinically relevant amastigote form of the parasite [151], though more recent studies suggest that this issue may be resolved through the use of nitrogenous analogs [154]. Further in vivo studies are required to benchmark the activity of these compounds in the context of other agents; for example, while risedronate is the best-characterized compound of the class, it has not been tested orally and has not been evaluated in any chronic disease model. In addition, there has been little effort to optimize the activity against trypanosomal FPPS while minimizing cross-reactivity against host targets; given the bisphosphonate warhead-driven activity of this class, species selectivity is likely to present a substantial challenge.

DNDi has joined a number of other research teams in focusing its attention on the sterol biosynthesis pathway as a source of novel targets for the treatment of Chagas disease [155]. This effort has proceeded along two main fronts:

1. Target/mechanism validation, using existing (non-optimized) agents
2. The discovery of novel optimized SBIs that meet the Chagas TPP

A clinical study initiated in conjunction with Eisai Pharmaceuticals (Japan) will test the efficacy of Eisai's potent antifungal triazole E-1224 (**42**, Scheme 11), a prodrug of ravuconazole (**41**) in treating the disease [156]. The pharmacokinetic

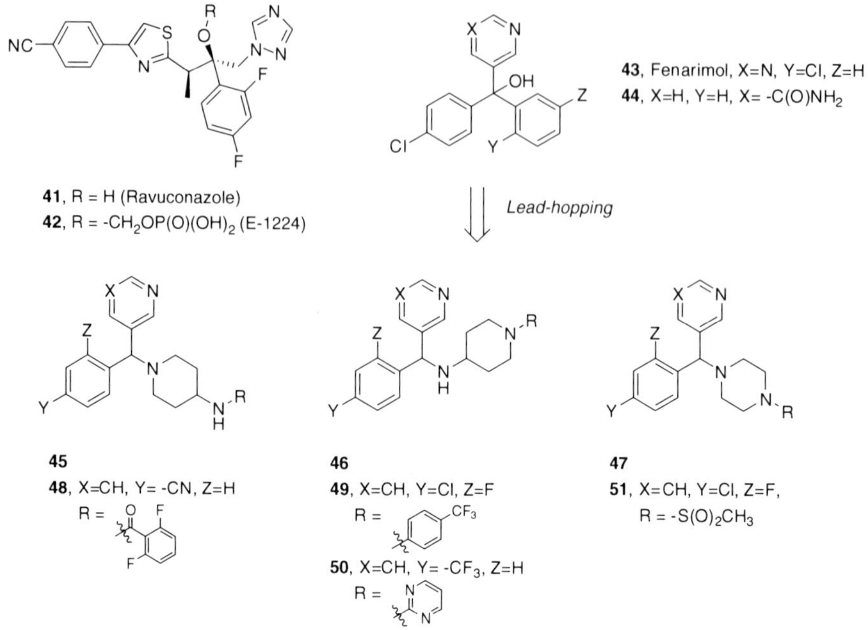

43, Fenarimol, X=N, Y=Cl, Z=H
44, X=H, Y=H, X= -C(O)NH_2

41, R = H (Ravuconazole)
42, R = -$CH_2OP(O)(OH)_2$ (E-1224)

Lead-hopping

45
48, X=CH, Y= -CN, Z=H

46
49, X=CH, Y=Cl, Z=F
50, X=CH, Y= -CF_3, Z=H

47
51, X=CH, Y=Cl, Z=F, R = -$S(O)_2CH_3$

Scheme 11 Sterol biosynthesis inhibitors for Chagas disease

profile of E-1224 is characterized by a high volume of distribution and long terminal half-life; both of these elements are expected to provide superiority over the first-generation SBIs that have been evaluated in the past. The results of this proof-of-concept study should help to clarify the role of SBIs as potential novel agents in the Chagas chemotherapeutic armamentatium. It is anticipated that a drug of this class might find utility as monotherapy, or in combination with existing agents [157].

While these validation-of-mechanism studies are under way, DNDi has assembled a team that is charged with the goal of developing a superior candidate based on this strategy. This team, based largely in Australia, combines experts from Epichem Ltd (medicinal chemistry), Murdoch University (parasitology), and the Center for Drug Candidate Optimization at Monash University (DMPK evaluation) with disease-specific expertise available through the Federal University of Ouro Preto in Brazil.

One key element of the Chagas TPP that is not appropriately addressed by the existing SBIs is "drug cost." Typical azole antifungal SBIs such as posaconazole, ravuconazole (**41**, Scheme 11), and its prodrug E-1224 (isavuconazole; **42**) have complex chemical structures, requiring lengthy chemical syntheses that make these drugs expensive to produce. The Chagas Lead Optimization Consortium initiated a search for alternative, simpler chemical scaffolds that might act through a similar mechanism. This effort led to the identification of fenarimol (compound **43**, Scheme 11), a fungicide that has recently been demonstrated to have activity against *Leishmania* [158], as a lead compound. **43** demonstrated activity against *T cruzi* in an in vitro assay measuring parasite viability [159]; and it was even active (though inferior to benznidazole) when given orally in an acute mouse model of Chagas disease (Chatelain, personal communication).

Evaluating fenarimol against the Chagas TPP, three major issues in particular will be the focus of SAR optimization studies:

1. Potency/PK profile: Robust antiparasitic activity will probably require the maintenance of chronic drug levels in excess of the *T. cruzi* MIC. Ideally, this profile can be achieved with once-daily dosing.
2. Safety: The target of this class of SBIs is CyP51, the sterol-C14-demethylase. It is anticipated that selectivity against other CyPs will be a critical SAR challenge. CyP cross-reactivity can be a source of undesired drug–drug interactions, and many agents of this class are susceptible; for example, posaconazole is a potent inhibitor of CyP3A4, while fenarimol itself blocks both CyP2D6 and CyP3A4.
3. Cost-of-goods: Structure–activity optimization should be accomplished with minimal increases in structural complexity.

Synthetic analogs are evaluated in a rigorous testing scheme that progresses from high-throughput in vitro through disease-relevant in vivo characterization (Fig. 1). Along the way, staged DMPK and safety/selectivity analyses ensure that the progressed compounds have an optimal chance to achieve key elements of the Chagas TPP. Initial structure–activity optimization focused on modifying the substitution pattern of **43**. Through adjustment of the heterocyclic moiety and alternative functionalization of one aromatic ring, a fenarimol analog was developed

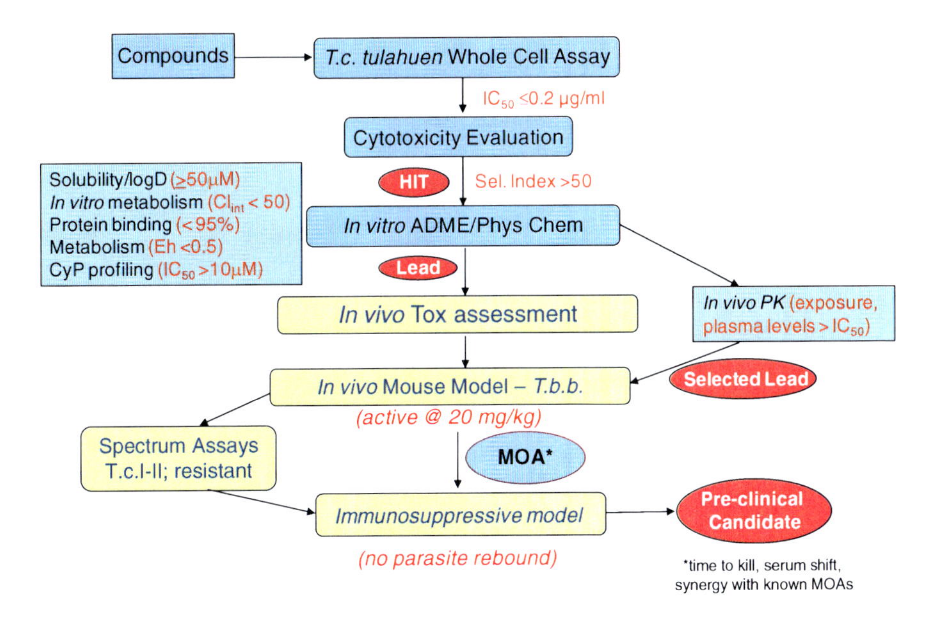

Fig. 1 Chagas testing cascade

(compound **44**, Scheme 11), which has increased antiparasitic activity and an improved pharmacokinetic profile [159]. **44** provides robust suppression of parasitemia in a mouse model of established *T. cruzi* infection, when dosed orally QD for 5 days.

In order to increase the opportunities for success, the Chagas chemistry team initiated a thorough lead-hopping exercise to identify alternative scaffolds related to the fenarimol core [159, 160]. The goal of this effort was to identify one or more core structures which better address the key challenges noted above. In particular, it was noted that one of the aromatic rings in the triaryl-core of **43** (the one most readily modified during SAR studies) can be replaced by a variety of linked heterocycles. The resultant novel cores **45**, **46**, and **47** have all been targeted for analog synthesis. In each case [159, 160], substantial improvements in potency have been noted (e.g., compounds **48–51**; compounds from each series have in vitro potencies in the single-digit nanomolar range). Preferred analogs (e.g., **50**) also show increased selectivity over other CyP isoforms [160]. Additional SAR studies across these fenarimol-based scaffolds have focused on reducing hepatic metabolism, with the consequence that compounds such as **49** and **51** have superior pharmacokinetic profiles [160], predictive of once-daily dosing. Compound **49**, dosed orally once-daily at 20 mg/kg for 10 days, is able to reduce parasite load to undetectable levels in a mouse model of established *T. cruzi* infection [160].

Compounds that are effective at clearing parasites in an acute experiment are next stress-tested by subjecting the test animals to multiple rounds of immunosuppression. This additional step is designed to reveal any parasites that avoid drug action by "hiding out" in sanctuary sites. To date, none of the fenarimol-based lead compounds has reproducibly blocked immunosuppressive rebound in this stringent model; and so work continues to optimize their profiles in search of an optimal candidate for treating established Chagas disease.

3.2 Isoprenoid Pathway Targets

Oxidosqualene cyclase, the enzyme responsible for synthesizing lanosterol from acyclic precursors, has also been validated as a target for *T. cruzi*, though the charged inhibitors prepared to date (e.g., **40**, Scheme 10) lack drug-like properties [161].

Squalene synthase (SQS) is another enzyme on the isoprenoid pathway that has received substantial attention as an antitrypanosomal target [142]. SQS occupies a critical branch point in the terpenoid pathway; blockade of this enzyme disrupts steroid synthesis without impacting the production of other isoprenoid products. Quinuclidine-based (e.g., E-5700, compound **52**, Scheme 12) [162] and phosphonate-based (ER27856, **53**) inhibitors of SQS block parasite replication and alter parasitic morphology when applied to *T. cruzi* cultures in vitro [163]. Other related quinuclidines are inhibitors of *Leishmanial* SQS, and inhibit the growth of *L. major* in vitro, though only at much higher concentrations [164]. The quinuclidines in

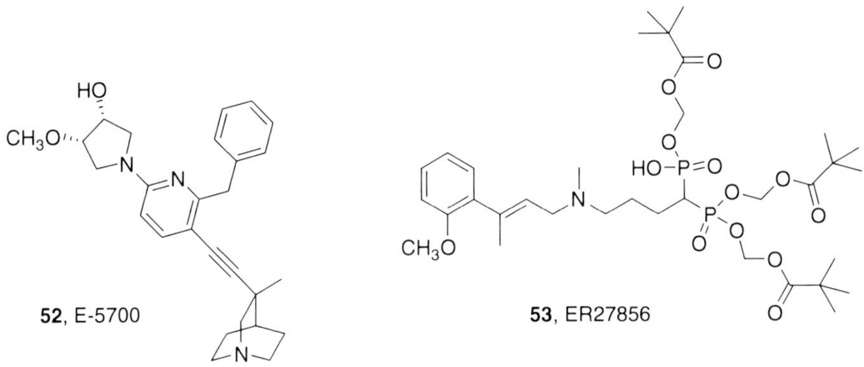

52, E-5700

53, ER27856

Scheme 12 Squalene synthetase inhibitors

particular appear to have drug-like properties, and antiparasitic activity is observed at concentrations similar to their in vitro potencies against trypanosomal SQS [165], suggesting that the compounds are cell-permeable. MIC values against extracellular epimastigote and intracellular amastigote forms of the parasite are similar, which is a desirable profile for a potential anti-Chagas agent. In an acute murine model of Chagas disease, E-5700 given at a high dose orally for 30 days controlled parasitemia as well as a nifurtimox control [165]. More recently it has been noted that these compounds (not surprisingly, given their origin as human lipid-lowering agents) lack selectivity for *Tc*SQS versus *h*SQS. The work to date has validated SQS as an interesting target, but further work is necessary to target the parasitic isoform. It is also a concern that a compound like E-5700, which is so potent in vitro, requires such large doses to be effective in a low-stringency in vivo model; this result may indicate some problem related to its pharmacokinetic and/or pharmacodynamic profile, but no data are available to help clarify this issue.

Attempts to target the prenylation enzyme farnesyltransferase (FTase) have led in a fascinating and unexpected direction. The FTase inhibitor tipifarnib (compound **36**, Scheme 9) is a nanomolar inhibitor of the growth of *T cruzi* epimastigotes; interestingly the antitrypanosomal activity of **36** is superior to its FTase activity, suggesting that it might be acting through an alternative mechanism. Additional studies with this compound have shown that, while it does act via inhibition of sterol biosynthesis, this intervention takes place at a different step in the pathway. The imidazole-containing **36** appears [166] to mimic the activity of the azole antifungals by inhibiting the trypanosomal sterol C-14α demethylase (also known as CyP51). Optimization of the tipifarnib lead [167] has produced analogs with superior activity against the amastigote form of the Chagas parasite, and with vastly improved selectivity against the original target *h*FTase [168]. The identification of CyP51 as the true target of these agents creates an additional selectivity concern; it must be established that these compounds do not inhibit critical host CyPs. While the activity of **36** and congeners against this family of key metabolic enzymes has not been explored in detail at this time, CyP3A4 inhibition has been measured for selected compounds in the series; these generally show good selectivity. An optimized analog **37** (Scheme 9), given orally at a high dose (50 mg/kg twice daily for 20 days), is effective at suppressing parasitemia and increasing survival in mice with an established *T. cruzi* infection [168], an in vivo model that is rather stringent. This exciting efficacy result is driven by and consistent with the high drug exposures (concentrations in the μM range, persisting for at least 5 h post-dose) recorded in an oral pharmacokinetic study [168]. While these high-dose results must be extrapolated with some caution (as PK profiles of CyP-inhibitory compounds are notoriously nonlinear), further evaluation of this class of molecules is definitely warranted.

3.3 Protein Myristoylation

Reinforcing the premise that a properly functioning membrane is critical for maintaining kinetoplastid viability, enzymes responsible for protein mobilization have also proved to be important antiparasitic targets. Of note is the trypanosomal *N*-myristoyltransferase (NMT), which is responsible for the myristoylation (and subsequent membrane localization) of a variety of critical proteins. RNA-interference studies have indicated that *T. brucei* NMT is essential for parasite survival [169]; while blockade of the host ortholog of this enzyme might be undesirable, a relatively low interspecies homology suggests the possibility that selective inhibitors might be designed.

A recent report [170] describes a novel series of drug-like NMT inhibitors based on a pyrazole-4-benzenesulfonamide scaffold (Scheme 13). First identified during a high-throughput screen of a diverse compound collection, the lead compound **54** was optimized for potency against TbNMT, with an improvement of three logs of activity; selectivity vs. human NMTs is modest at best. Analog **55** has a pharmacokinetic profile (20% bioavailability) that allows the compound to be tested in an in vivo mouse model of acute HAT. **55** is active in this model, suppressing parasitemia while increasing survival, thus validating NMT inhibition as a new mechanism for treating trypanosomiasis. Due to a relatively short half-life, the compound is dosed twice-daily in this experiment; an oral dose of >10 mg/kg, given BID for 4 days, is sufficient to protect all animals for at least 30 days. Despite

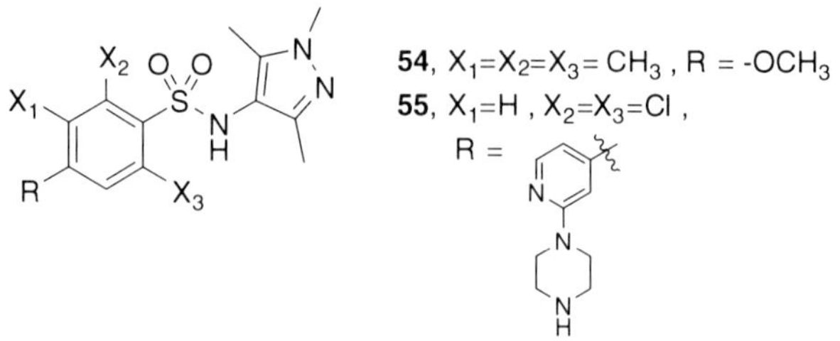

54, $X_1=X_2=X_3=CH_3$, R = $-OCH_3$

55, $X_1=H$, $X_2=X_3=Cl$,

R =

Scheme 13 *N*-myristoyltransferase (NMT) inhibitors

a high level of activity against mammalian NMT orthologs, no overt toxicity was observed during these experiments.

While these results suggest the likelihood that **55** or a further-optimized analog could be useful as a treatment for stage 1 HAT, issues with CNS penetration in this series create a barrier to activity in stage 2 disease. Further studies are under way to attempt to improve this aspect of the overall profile of this new class of antitrypanosomal agents.

3.4 Dihydrofolate Reductase and Pteridine Reductase

Trypanosomes have unique strategies for ensuring their supply of critical folate cofactors. These organisms rely on their hosts to supply folate and pterin precursors, but are surprisingly insensitive to blockade of the traditional salvage pathway, for example through inhibition of dihydrofolate reductase (DHFR) with methotrexate [171]. Studies have revealed an alternative, bypass mechanism for pterin scavenging [172, 173] that involves amplification of the trypansomal pterin reductase (PTR1). Knockout of PTR1 reduces parasitic viability and increases susceptibility to traditional antifolates, suggesting that PTR1 inhibitors might be useful antiparasitic agents, alone or in combination [174]. Medicinal chemistry efforts in this arena have generally recognized the functional homologies between DHFR and PTR1; agents designed to block one of these targets may exhibit cross-reactivity against the other, with the ultimate test being their ability to block replication of the parasite.

Screening of one collection of folate-like molecules led to the identification of a subset of inhibitors (including **56**, Scheme 14) with micromolar activity against *Leishmanial* DHFR and submicromolar activity against L-PTR1 [175]. Rescreening of the collection using these early leads as substructures led to modest improvements in potency; but selectivity against human DHFR remained generally low. Families of 2,4-diaminopyrimidines (including **57**) and 2,4-diaminoquinazolines (e.g., **58**) have been reported to be inhibitors of trypanosomal DHFR [176–179], with in vitro activity against both *Trypanosoma* and *Leishmania*. The success of these compounds as antiparasitic agents suggests the possibility that they may be cross-reactive against PTR1, though this question was not explored in the original studies.

Efforts to design inhibitors of PTR1 have been aided by reports of the crystal structures of *T. cruzi* [180], *T. brucei* [181], and *L. major* [182] enzymes, complexed with both substrates and inhibitors. In several cases, molecular docking studies using these crystal structures have been used to suggest or optimize new scaffolds. Using this approach, three distinct scaffolds (represented by **59**, **60**, and **61**, Scheme 14) were identified as potential starting point [183]. Early SAR studies indicate some ability to improve the PTR1 activity for each of these leads, but none of the reported compounds is as active as methotrexate. The most active agents in this group have antiparasitic activity against *T. brucei*, but ED_{50} values are in the hundreds of micromolar range.

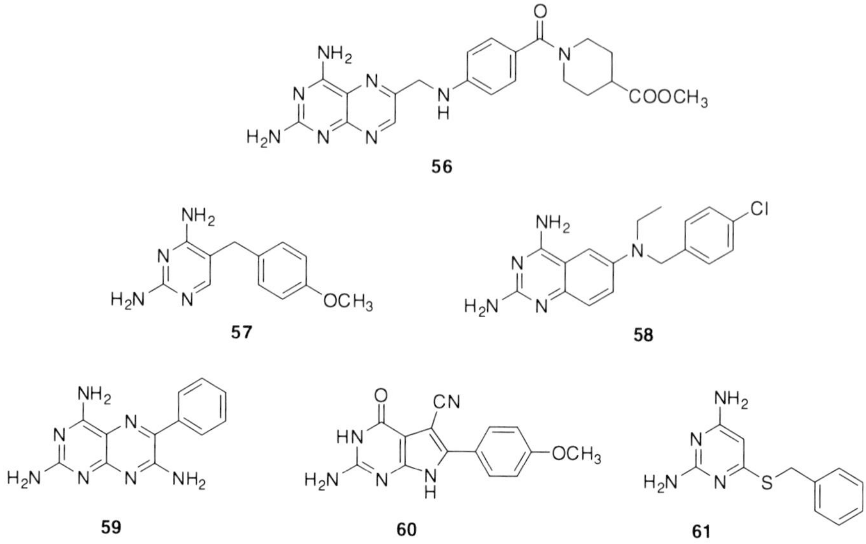

Scheme 14 DHFR and pteridine reductase inhibitors

The issue of species selectivity, and its impact on safety margins, hangs over this antifolate/antipterin approach. The toxicity profile of classic antifolates like methotrexate makes them unsuitable for use as antiparasitic agents; so selectivity against the host enzyme(s) will be a critical element of the target profile. While some preliminary work has been done to address this question, there is a long way to go before a potential candidate can be identified from this class.

3.5 *Protease Inhibitors*

A number of parasite-specific proteases, and in particular cysteine proteases, have been identified as potential targets for antiprotozoal drug therapy [184]. Members of cysteine protease group B (CPG-B), related to the papain family, play a critical role in the life cycle of *Leishmania mexicana*, the causative agent for cutaneous leishmaniasis [185, 186], while cruzipain (cruzain) is a validated target for *T. cruzi*. It is notable that the protozoal cysteine proteases tend to be localized on the surface of the parasite, making them particularly accessible, and thus attractive from a drug discovery perspective [187]. Various academic and industrial groups have targeted these proteases through screening and/or rational design-based approaches. The most advanced molecule to result from these efforts is the vinyl sulfone K777 (**62**, Scheme 15). K777 is a potent inhibitor of cruzipain [188], with significant levels of cross-reactivity against other cysteine proteases, including the cathepsins. It is orally deliverable, with an oral bioavailability (in rats) of ~20% [189].

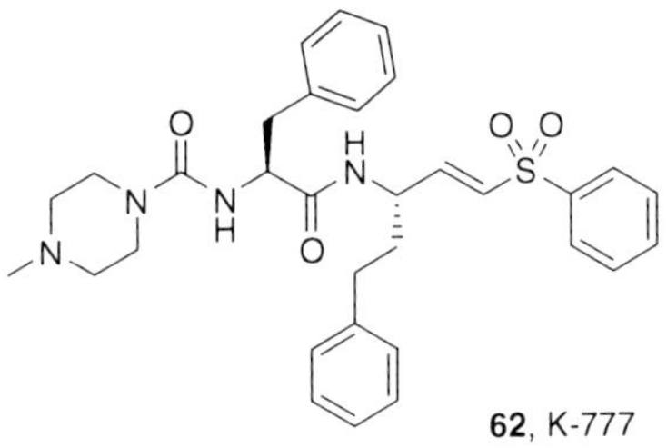

62, K-777

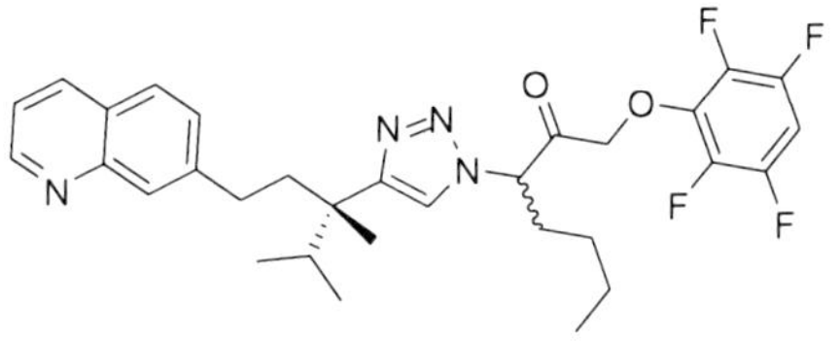

63

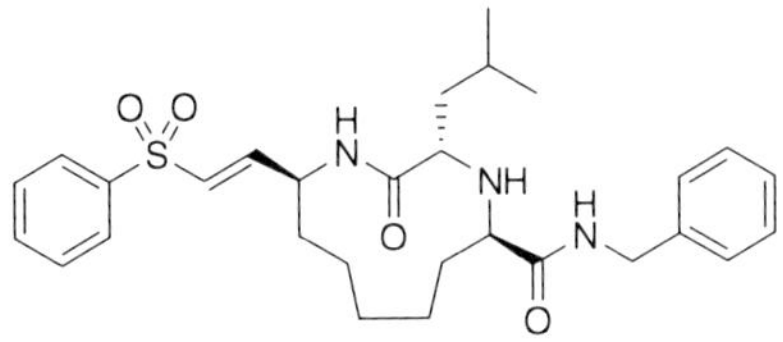

64

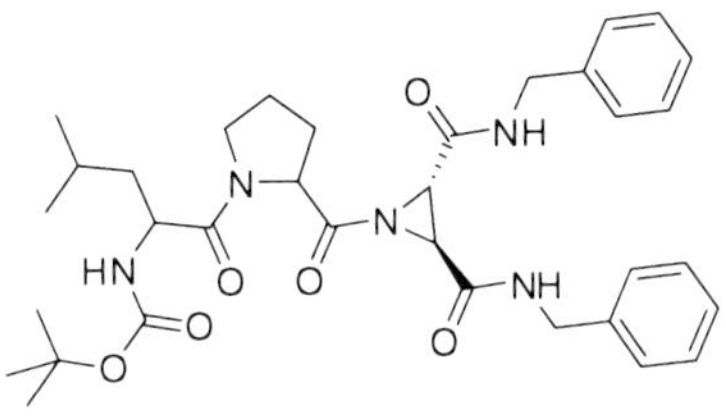

65

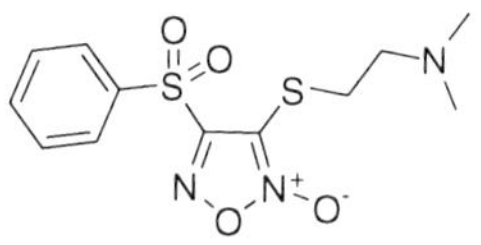

66, SNO-102

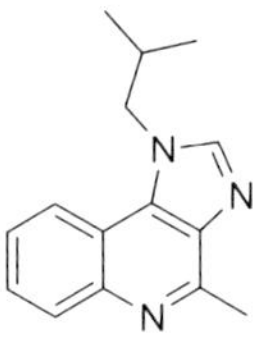

67, Imiquimod

Scheme 15 Protozoal protease inhibitors

The compound kills trypanosomes in vitro, and a 3–4-week course of K777, given i.p., twice daily at a dose of 1 mg per mouse, is sufficient to cure infected mice in an acute disease model [190]. Importantly, it has activity against a variety of *T. cruzi* strains, including those resistant to current therapies [188], and it is even effective (albeit at heroic doses) in a particularly aggressive (but nonetheless clinically relevant) model of *T. cruzi* infection involving an immunodeficient mouse strain [191]. A clinical development program for K777 was initiated by the Institute for OneWorld Health, and has more recently been taken over by the National Institute for Allergy and Infectious Diseases (NIAID). Some potential safety concerns have been noted. For example, the compound has been identified as a potent mechanism-based inhibitor of CyP3A4 [189], though the safety consequences of this off-target activity in a Chagas patient population has not been established. Preclinical toxicology studies (7-day dosing) indicate that the compound is generally safe, but causes significant elevations of the hepatic transaminases ALT and AST at doses of 100 mg/kg/day and higher [192]. This latter result originally led to the suspension of the K777 development effort. Subsequent experiments have indicated that the therapeutic safety margin for this adverse event, while not optimal, may be sufficient to allow a clinical program to proceed. Preclinical evaluation of K777 is ongoing; much more work is required to establish its place as a new addition to the Chagas armamentarium, or as a proof-of-concept for cysteine protease inhibition as a novel therapeutic strategy.

The promising early data from studies of K777 have stimulated additional research to identify cruzipain inhibitors with improved properties and/or using alternative chemical scaffolds. In an attempt to modulate the Michael reactivity of the vinyl sulfone moiety, vinyl sulfonates and vinyl sulfonamides have been developed and reported to show potent antitrypanosomal activity [193, 194]. An early report describes thiosemicarbazone-based inhibitors of cruzipain [195]; these compounds contain a thiocarbonyl moiety that is sometimes considered to be a red-flag for further development, and have not been extensively studied. Activity-based screening of several libraries of warhead-based cysteine protease inhibitors has resulted in the identification of the phenoxymethyl ketone **63** (Scheme 15), a potent inhibitor of cruzipain which is trypanocidal against *T. cruzi* in cell culture [196]. And a series of conformationally constrained analogs of K777 have been prepared using structural information derived from the co-crystal structure of a closely related analog bound to cruzipain [197]. Unfortunately, these macrocyclic compounds (e.g., **64**, Scheme 15) are dramatically less potent and selective than the parent.

Other kinetoplastid parasites have been targeted through inhibition of their cysteine proteases. A series of aziridine-2,3-dicarboxylates (e.g., **65**, Scheme 15) have been demonstrated to have anti-leishmanial activity [198], though the specific protease target(s) have not been identified, and evaluation of these agents has not progressed beyond the in vitro stage. Peptidomimetic inhibitors *of L. mexicana* CPG-B have been shown to have a modest effect on survival of the parasite, though these modified 8-mers are not expected to have drug-like properties [199]. Moreover, K777 itself has shown anti-leishmanial effects linked to the blockade of parasitic differentiation and autophagy [187].

3.6 NO Regulators

As an alternative to the use of substrate-competitive inhibitors to modulate the activity of the trypanosomal cysteine proteases, S-nitrosation of the catalytic nucleophile has also been shown to be an effective strategy for suppressing parasitic infections [200]. Thus nitric oxide has been demonstrated to have trypanocidal effects in vitro [201, 202], and NO-releasing patches have been evaluated for the treatment of cutaneous leishmaniasis [203–205]. These early results stimulated interest in this approach, and other researchers have explored the use of NO donors to treat kinetoplastid infections [206]. The furoxazan-based NO-donor SNO-102 (**66**, Scheme 15) is a time-dependent inactivator of both cruzipain and the *L. infantum* cysteine proteases [207]; consistent with a mechanism involving direct S-nitrosation, this inactivation is reversed by treatment with glutathione.

In principle, another approach to modulating cysteine protease activity via the nitric oxide system would be to stimulate parasitic nitric oxide synthase (NOS) enzyme activity. Imiquimod (Aldara®) is an immunomodulatory agent that is approved for the treatment of genital warts; its antiviral activity is due in part to its ability to activate the inducible NOS isoform (iNOS). This compound (**67**, Scheme 15) has subsequently been shown to be active in vitro (at a concentration of 1 μg/mL) against *Leishmania* infections in macrophages [208]. Unexpectedly, the compound does *not* show direct toxicity against the parasite in the absence of the macrophage host, suggesting that its leishmanicidal activity is driven primarily through stimulation of the macrophage immune response. This host-based immunostimulatory effect is nonetheless driven by NO, as addition of the iNOS inhibitor L-NMMA abrogates the antiparasitic effect of **67**. Pilot Phase II studies of an imiquimod-based cream found that it is highly effective in curing meglumine-resistant cutaneous leishmaniasis infections [209, 210], demonstrating superiority to the current standard-of-care, pentavalent antimonials [211] in naïve patients. Imiquimod has also been shown to combine well with more traditional therapies for cutaneous leishmaniasis [210–212], and its off-label use for this indication continues to expand.

In considering this strategy for the modulation of cysteine protease activity, it is important to recognize that nitric oxide is an important signaling molecule as well as a potent vasodilator. An ideal agent of this class should be selective for parasite targets vs. the host homologs, to provide parasite-specific protease inactivation. As a rule, the NO donors and NOS inhibitors explored to date were originally optimized for activity in man, and so this issue of selectivity remains to be addressed. And the data for imiquimod suggest that parasite-specific targeting may not always be sufficient (or even desirable), for example when the host immune response plays a critical role in clearing parasites.

3.7 *Isoprenoid Pathway Targets*

It has been well established that agents that block cell growth can have antiparasitic activity. In this arena, the key challenge is to establish selectivity for the parasitic target with minimal interruption of host processes. The specific examples note below represent the most advanced efforts reported to date in this arena.

3.7.1 Tubulin Binders

Tubulin binders are known cytotoxic agents that have been used as anticancer agents, but have also seen utility as anti-helminthics [213]. A screen of classical

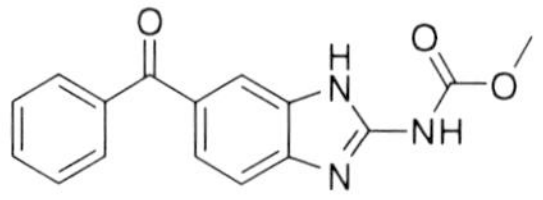

68, Mebendazole

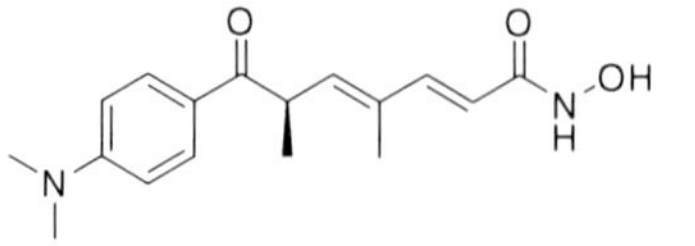

70, TrichostatinA

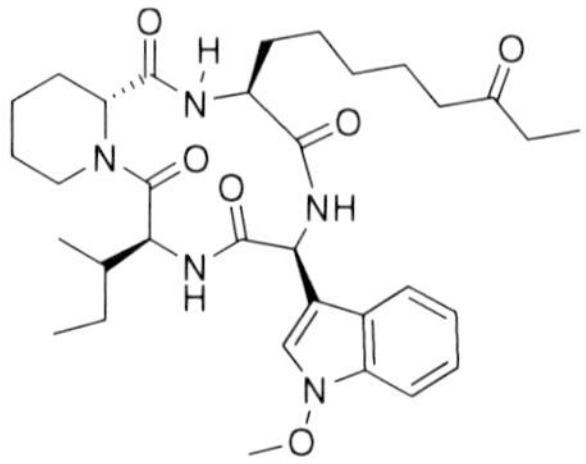

69, Apicidin

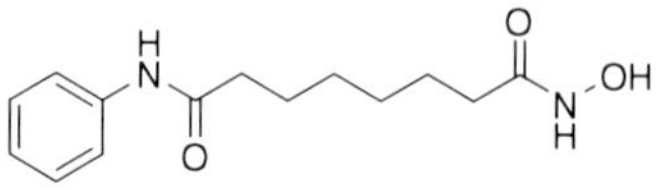

71, SAHA

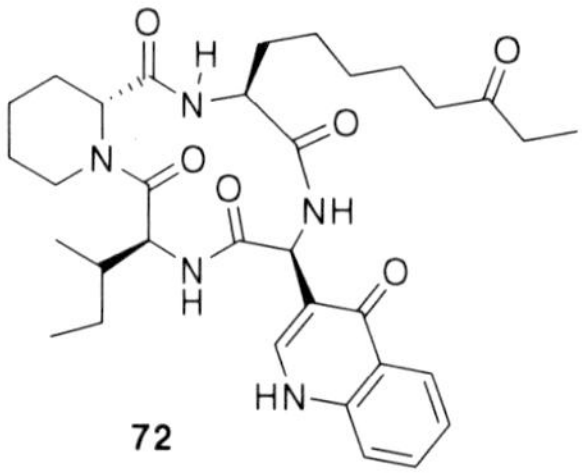

72

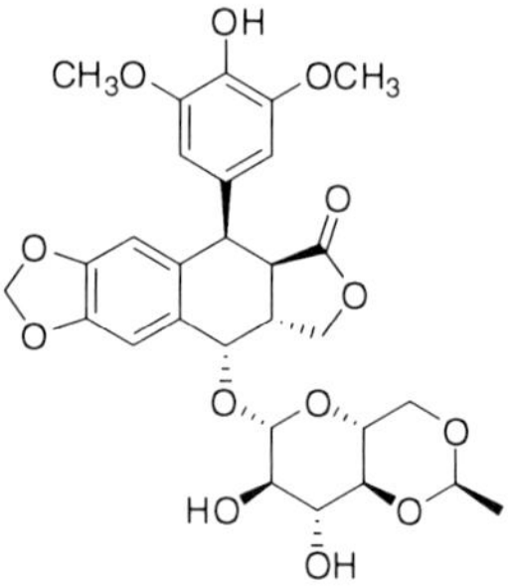

74, Etoposide

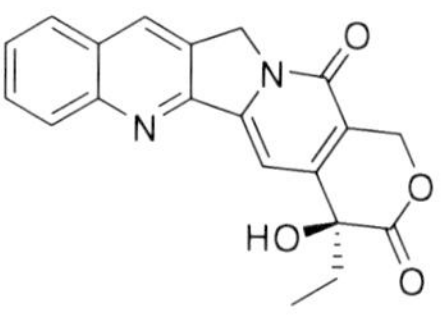

76, Camptothecin

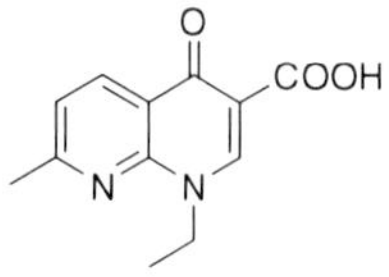

73, Nalidixic acid

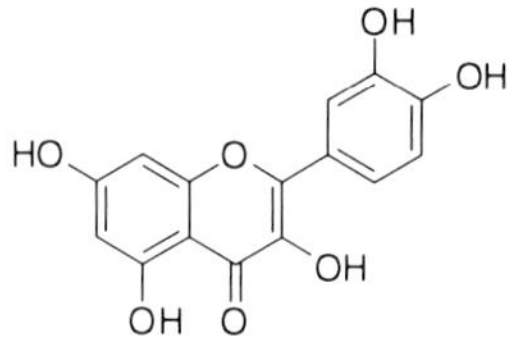

75, Quercetin

Scheme 16 Inhibitors of cellular replication

tubulin binders indicates that this class can block the proliferation of trypansomes as well [214]. Of note is mebendazole (**68**, Scheme 16), which shows some selectivity for binding to *T. brucei* tubulin vs. its mammalian counterpart. While these results suggest trypanosomal tubulin as a potential target, extensive work will be required to optimize selectivity and establish the safety profile for this mechanism.

3.7.2 HDAC Inhibitors

The observation that the antiprotozoal agent apicidin (**69**, Scheme 16) is also an inhibitor of histone deacetylase (HDAC) enzymes [215] suggested that other HDAC inhibitors might be useful for the treatment of parasitic diseases. More recently, the HDAC inhibitors trichostatin A (TSA, **70**) and (to a lesser degree) suberoylanilide hydroxamic acid (SAHA, **71**) have demonstrated good activity against *L. donovani*, with the former showing effects equivalent to pentamidine [216]. Each of these lead structures is also a potent inhibitor of human HDACs; so there are significant safety concerns with their therapeutic use. Researchers at Merck have prepared a series of apicidin analogs (e.g., **72**) with potent antiprotozoal activity and improved selectivity for parastic vs. human HDAC targets [217–219]. There is no information in these studies to indicate whether these modified apicidins have pharmacokinetic properties superior to the marginal profile of the parent [220].

3.7.3 Kinetoplastid Topoisomerase Inhibitors

Trypanosomal topoisomerases play a key role in organizing the parasite's DNA, and are thus expected to represent potential drug targets. Inhibitors of human topoisomerase II, such as nalidixic acid (**73**) and etoposide (**74**), have been shown also to inhibit the growth of *T. cruzi* and *L. amazonensis*; this trypanocidal effect is associated with ultrastructural changes that are consistent with this mode of action [221]. Several recent observations regarding the distinct structure and characteristics of kinetoplastid topoisomerases [222] provide a rationale for the development of trypanosome-selective topo-inhibitors, including a family of flavones [223] related to quercetin (**75**). The unique sensitivity of *L. donovani* to camptothecin (**76**) suggests a starting point for the development of improved agents targeting this mechanism [224].

It is worth noting that, until very recently, the approaches described above have been pursued largely in the academic domain. There is little motivation, from a purely financial perspective, for pharmaceutical companies to engage in efforts to target these "neglected diseases," which generally attack populations that lack economic clout. The academic groups, while often highly motivated, generally lack the drug discovery expertise and/or broad-based testing capabilities required to develop a safe and effective medication with a desirable pharmaceutics profile. More recently, however, this situation is starting to change. Pharma has begun to acknowledge its social responsibility to contribute in this domain, as well as its unique ability to assemble the resources necessary for a successful drug discovery effort. In addition to initiating in-house screening and optimization efforts targeting the neglected tropical diseases, pharmaceutical houses are now collaborating with disease-specific product development partnerships (PDPs) by providing compound screening libraries, and are offering their expertise and capacity to help prosecute interesting leads evolving from academic laboratories.

4 Optimizing the Utility of Existing Agents ("the Low-Hanging Fruit")

4.1 Nifurtimox–Eflornithine Combination Therapy for HAT

The ODC inhibitor eflornithine is currently the safest and most effective treatment for stage 2 HAT infections, but several practical considerations limit its utility in the field. When used as monotherapy, its ability to kill parasites in the brain is linked to the maintenance of consistently high drug levels in the CNS compartment. Limitations in the drug's PK profile translate into a dosing protocol in which eflornithine is given *intravenously, via slow infusion, every 6 h for 14 days*. These requirements place constraints on patients and caregivers alike. They also place constraints on the health care-delivery capacity in the rural settings where HAT outbreaks typically occur. In particular, the drug supply required for a single course of patient therapy occupies a volume of 1 m^3, which volume must often be transferred to remote regions with minimal infrastructural support, and stored once in place.

These practical concerns led DNDi to explore the possibility of using eflornithine as one component in a multidrug combination with the goal of reducing drug burden and improving safety and efficacy [225]. In practice, it was discovered that nifurtimox, a standard of care for Chagas' disease and occasionally used for the treatment of HAT, combines well with eflornithine, producing a substantial decrease in the dosing requirement for each component. Clinical trials in Congo [226, 227] and in Uganda [228] each confirmed that a course of treatment involving eflornithine infusion (every 12 h, 400 mg/kg each dose) coupled with nifurtimox given orally (every 8 h, 15 mg/kg each dose) for a period of 7 days, is non-inferior to the classic course of eflornithine-only (every 6 h, 400 mg/kg each dose, for 14 days) for the treatment of stage 2 HAT. The number of required eflornithine infusions is cut in half; the overall drug requirement (and volume requirement) per patient is reduced by 75%. *The number of patients who can be treated with a truckload of supplies increases fourfold.*

On the basis of the above studies, nifurtimox–eflornithine combination therapy (NECT) has been approved for the treatment of stage 2 HAT [229], and has been added to the WHO Model List of Essential Medicines [230, 231]. Current efforts focus on the provision of drug supplies to endemic areas, and on combating the inertia that allows the toxic drug melarsoprol to remain a first-line therapy for stage 2 HAT in many regions [232].

4.2 Pediatric Benznidazole for Chagas

The nitroheterocycle benznidazole (Bzn) is one of only two drugs approved for the treatment of Chagas' disease. While the drug is generally effective in controlling parasite burden, its safety profile limits patient compliance with the required extended courses of therapy. Observers have noted that these adverse events appear to be minimized in children, suggesting that benznidazole might be more effective when targeted to a pediatric population. Ironically, this population has historically had the least access to Bzn treatment for the simple reason that dosing modalities have not been scaled to accommodate smaller/lower body weight patients. To address this therapeutic gap, DNDi has formed a partnership with the Pharmaceutical Laboratory of Pernambuco (LAFEPE) to produce and distribute a pediatric formulation of Bzn [233]. By analyzing the current practices for the treatment of pediatric patients, an expert panel determined that a dose size of 12.5 mg Bzn was optimal, with the youngest patients receiving a single tablet twice daily, while older children are treated with two tablets at each dosing interval [234]. This pediatric formulation is expected to be available for use in endemic countries in 2010. At the same time, an ongoing study on the population pharmacokinetics of Bzn in patients 2–12 years of age should provide valuable guidance with regard to safety, efficacy, and optimal dosing ranges to guide future formulation efforts [235].

4.3 *Combination Therapy for Visceral Leishmaniasis (Kala-Azar)*

While a number of therapeutic options exist for treating visceral leishmaniasis (liposomal amphotericin, miltefosine, and paromomycin in addition to the traditional antimonials), the standard of care for this debilitating infection remains unsatisfactory [236, 237]. All of the available treatments are associated with significant adverse reactions; a problem that is exacerbated by the extended courses of therapy (up to 30 days) which are the current practice. Very little work has been done to optimize the use of these agents, and in fact many of these are not available in all endemic regions.

To address these issues, DNDi has created a series of VL Combination Therapy Consortia that will explore the potential of drug combinations to reduce side effects while maintaining/improving efficacy. It is also anticipated that combination therapy may offer the opportunity to suppress the drug resistance that is already evolving against many of these agents, and in particular against the antimonials. The first of these consortia is based in India, where each of the drugs listed above has already been approved for first-line therapy. A Phase 2 exploratory study tested the relative effectiveness of three different combinations (none of which includes an antimonial), simultaneously exploring a short-course strategy in which the drugs are given for periods of 7–14 days [238]. The combinations tested include

1. Miltefosine plus Paromomycin
2. Amphotericin B liposomes (Ambisome®) plus miltefosine
3. Ambisome® plus Paromomycin

Early evaluation of trial results indicates that the combination of Ambisome with miltefosine provides excellent suppression of an established *Leishmania* infection; substantial benefit is seen with treatment regimens as short as 1 week [238]. Data from other ongoing combination arms will be used to select optimal combinations for more extensive clinical evaluation. More than one of these combinations may prove to have therapeutic utility, allowing for a new treatment paradigm in which local conditions (including drug susceptibilities and resistance profiles) will drive the selection of locally optimal treatment strategies. In any case, "it is anticipated that these combination treatments will be shorter, safer and cheaper than the current standard monotherapy treatments in the region" [239].

The lessons of this first VL Combination Therapy Consortium are being extended into Africa [240] and Central America [241]. In these latter cases, a critical first step is to register the more modern (non-antimony-based) agents as monotherapy. In Africa, these efforts are coordinated through the DNDi-sponsored Leishmania East Africa Platform (LEAP). LEAP has initiated a Phase 2 study comparing the efficacy of paromomycin with the current standard of care sodium stibogluconate, with a third treatment arm receiving a combination of these agents [242]. As part of these trials, the doses of each drug will also be optimized to achieve maximal effect with minimal adverse outcomes, both singly and in combination. Recently completed study at four sites in Ethiopia, Kenya, and Uganda have demonstrated that paromomycin alone is not an optimal therapeutic approach [243], but can be improved through the use of higher doses or longer treatment regimens [244]; combining these two drugs allows for simultaneous reduction in dose (of paromomycin) and dosing duration (for both agents) while retaining maximal efficacy [245, 246].

5 Chemical Platform-Based Approaches to Developing New Antiprotozoals

In addition to the target-based approaches described above, the DNDi-sponsored Lead Optimization Consortia have focused their attention of specific classes of molecules that have demonstrated broad-based antimicrobial activity ("antiparasitic platforms"). These platforms are typically endowed with a cluster of desirable properties, including robust in vitro and in vivo activity; the challenge for the lead-optimization teams is to identify the subset of analogs within the chemotype that are best aligned with the TPP for the specific disease that is being targeted. Some examples of these efforts are highlighted in the following section.

5.1 Fexinidazole for HAT

The success of NECT for the treatment of HAT has spurred interest in a re-examination of the nitroheterocycle platform to identify improved agents for the treatment of the kinetoplastid diseases. DNDi performed a thorough review of the historic literature in this field, with the primary goal of finding compounds with

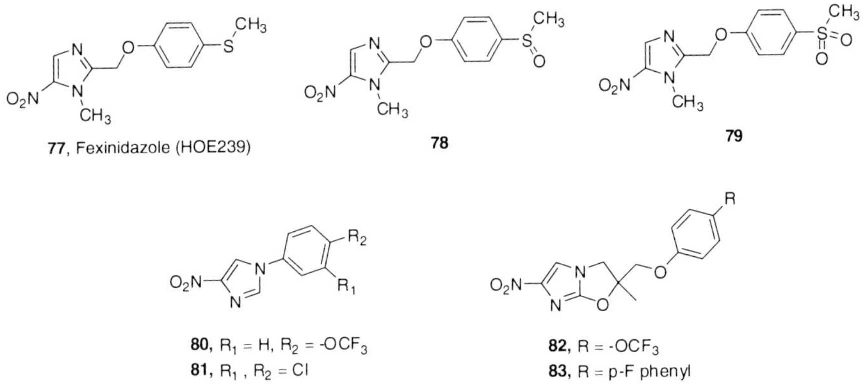

77, Fexinidazole (HOE239)

78

79

80, R_1 = H, R_2 = -OCF_3
81, R_1 , R_2 = Cl

82, R = -OCF_3
83, R = p-F phenyl

Scheme 17 Nitroheterocycles for kinetoplastid diseases

a superior safety profile [247]. This effort led to the identification of fexinidazole (HOE 239; compound **77**, Scheme 17). Of particular note is the observation that while **O1**, like other nitroheterocycles, gives a positive Ames test [248], the compound is not genotoxic in mammalian cells [249]. **77** is trypanocidal against both *T. b. gambiense* and *T. b. rhodesiense* in in vitro assays, with IC_{50} values near 1 μM. Time-kill studies indicate that the drug acts relatively rapidly, with in vitro potency reaching its maximum after an exposure of 24–48 h [249].

Compound **77**, given orally, cures both acute and established CNS infections in mice [249], and is particularly effective in conjunction with a single priming dose of suramin [250]. For the treatment of chronic infections, substantial doses of **77** (5 days of 50 mg/kg BID IP, 5 days of 200 mg/kg/day orally) are required to achieve a complete cure. The compound also has activity (superior to the benchmark agents benznidazole and nifurtimox) in suppressing *T. cruzi* infection; though prolonged treatment is required to effectively clear the parasite from the mice in this model [251].

Studies of fexinidazole are complicated by its complex metabolic fate [252]. Two primary metabolites are formed in preclinical species as well as in man, the sulfoxide (**78**) and sulfone (**79**). Both of these are active antitrypanosomal agents in their own right, with potencies as good as or superior to the parent **77**. Thus, fexinidazole presents the interesting scenario of being both active principle and also prodrug, with the relevance of each of these roles potentially dependent upon both time course and site of action.

Pharmacokinetic evaluations in several species confirm the conclusions gleaned from the in vivo efficacy studies [253]. Parent drug **77** is well absorbed from the gastrointestinal tract, but is rapidly converted to metabolites **78** and **79**, which persist longer. All three entities appear to penetrate the brain efficiently [254]. Since each has similar antitrypanosomal activity, together they contribute to the in vivo activity of fexinidazole against both stage 1 and stage 2 disease.

Preclinical evaluation suggests that the safety profile of fexinidazole is superior to existing agents for the treatment of HAT. When tested against a selectivity panel of enzymes, receptors, and ion channels, **77** was only active against the muscarinic

(M1, M2) receptor and a monoamine oxidase isoform (MAO-B); these interactions occurred at drug concentrations that make a relevant therapeutic interaction unlikely [252]. Neither of the primary metabolites shows any activity against the same panel of targets. Neither **77** nor the sulfoxide metabolite **78** shows any inhibition of hERG channel activity; sulfone metabolite **O3** does significantly decrease hERG peak tail current, but only at a dose of 30 μM [253]. A No-Observed-Effect-Level (NOEL) of 1,000 mg/kg (the highest dose tested) was recorded in an Irwin study of the behavioral effects of **77** in mice, and a similar NOEL was observed in a cardiovascular safety study in dogs [253]. In the latter study, modest increases in body temperature were noted at lower drug doses.

Nitroheterocycles have historically not been used for the treatment of HAT, and so the issue of resistance to this class of drugs is not yet a concern. However, the recent success of NECT is likely to lead to a dramatic increase in the exposure of *T. brucei* to nifurtimox (Nfx) therapy. In this light, a recent study [255] of cross-resistance is worthy of note. In vitro selection of *T. b. brucei* using Nfx leads to the identification of several resistant mutants. These Nfx-resistant parasites retain a high level of virulence and are significantly cross-resistant against both fexinidazole and its metabolites. These results reinforce the conclusion that **77** may find its greatest utility as part of a novel drug combination therapy.

Fexinidazole entered Phase I clinical evaluation in September 2009, with a target population that includes both stage 1 and stage 2 HAT patients.

The search for optimized nitroheterocycles to treat African trypanosomiasis continues. More recently [256], a DNDi-supported multinational consortium has reported on their studies of 1-aryl-4-nitroimidazoles as antitrypanosomals. Noting that the kinetoplastid nitroreductases tend to be more active than their mammalian counterparts, the team focused their attention on developing agents with low redox potentials. The resultant compounds **80** and **81** (Scheme 17) are active in both acute and chronic HAT models and are non-mutagenic in mammalian cells. Further evaluation is under way to determine their suitability as backup candidates to fexinidazole. In a similar vein, a family of 2,4-dinitro-6-trifluoromethylbenzene derivatives (**82** and **83**, Scheme 17) have demonstrated in vitro and (in one case) in vivo activity against *T. brucei* and other parasites [248, 254], but have not been explored further. Preliminary evidence suggests, intriguingly, that the antiparasitic activity of this class of compounds might *not* be linked to tubulin binding (W. Best, personal communication).

5.2 Cell Cycle Inhibitors for HAT

Two independent lines of inquiry focused the attention of researchers on a class of cell cycle inhibitors that have antitrypanosomal properties [257]. A targeted-screening effort performed by the Drug Discovery Unit of the University of Dundee [258], using a screening library of likely kinase inhibitors, identified a cluster of 2,4-diaminothiazole-5-ketones (e.g., compound **84**, Scheme 18) as inhibitors of

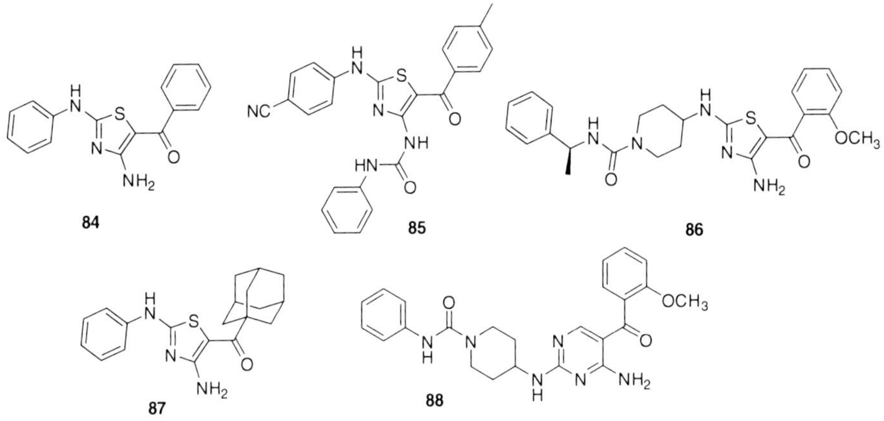

84 85 86 87 88

Scheme 18 Cell cycle inhibitors for HAT

T. brucei glycogen synthase kinase-3 (GSK-3). Concurrently, a broad-based screen of an internal compound collection for antitrypanosomal agents, performed at Scynexis (North Carolina, USA) under agreement with DNDi, identified a very similar structural motif (compound **85**). Recognizing this convergence of effort, these teams joined together to optimize the properties of their lead structures.

Modification of the side-chain functionality of **84/85** has led to improvements in potency, solubility, and metabolic stability [257]. Optimized diaminothiazoles like **86** are potent in vitro antiparasitics (IC_{50} values in the tens of nanomolar against *T. brucei*) and are active in an acute mouse model of HAT. Of particular note is compound **87**, which is highly active against trypanosomes in vitro and demonstrates a high brain/plasma ratio (albeit with modest exposures in both compartments). But the class suffers significant liabilities as well. Their generally poor aqueous solubility sometimes limits drug exposure when the compounds are given orally; and it has proved difficult to identify an analog that combines metabolic stability with good CNS permeability. As a consequence, minimal progress has been made toward the identification of a stage 2-active agent from this class.

Some of the liabilities of the diaminothiazole series have been addressed through a core-hopping exercise leading to a 2,4-diaminopyrimidine-5-ketone (e.g., compound **88**, Scheme 18). Optimized pyrimidine analogs like **88** retain the potency of their thiazole counterparts and are generally more soluble [257]. They also exhibit improved microsomal stability and (presumably as a result) are active at lower doses in the acute mouse model. Despite these advantages, even the best of these analogs fails in the chronic CNS model, and so this approach has been de-emphasized by the partners.

5.3 *Nitroheterocycles for Leishmaniasis*

The success of the HAT team in identifying fexinidazole has stimulated interest in the potential of finding optimized nitroheterocycles to target other kinetoplastid diseases. The VL Lead Optimization Consortium, with medicinal chemistry and pharmacology studies performed at Advinus and Central Drug Research Institute (CDRI), initiated a screen of this class against *L. donovani* in macrophages. Compounds that were active in vitro with minimal cytotoxicity against a mammalian cell line were next evaluated in a hamster model of visceral leishmaniasis [259]. These studies have led to the identification of two potential lead compounds, DNDi-VL-2001 (**82**, Scheme 17) and DNDi-VL-2075 (**83**). Each of these compounds provides substantial suppression of parasitemia at higher doses in the in vivo model, though neither appears to be as effective as miltefosine. Further profiling of these agents will explore optimal dosing regimens and determine whether their efficacy/safety profiles warrant further evaluation in the clinic. In the meantime, the preparation and evaluation of more potent and more stable analogs are underway.

5.4 *Oxaboroles for Kinetoplastid Diseases*

Compounds containing an oxaborole functionality have demonstrated activity against a variety of infectious agents [260, 261]; this scaffold forms one key core technology for the biopharmaceutical company Anacor (California, USA). As part of a broader program to fully explore the potential of this novel pharmacophore, Anacor collaborated with the Sandler Center at UCSF (California, USA) to test a set of representative oxaboroles against *T. brucei*, the causative agent for HAT [262]. This initial screen indicated that the oxaborole platform was capable of delivering potent antitrypanosomals; and, in fact, some of these early leads demonstrated in vivo activity in a murine model of stage 1 HAT. In order to pursue this opportunity, Anacor and DNDi formed a partnership that allows DNDi to pursue the use of this class of agents for the treatment of kinetoplastid diseases. This collaboration has spawned several research programs, as described below.

5.4.1 The Development of SCYX-7158 for HAT

In order to identify an optimized oxaborole suitable for the treatment of HAT, DNDi has assembled a consortium that, together, can provide most of the capabilities required for a drug discovery/candidate selection program. In addition to Anacor, who provided access to their compound collection to establish early structure–activity relationships, this team includes Scynexis Inc. (North Carolina, USA), who take responsibility for all chemistry, in vitro pharmacology, and DMPK activities, and the Bacchi/Yarlett group at Pace University (New York, USA), who evaluate key compounds in acute and chronic murine models of HAT.

The HAT consortium established a staged testing cascade (Fig. 2) to allow for the efficient comparison and triage of project compounds, and for the steady advancement of analogs with superior properties. The progression of the oxaborole

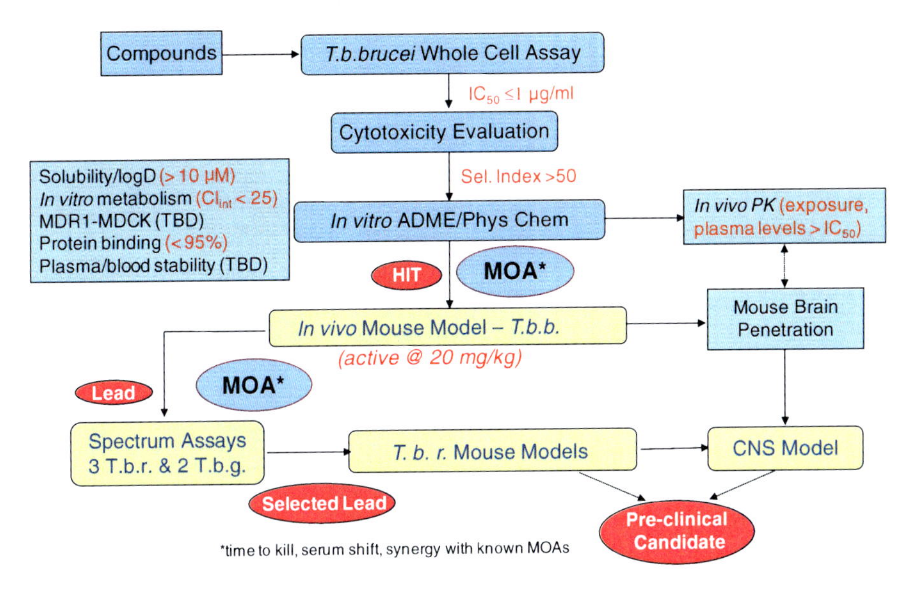

Fig. 2 HAT testing cascade

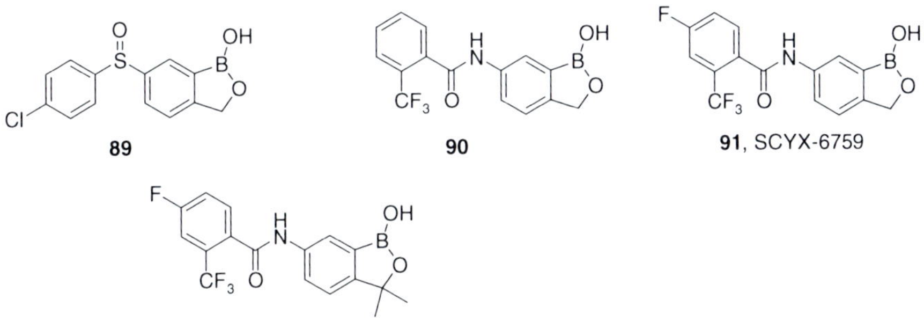

89

90

91, SCYX-6759

92,SCYX-7158

Scheme 19 Oxaboroles for treating kinetoplastid diseases

optimization effort has been marked by several notable breakthroughs. Members of the original test-set [262] evaluated by the UCSF team (e.g., **89**, Scheme 19) already met the team's potency goals (in vitro IC_{50} values in the tens of nanomolar), and were minimally cytotoxic; so the program was able to begin at a relatively advanced stage. Early medicinal chemistry efforts focused on improving the compounds' pharmacokinetic profiles, with the goal of demonstrating oral activity in an acute disease model.

It was noted [263] that a subset of analogs having a benzamide moiety at the 6-position of the benzoxaborole system were superior to the corresponding 6-sulfides and -sulfoxides, when given i.p. in the acute in vivo model. This result was believed (and later proven [264]) to reflect an improvement in drug exposure levels, leading to prioritization of the 6-benzamide subseries for further optimization. In vitro metabolism studies (measuring compound half-lives in microsomal incubations) suggested that first-generation benzamides such as **90** were susceptible to Phase I (oxidative) metabolism. In order to suppress this potential metabolic liability, a series of analogs were prepared in which putative metabolic sites were altered through substitution [265]. Replacement of the hydrogen atom at the 4-position of the benzamide group with F (giving SCYX-6759/AN-4169, compound **91**) resulted in a substantial increase in microsomal half-life without a loss of in vitro antitrypanosomal potency. In vivo activity also improved, presumably as a reflection of an improved pharmacokinetic profile; compound **91** was the first oxaborole to show robust activity in the acute mouse model when dosed orally [266]. This result is readily explained by the pharmacokinetic profile of the compound, which shows that oral administration of **91** (at 9 mg/kg, in mice) provides elevated ($C_{max} > 5$ μg/mL) and sustained (half-life = 20 h) drug levels in the systemic circulation [267]. And **91** is also active in a more stringent mouse model in which parasites are allowed to infect the CNS before initiation of drug therapy, suggesting the exciting possibility that agents of this class might find utility for the treatment of stage 2 HAT [268]. But activity in the stage 2 model requires substantially higher drug doses than were used in the acute model, suggesting that CNS levels of **91** might be lower than circulating levels. This hypothesis is confirmed by pharmacokinetic analysis, which indicates that the initially high brain levels of **91** are rapidly cleared from the CNS compartment.

The Scynexis team was eventually able to balance brain and plasma drug levels through modification of the oxaborole core [265]. Placement of a pair of methyl groups at C-3 of the boron-containing ring leads to SCYX-7158 (compound **92**, Scheme 15). This modification triggers a half-log decrease in in vitro potency, but concomitant improvements in the pharmacokinetic profile more than compensate for this relatively modest effect. The plasma profile **92** is similar to that of **91**, and, in fact, the two compounds display similar potency and efficacy in the acute mouse model. But brain levels of **92** remain high throughout the 24-h time course of the pharmacokinetic experiment, resulting in an overall brain exposure that is six to eight times higher (on a per dose basis) than that achieved with **91**. As a consequence, **92** is dramatically more potent in the chronic HAT model than its precursor, and most notably is able to affect a complete cure at a modest dose, given orally, once daily, for 7 days.

In light of these results, SCYX-7158 is currently under evaluation as a potential clinical candidate for the treatment of both stage 1 and stage 2 HAT [269]. Pharmacokinetic studies in multiple species confirm the exceptional dose-exposure profile originally observed in the mouse and predict a desirable profile in man. Mode of action studies have not yet revealed a mechanism for the antiparasitic properties of the compound, though studies of an earlier oxaborole suggest that its antifungal activity is driven through inhibition of an aminoacyl tRNA synthetase [260]. Preclinical safety studies indicate that P4 is unreactive against a "selectivity panel" of relevant enzymes, receptors, and ion channels; it has no affinity for the hERG channel and is negative in an Ames assay. The compound is moving actively through toxicology studies and pharmaceutics evaluation, with a goal (results permitting) of entering Phase I trials in 2011.

5.4.2 The Development of AN-4169 for Leishmaniasis

In the course of optimizing the oxaborole scaffold against the elements of the HAT TPP, compounds from this class were tested by Anacor and by Scynexis for activity against other kinetoplastids. This testing revealed the broad-based antiprotozoal properties of this series, suggesting their potential use in other neglected diseases. In particular, the search for novel anti-leishmanials has been full of roadblocks [259], making a new series particularly welcome.

SCYX-6759/AN-4169 (**91**), which was previously shown to have a robust pharmacokinetic profile with minimal CNS exposure during the HAT program, is active against *L. donovani* infection in macrophages with an IC_{50} value in the single-digit micromolar range [259]. As a result of this attractive combination of properties, the compound was rapidly moved to in vivo evaluation in the benchmark hamster model of visceral leishmaniasis. **91** suppresses the proliferation of parasites in this model; though this benefit is only observed at doses near the compound's MTD, and the compound is clearly inferior to miltefosine in this model. These early efforts highlight the potential of oxaboroles for this indication, and studies are under way to identify more active analogs from this class.

5.4.3 The Development of Oxaboroles for Chagas Disease

To further probe the anti-kinetoplastid activity of the oxaborole core, a subset of compounds from the Anacor collection was screened for activity against *T. cruzi*, the causative agent for Chagas disease. Preliminary results suggest that, once again, it is possible to identify potent antitrypanosomal agents containing this scaffold. Of particular note is the observation that SCYX-6759/AN-4169 (compound **91**, Scheme 15), the same compound selected for anti-leishmanial testing, is active in this in vitro screen, with potency in the low-micromolar range. The pharmacokinetic profile of this compound has been examined in some detail [270]; it is known to give excellent systemic exposures upon oral dosing in multiple species, with a long half-life. While modest brain-to-plasma ratios made **91** an unsuitable candidate for the treatment of stage 2 HAT, this does not present a difficulty (and might be considered an advantage) for its use in Chagas disease.

To examine this point further, **91** has been evaluated in several in vivo models of *T. cruzi* infection [271, 272]. Oral dosing of **91**, once daily for 5 days, reduces disease burden to below-measurable levels in a mouse model of established *T. cruzi* infection. Mechanistic studies indicate that the activity of **P3** is trypanocidal. Despite this robust response, parasites rebound after the test animals are subjected to multiple rounds of immunosuppression, suggesting that further optimization of compound properties, or of dosing strategies, will be required.

SCYX-6759 thus represents an intriguing lead for the treatment of Chagas disease and validates the utility of the oxaborole scaffold in this space. Further optimization of this series is underway, beginning with a more complete screen of the Anacor collection, and continuing with the synthesis of new molecules designed to test specific SAR hypotheses that have developed.

6 Summary and Conclusions

The state of therapeutic practice for treating the kinetoplastid diseases has remained largely unchanged for decades. The greatest progress has been made with regard to the treatment of leishmaniasis, where relatively recent introductions of paromomycin, miltefosine, and liposomal amphotericin have provided improved efficacy (particularly against drug-resistant forms of the parasite) with reduced side effects. For HAT, the most a considerable improvement of available treatments was the introduction of DFMO. This was the first alternative to the use of Mlerasoprol which is lethal in 5% of cases, affected by the second stage of the diseases. More recently, introduction of the NECT combination therapy allowed use of DFMO and Nifurtimox in field conditions. For Chagas disease, therapeutic practice has changed little since the 1970s, and no truly effective treatment exists. Despite the high burden of mortality and morbidity associated with these diseases, they remain among the "most neglected." These outcomes are the direct result of a lack of interest in NTD research, particularly during the "glory decades" of evolving modern pharmaceutical research in the 1970s to 1990s.

The new millennium has brought a resurgence in interest in the development of novel drugs to treat these diseases of the disenfranchised. Of note, the pharmaceutical industry has committed to support this effort, donating their expertise and in some cases making their compound collections available as a source of "drug-like" leads. Major funding initiatives from the World Health Organization, the Gates Foundation, and the Wellcome Trust, among others, have stimulated new research in the field, while the development of the PDP model has helped to ensure that the best candidates are prioritized for clinical evaluation. DNDi in particular has demonstrated a pragmatic approach to improving the lives of patients, focusing first on "low-hanging fruit" like improved dosage forms and well-characterized drug combinations. These early successes are already making a difference in the field, while the preliminary results of new research suggest the possibility for transformational improvements in disease control and prevention. Drug development is always a painfully slow process, and committed support will be required to fully evaluate these new approaches. But for those in the field, the current nexus of evolving science and increased commitment suggests a unique opportunity to finally impact the lives of those suffering from these maladies.

Acknowledgments The authors would like to thank Federica Givannini for critical reading of the manuscript. DNDi is grateful to its donors, public, and private, who have provided funding to DNDi since its inception in 2003. With the support of these donors, DNDi is well on its way to achieving the objectives of a robust pipeline with the aim to deliver six to eight new treatments by 2014. A full list of DNDi's donors can be found at: http://www.dndi.org/index.php/donors.html?ids=8. The donors had no role in study design, data collection and analysis, decision to publish, or preparation of the manuscript.

References

1. Adhikari SR, Maskay NM, Sharma BP (2009) Paying for hospital-based care of Kala azar in Nepal: assessing catastrophic, impoverishment and economic consequences. Health Policy Plan 24:129–139
2. Anoopa Sharma D, Bern C, Varghese B, Chowdhury R, Haque R, Ali M, Amann J, Ahluwalia IB, Wagatsuma Y, Breiman RF, Maguire JH, McFarland DA (2006) The economic impact of visceral leishmaniasis on households in Bangladesh. Trop Med Int Health 11:757–764
3. Boelaert M, Meheus F, Sanchez A, Singh SP, Vanlerberghe V, Picado A, Meessen B, Sundar S (2009) The poorest of the poor: a poverty appraisal of households affected by visceral leishmaniasis in Bihar, India. Trop Med Int Health 14:639–644
4. Remme JHF, Feenstra P, Lever PR, Médici A, Morel C, Noma M, Ramaiah KD, Richards F, Seketeli A, Schmunis G, van Brakel WH, Vassall A (2006) Tropical diseases targeted for elimination: Chagas disease, lymphatic filariasis, onchocerciasis, and leprosy. In: Disease control priorities in developing countries, 2nd edn. Oxford University Press, New York, pp 433–450. doi: 10.1596/978-0-821-36179-5/Chpt-22
5. Bern C, Montgomery SP, Katz L, Caglioti S, Stramer SL (2008) Chagas disease and the US blood supply. Curr Opin Infect Dis 21(5):476–482
6. Gascon J, Bern C, Pinazo MJ (2010) Chagas disease in Spain, the United States and other non-endemic countries. Acta Trop 115(1–2):22–27
7. Swallow BM (2000) Impacts of trypanosomiasis on African agriculture. PAAT Technical and Scientific Series, No. 2. FAO, Rome
8. Brun R et al (2010) Human African trypanosomiasis. Lancet 375(9709):148–159
9. Bonds MH, Keenan DC, Rohani P, Sachs JD (2010) Poverty trap formed by the ecology of infectious diseases. Proc Biol Sci 277(1685):1185–1192
10. Sachs JD (2007) Breaking the poverty trap. Targeted investments can trump a region's geographic disadvantages. Sci Am 297(3):40, 42
11. Nussbaum K, Honek J, Cadmus CM, Efferth T (2010) Trypanosomatid parasites causing neglected diseases. Curr Med Chem 17(15):1594–1617
12. Chappuis F et al (2007) Visceral leishmaniasis: what are the needs for diagnosis, treatment and control? Nat Rev Microbiol 5:873–882. doi:10.1038/nrmicro1748
13. Costa CH (2008) Characterization and speculations on the urbanization of visceral leishmaniasis in Brazil. Cad Saude Publica 24:2959–2963
14. David CV, Craft N (2009) Cutaneous and mucocutaneous leishmaniasis. Dermatol Ther 22 (6):491–502
15. Shaw JJ (1994) Taxonomy of the genus Leishmania: present and future trends and their implications. Mem Inst Oswaldo Cruz 89:471–478
16. Desjeux P (2004) Leishmaniasis: current situation and new perspectives. Comp Immunol Microbiol Infect Dis 27:305–318
17. Bern C, Maguire JH, Alvar J (2008) Complexities of assessing the disease burden attributable to leishmaniasis. PLoS Negl Trop Dis 2:e313

18. Mubayi A, Castillo-Chavez C, Chowell G, Kribs-Zaleta C, Ali Siddiqui N, Kumar N, Das P (2009) Transmission dynamics and underreporting of Kala-azar in the Indian state of Bihar. J Theor Biol 262:177–185
19. Herwaldt BL (1999) Leishmaniasis. Lancet 354(9185):1191–1199
20. Murray HW, Berman JD, Davies CR, Saravia NG (2005) Advances in leishmaniasis. Lancet 366(9496):1561–1577
21. Rittig MG, Bogdan C (2000) Leishmania–host-cell interaction: complexities and alternative views. Parasitol Today 16:292–297
22. van Griensven J, Balasegaram M, Meheus F, Alvar J, Lynen L, Boelaert M (2010) Combination therapy for visceral leishmaniasis. Lancet Infect Dis 10:184–194
23. Shankar EM, Vignesh R, Murugavel KG, Balakrishnan P, Ponmalar E, Rao UA, Velu V, Solomon S (2009) Common protozoans as an uncommon cause of respiratory ailments in HIV-associated immunodeficiency. FEMS Immunol Med Microbiol 57(2):93–103
24. Chappuis F, Rijal S, Soto A, Menten J, Boelaert M (2006) A meta-analysis of the diagnostic performance of the direct agglutination test and rK39 dipstick for visceral leishmaniasis. BMJ 333(7571):723
25. Goto H, Lindoso JA (2010) Current diagnosis and treatment of cutaneous and mucocutaneous leishmaniasis. Expert Rev Anti Infect Ther 8(4):419–433
26. Crum NF, Aronson NE, Lederman ER, Rusnak JM, Cross JH (2005) History of U.S. military contributions to the study of parasitic diseases. Mil Med 170(4 Suppl):17–29
27. Kenner JR, Aronson NE, Benson PM (1999) The United States military and leishmaniasis. Dermatol Clin 17:77–92
28. Ross R (1903) Note on the bodies recently described by Leishman and Donovan. BMJ 2:1261–1262
29. Kinnamon KE, Steck EA, Loizeaux PS et al (1979) Leishmaniasis: military significance and new hope for treatment. Milit Med 144:660–664
30. Coleman RE, Hochberg LP, Putnam JL, Swanson KI, Lee JS, McAvin JC, Chan AS, Oguinn ML, Ryan JR, Wirtz RA, Moulton JK, Dave K, Faulde MK (2009) Use of vector diagnostics during military deployments: recent experience in Iraq and Afghanistan. Mil Med 174(9):904–920
31. Coleman RE, Burkett DA, Putnam JL, Sherwood V, Caci JB, Jennings BT, Hochberg LP, Spradling SL, Rowton ED, Blount K, Ploch J, Hopkins G, Raymond JL, O'Guinn ML, Lee JS, Weina PJ (2006) Impact of phlebotomine sand flies on U.S. Military operations at Tallil Air Base, Iraq: 1. background, military situation, and development of a "Leishmaniasis Control Program". J Med Entomol 43(4):647–662
32. Aronson NE, Sanders JW, Moran KA (2006) In harm's way: infections in deployed American military forces. Clin Infect Dis 43(8):1045–1051
33. Murray CK (2008) Infectious disease complications of combat-related injuries. Crit Care Med 36(7 Suppl):S358–S364
34. Murray CK (2008) Epidemiology of infections associated with combat-related injuries in Iraq and Afghanistan. J Trauma 64(3 Suppl):S232–S238
35. World Health Organization (2005) Regional strategic framework for elimination of kala-azar from the South-East Asia region (2005–2015). WHO Regional Office for South-East Asia, New Delhi
36. Seaman J, Mercer AJ, Sondorp E (1996) The epidemic of visceral leishmaniasis in western Upper Nile, southern Sudan: course and impact from 1984 to 1994. Int J Epidemiol 25:862–871
37. WHO Technical Report Series 946 (2010) Control of the Leishmaniases, Report of a meeting of the WHO Expert Committee on the Control of Leishmaniases, Geneva, 22–26 March 2010
38. Buscaglia CA et al (2006) Trypanosoma cruzi surface mucins: host-dependent coat diversity. Nat Rev Micro 4(3):229–236
39. Hotez PJ, Bottazzi ME, Franco-Paredes C, Ault SK, Periago MR (2008) The neglected tropical diseases of Latin America and the Caribbean: a review of disease burden and

distribution and a roadmap for control and elimination. PLoS Negl Trop Dis 2:e300. doi:10.1371/journal.pntd.0000300
40. World Health Organization (2004) World health report 2004. Changing history. World Health Organization, Geneva
41. Mathers CD, Ezzati M, Lopez AD (2007) Measuring the burden of neglected tropical diseases: the global burden of disease framework. PLoS Negl Trop Dis 1:e114. doi:10.1371/journal.pntd.0000149
42. Moncayo A, Ortiz Yanine MI (2006) An update on Chagas disease (human American trypanosomiasis). Ann Trop Med Parasitol 100:663–677
43. Tibayrenc M, Ayala FJ (2002) The clonal theory of parasitic protozoa: 12 years on. Trends Parasitol 18:405–410
44. Andrade LO, Andrews NW (2005) The Trypanosoma cruzi-host-cell interplay: location, invasion, retention. Nat Rev Micro 3(10):819–823
45. Andrade LO, Andrews NW (2005) The Trypanosoma cruzi-host-cell interplay: location, invasion, retention. Nat Rev Microbiol 3(10):819–823
46. Barrett MP et al (2003) The trypanosomiases. Lancet 362:1469–1480
47. Ribeiro I, Sevcsik A-M, Alves F, Diap G, Don R et al (2009) New, improved treatments for Chagas disease: from the R&D pipeline to the patients. PLoS Negl Trop Dis 3(7):e484. doi:10.1371/journal.pntd.0000484
48. Zhang L, Tarleton RL (1999) Parasite persistence correlates with disease severity and localization in chronic Chagas' disease. J Infect Dis 180:480–486
49. Rassi A Jr, Rassi A, Marin-Neto JA (2010) Chagas disease. Lancet 375(9723):1388–1402
50. Schofield CJ, Dias JC (1999) The Southern Cone Initiative against Chagas disease. Adv Parasitol 42:1–27
51. Coura JR, Dias JCP (2009) Epidemiology, control and surveillance of Chagas disease - 100 years after its discovery. Mem Inst Oswaldo Cruz 104(Suppl):I31–I40
52. Ramsey JM, Schofield CJ (2003) Control of Chagas disease vectors. Salud pública Méx (online) 45(2):123–128. ISSN 0036–3634. doi: 10.1590/S0036-36342003000200010
53. Franco-Paredes C, Von A, Hidron A, Rodríguez-Morales AJ, Tellez I et al (2007) Chagas disease: an impediment in achieving the Millennium Development Goals in Latin America. BMC Int Health Hum Rights 7:7
54. Prata A (2001) Clinical and epidemiological aspects of Chagas disease. Lancet Infect Dis 1:92–100
55. Schmunis GA (2007) Epidemiology of Chagas disease in non-endemic countries: the role of international migration. Mem Inst Oswaldo Cruz 102(Suppl 1):75–85
56. Paulinoa M, Iribarnea F, Dubinb M, Aguilera-Moralesc S, Tapiad O, Stoppani AO (2005) Mini Rev Med Chem 5:499–519
57. Simarro PP, Jannin J, Cattand P (2008) Eliminating human African trypanosomiasis: where do we stand and what comes next? PLoS Med 5(2):e55
58. Kennedy PG (2008) The continuing problem of human African trypanosomiasis (sleeping sickness). Ann Neurol 64(2):116–126
59. Kennedy PG (2008) Diagnosing central nervous system trypanosomiasis: two stage or not to stage? Trans R Soc Trop Med Hyg 102(4):306–307
60. WHO (2004) The World Health Report 2004. Shaping history. Annex Table 3: Burden of Disease in DALYs by cause, sex and mortality stratum in WHO regions, estimates for 2002, World Health Organization, Geneva (2004)
61. Lutumba P, Makieya E, Shaw A, Meheus F, Boelaert M (2007) Human African trypanosomiasis in a rural community, Democratic Republic of Congo. Emerg Infect Dis 13 (2):248–254
62. Field MC, Carrington M (2009) The trypanosome flagellar pocket. Nat Rev Micro 7(11):775–786
63. Kennedy PG (2006) Diagnostic and neuropathogenesis issues in human African trypanosomiasis. Int J Parasitol 36(5):505–512

64. Enanga B, Burchmore RJ, Stewart ML, Barrett MP (2002) Sleeping sickness and the brain. Cell Mol Life Sci 59(5):845–858
65. Kennedy PG (2004) Human African trypanosomiasis of the CNS: current issues and challenges. J Clin Invest 113(4):496–504
66. Blum J, Schmid C, Burri C (2006) Clinical aspects of 2541 patients with second stage human African trypanosomiasis. Acta Trop 97(1):55–64
67. Checchi F, Filipe JA, Haydon DT, Chandramohan D, Chappuis F (2008) Estimates of the duration of the early and late stage of gambiense sleeping sickness. BMC Infect Dis 8:16
68. Checchi F, Filipe JAN, Barrett MP, Chandramohan D (2008) The natural progression of Gambiense sleeping sickness: what is the evidence? PLoS Negl Trop Dis 2:e303
69. Chappuis F, Loutan L, Simarro P, Lejon V, Büscher P (2005) Options for field diagnosis of human african trypanosomiasis. Clin Microbiol Rev 18(1):133–146
70. Steverding D (2008) The history of African trypanosomiasis. Parasit Vectors 1(1):3
71. de Raadt P (2005) The history of sleeping sickness. http://www.who.int/trypanosomiasis_african/country/history/en/print.html
72. Berrang Ford L (2007) Civil conflict and sleeping sickness in Africa in general and Uganda in particular. Confl Health 1:6
73. Kaba D, Dje NN, Courtin F, Oke E, Koffi M, Garcia A, Jamonneau V, Solano P (2006) The impact of war on the evolution of sleeping sickness in west-central Côte d'Ivoire. Trop Med Int Health 11(2):136–143
74. Stanghellini A, Josenando T (2001) The situation of sleeping sickness in Angola: a calamity. Trop Med Int Health 6(5):330–334
75. Fèvre EM, Picozzi K, Fyfe J, Waiswa C, Odiit M, Coleman PG, Welburn SC (2005) A burgeoning epidemic of sleeping sickness in Uganda. Lancet 366(9487):745–747
76. Legros D, Ollivier G, Etchegorry M, Paquet C, Burri B, Jannin J, Buscher P (2002) Treatment of human African trypanosomiasis—present situation and needs for research and development. Lancet Infect Dis 2:437–440
77. Fèvre EM, Wissmann BV, Welburn SC, Lutumba P (2008) The burden of human african trypanosomiasis. PLoS Negl Trop Dis 2(12):e333
78. Cecchi G, Paone M, Franco JR, Fevre EM, Diarra A, Ruiz JA, Mattioli RC, Simarro PP (2009) Towards the Atlas of human African trypanosomiasis. Int J Health Geogr 8:15
79. Jenh AM, Pham PA (2008) Zambia HIV National Guidelines, Johns Hopkins Point-of-Care IT Center. http://www.zambiahivguide.org/drugs/antimicrobial_agents/suramin.html?contentInstanceId=438701
80. World Health Organization, Media Centre (2006) Fact sheet no. 259, "African Trypanosomiasis"
81. de Nazaré C, Soeiro M, de Souza EM, Boykin DW (2007) Antiparasitic activity of aromatic diamidines and their patented literature. Expert Opin Ther Pat 17(8):927–939
82. Bacchi C (2009) Chemotherapy of human African trypanosomiasis. Interdisciplinary Perspectives on Infectious Diseases (Hindawi Publishing Corporation) 1–5
83. Heby O, Persson L, Rentala M (2007) Targeting the polyamine biosynthetic enzymes: a promising approach to therapy of African sleeping sickness, Chagas' disease, and leishmaniasis. Amino Acids 33:359–366, PMID: 17610127
84. Jacobs RT, Ding C (2010) Recent advances in drug discovery for neglected tropical diseases caused by infective kinetoplastid parasites. Annual Reports in Medicinal Chemistry, Volume 45, 2010 Elsevier Inc.
85. Dohn M, Weinberg WG, Torres RA, Follansbee SE, Caldwell PT, Scott JD, Gathe JC Jr, Haghighat DP, Sampson JH, Spotkov J, Deresinski SC, Meyer RD, Lancaster DJ (1994) Oral atovaquone compared with intravenous pentamidine for Pneumocystis carinii pneumonia in patients with AIDS. Ann Int Med 121(3):174–180, PMID: 7880228
86. Feddersen A, Sack K (1991) Experimental studies on the nephrotoxicity of pentamidine in rats. J Antimicrob Chemother 28(3):437–446, PMID: 1960124

87. Assan R, Perronne C, Assan D, Chotard L, Mayaud C, Matheron S, Zucman D (1995) Pentamidine-induced derangements of glucose homeostasis. Determinant roles of renal failure and drug accumulation. A study of 128 patients. Diabetes Care 18(1):47–55, PMID: 7698047
88. Antoniou T, Gough KA (2005) Early-onset pentamidine-associated second degree heart block and sinus bradycardia: Case report and review of the literature. Pharmacotherapy 25(6):899–903, PMID: 15927910
89. Denise H, Barrett MP (2001) Uptake and mode of action of drugs used against sleeping sickness. Biochem Pharmacol 61(1):1–5, PMID: 11137702
90. Werbovetz K (2006) Diamidines as antitrypanosomal, antileishmanial and antimalarial agents. Curr Opin Investig Drugs 7(2):147–157, PMID: 16499285
91. Soeiro MN, de Castro SL, de Souza EM, Batista DG, Silva CF, Boykin DW (2008) Diamidine activity against trypanosomes: the state of the art. Curr Mol Pharmacol 1(2): 151–161, PMID: 20021429
92. de Nazaré M, Soeiro C, de Souza EM, Boykin DW (2007) Antiparasitic activity of aromatic diamidines and their patented literature. Exp Opin Ther Pat 17(8):927–939
93. Ansede JH, Anbazhagan M, Brun R, Easterbrook JD, Hall JE, Boykin DW (2004) O-alkoxyamidine prodrugs of furamidine: In vitro transport and microsomal metabolism as indicators of in vivo efficacy in a mouse model of Trypanosoma brucei rhodesiense infection. J Med Chem 47(17):4335–4338, PMID: 15294005
94. Das BP, Boykin DW (1977) Synthesis and antiprotozoal activity of 2,5-bis(4-guanylphenyl) furans. J Med Chem 20(4):531–536, PMID: 321783
95. Clement B, Raether W (1985) Amidoximes of pentamidine: synthesis, trypanocidal and leishmanicidal activity. Arzneimittelforschung 35(7):1009–1014, PMID: 4052136
96. Yeates C (2003) DB-289. IDrugs 6(11):1086–1093, PMID: 14600842
97. Midgley I, Fitzpatrick K, Taylor LM, Houchen TL, Henderson SJ, Wright SJ, Cybulski ZR, John BA, McBurney A, Boykin DW, Trendler KL (2007) Pharmacokinetics and metabolism of the prodrug DB289 (2,5-bis[4-(N-methoxyamidino)phenyl]furan monomaleate) in rat and monkey and its conversion to the antiprotozoal/antifungal drug DB75 (2,5-bis(4-guanylphenyl) furan dihydrochloride). Drug Metab Dispos 35(6):955–967, PMID: 17360833
98. Zhou L, Thakker DR, Voyksner RD, Anbazhagan M, Boykin DW, Hall JE, Tidwell RR (2004) Metabolites of an orally active antimicrobial prodrug, 2,5-bis(4-amidinophenyl) furan-bis-O-methylamidoxime, identified by liquid chromatography/tandem mass spectrometry. J Mass Spectrom 39(4):351–360, PMID: 15103648
99. Wang MZ, Saulter JY, Usuki E, Cheung YL, Hall M, Bridges AS, Loewen G, Parkinson OT, Stephens CE, Allen JL, Zeldin DC, Boykin DW, Tidwell RR, Parkinson A, Paine MF, Hall JE (2006) CYP4F enzymes are the major enzymes in human liver microsomes that catalyze the O-demethylation of the antiparasitic prodrug DB289 [2,5-bis(4-amidinophenyl) furan-bis-O-methylamidoxime]. Drug Metab Dispos 34(12):1985–1994, PMID: 16997912
100. Saulter JY, Kurian JR, Trepanier LA, Tidwell RR, Bridges AS, Boykin DW, Stephens CE, Anbazhagan M, Hall JE (2005) Unusual dehydroxylation of antimicrobial amidoxime prodrugsby cytochrome b5 and NADH cytochrome b5 reductase. Drug Metab Dispos 33:1886–1893, PMID: 16131524
101. Thuita JK, Karanja SM, Wenzler T, Mdachi RE, Ngotho JM, Kagira JM, Tidwell R, Brun R (2008) Efficacy of the diamidine DB75 and its prodrug DB289, against murine models of human African trypanosomiasis. Acta Trop 108(1):6–10, PMID: 18722336
102. Mdachi RE, Thuita JK, Kagira JM, Ngotho JM, Murilla GA, Ndung'u JM, Tidwell RR, Hall JE, Brun R (2009) Efficacy of the novel diamidine compound 2,5-Bis(4-amidinophenyl)-furan-bis-O-Methlylamidoxime (Pafuramidine, DB289) against Trypanosoma brucei rhodesiense infection in vervet monkeys after oral administration. Antimicrob Agents Chemother 53(3):953–957, PMID: 19064893
103. Sturk LM, Brock JL, Bagnell CR, Hall JE, Tidwell RR (2004) Distribution and quantitation of the anti-trypanosomal diamidine 2,5-bis(4-amidinophenyl)furan (DB75) and its N-methoxy prodrug DB289 in murine brain tissue. Acta Trop 91:131–143, PMID: 15234662

104. Wenzler T, Boykin DW, Ismail MA, Hall JE, Tidwell RR, Brun R (2009) New treatment option for second-stage African sleeping sickness: in vitro and in vivo efficacy of aza analogs of DB289. Antimicrob Agents Chemother 53(10):4185–4192, PMID: 19620327
105. Pholig G, Bernhard S, Blum J, Burri C, Mpanya Kabeya A, Fina Lubaki J-P, Mpoo Mpoto A, Fungula Munungu B, Kambau Manesa Deo G, Nsele Mutantu P, Mbo Kuikumbi F, Fukinsia Mintwo A, Kayeye Munungi A, Dala A, Macharia S, Miaka Mia Bilenge C, Kande Betu Ku Mesu V, Ramon Franco J, Dieyi Dituvanga N, Olson C (2008) Phase 3 trial of pafuramidine maleate (DB289), a novel, oral drug, for treatment of first stage sleeping sickness: safety and efficacy. In: 57th Meeting of the American Society of Tropical Medicine & Hygiene, Abstract No. 542, New Orleans
106. Nyunt MM, Hendrix CW, Bakshi RP, Kumar N, Shapiro TA (2009) Phase I/II evaluation of the prophylactic antimalarial activity of pafuramidine in healthy volunteers challenged with Plasmodium falciparum sporozoites. Am J Trop Med Hyg 80(4):528–535, PMID: 19346370
107. De Souza EM, Lansiaux A, Bailly C, Wilson WD, Hu Q, Boykin DW, Batista MM, Araújo-Jorge TC, Soeiro MN (2004) Phenyl substitution of furamidine markedly potentiates its antiparasitic activity against Trypanosoma cruzi and Leishmania amazonensis. Biochem Pharmacol 68:593–600, PMID: 15276066
108. De Souza EM, Melo G, Boykin DW, Kumar A, Hu Q, De Nazaré C, Soeiro M (2006) Trypanocidal activity of the phenyl-substituted analogue of furamidine DB569 against T. cruzi infection in vivo. J Antimicrob Chemother 58(3):610–614, PMID: 16854954
109. Brendle JJ, Outlaw A, Kumar A, Boykin DW, Patrick DA, Tidwell RR, Werbovetz KA (2002) Antileishmanial activities of several classes of aromatic dications. Antimicrob Agents Chemother 46:797–807, PMID: 11850264
110. Stephens CE, Brun R, Salem MM, Werbovetz KA, Tanious F, Wilson WD, Boykin DW (2003) The activity of diguanidino and 'reversed' diamidino 2,5-diarylfurans versus Trypanosoma cruzi and Leishmania donovani. Bioorg Med Chem Lett 13(12):2065–2069, PMID: 12781196
111. Seiler N (2003) Thirty years of polyamine-related approaches to cancer therapy. Retrospect and prospect. Part 1. Selective enzyme inhibitors. Curr Drug Targets 4(7):537–564, PMID: 14535654
112. Wu F, Grossenbacher D, Gehring H (2007) New transition state–based inhibitor for human ornithine decarboxylase inhibits growth of tumor cells. Mol Cancer Ther 6(6):1831–1839, PMID: 17575112
113. Coward JK, Pegg AE (1987) Specific multisubstrate adduct inhibitors of aminopropyltransferases and their effect on polyamine biosynthesis in cultured cells. Adv Enzyme Regul 26:107–113, PMID: 3673702
114. Singh S, Mukherjee A, Khomutov AR, Persson L, Heby O, Chatterjee M, Madhubala R (2007) Antileishmanial effect of 3-aminooxy-1-aminopropane is due to polyamine depletion. Antimicrob Agents Chemother 51(2):528–534, PMID: 17101681
115. Bitonti AJ, Byers TL, Bush TL, Casara PJ, Bacchi CJ, Clarkson AB Jr, McCann PP, Sjoerdsma A (1990) Cure of Trypanosoma brucei brucei and Trypanosoma brucei rhodesiense infections in mice with an irreversible inhibitor of S-adenosylmethionine decarboxylase. Antimicrob Agents Chemother 34:1485–1490, PMID: 1977366
116. Byers T, Casarat T, Bitonti A (1992) Uptake of the antitrypanosomal drug 5'-{[(Z)-4-amino-2-butenyljmethylamino}-5'-deoxyadenosine (MDL 73811) by the purine transport system of Trypanosoma brucei brucei. Biochem J 283:755–758, PMID: 1590765
117. Bacchi CJ, Barker RH Jr, Rodriguez A, Hirth B, Rattendi D, Yarlett N, Hendrick CL, Sybertz E (2009) Trypanocidal activity of 8-methyl-5'-{[(Z)-4-aminobut-2-enyl]-(methylamino)}adenosine (Genz-644131), an adenosylmethionine decarboxylase inhibitor. Antimicrob Agents Chemother 53(8):3269–3272, PMID: 19451291
118. Krauth-Siegel RL, Bauer H, Schirmer RH (2005) Dithiol proteins as guardians of the intracellular redox milieu in parasites: old and new drug targets in trypanosomes and malaria-causing plasmodia. Angew Chem Int Ed 44:690–715, PMID: 15657967

119. Krauth-Siegel RL, Comini MA (2008) Redox control in trypanosomatids, parasitic protozoa with trypanothione-based thiol metabolism. Biochim Biophys Acta 1780:1236–1248, PMID: 18395526
120. Rivera G, Bocanegra-García V, Ordaz-Pichardo C, Nogueda-Torres B, Monge A (2009) New therapeutic targets for drug design against Trypanosoma cruzi, advances and perspectives. Curr Med Chem 16(25):3286–3293, PMID: 19548870
121. Soeiro MN, de Castro SL (2009) Trypanosoma cruzi targets for new chemotherapeutic approaches. Expert Opin Ther Targets 13(1):105–121, PMID: 19063710
122. Galarreta BC, Sifuentes R, Carrillo AK, Sanchez L, Amado Mdel R, Maruenda H (2008) The use of natural product scaffolds as leads in the search for trypanothione reductase inhibitors. Bioorg Med Chem 16:6689–6695, PMID: 18558492
123. Bitonti AJ, Dumont JA, Bush TL, Edwards ML, Stemerick DM, McCann PP, Sjoerdsma A (1989) Bis(benzyl)polyamine analogs inhibit the growth of chloroquine-resistant human malaria parasites (Plasmodium falciparum) in vitro and in combination with alpha-difluoromethylornithine cure murine malaria. Proc Natl Acad Sci USA 86:651–655, PMID: 2463635
124. Bonnet B, Soullez D, Davioud-Charvet E, Landry V, Horvath D, Sergheraert C (1997) New spermine and spermidine derivatives as potent inhibitors of Trypanosoma cruzi trypanothione reductase. Bioorg Med Chem 5:1249–1256, PMID: 9377084
125. Baumann RJ, Hanson WL, McCann PP, Sjoerdsma A, Bitonti AJ (1990) Suppression of both antimony-susceptible and antimony-resistant Leishmania donovani by a bis(benzyl)polyamine analog. Antimicrob Agents Chemother 34:722–727, PMID: 2360812
126. Baumann RJ, McCann PP, Bitonti AJ (1991) Suppression of Leishmania donovani by oral administration of a bis(benzyl)polyamine analog. Antimicrob Agents Chemother 35:1403–1407, PMID: 1929300
127. Obexer W, Schmid C, Barbe J, Galy JP, Brun R (1995) (1995) Activity and structure relationship of acridine derivatives against African trypanosomes. Trop Med Parasitol 46 (1):49–53, PMID: 7631129
128. Khan MO, Austin SE, Chan C, Yin H, Marks D, Vaghjiani SN, Kendrick H, Yardley V, Croft SL, Douglas KT (2000) Use of an additional hydrophobic binding site, the Z site, in the rational drug design of a new class of stronger trypanothione reductase inhibitor, quaternary alkylammonium phenothiazines. J Med Chem 43:3148–3156, PMID: 10956223
129. Eberle C, Burkhard JA, Stump B, Kaiser M, Brun R, Krauth-Siegel RL, Diederich F (2009) Synthesis, inhibition potency, binding mode, and antiprotozoal activities of fluorescent inhibitors of trypanothione reductase based on mepacrine-conjugated diaryl sulfide scaffolds. ChemMedChem 4(12):2034–2044, PMID: 19847846
130. Spinks D, Shanks EJ, Cleghorn LAT, McElroy S, Jones D, James D, Fairlamb AH, Frearson JA, Wyatt PG, Gilbert IH (2009) Investigation of trypanothione reductase as a drug target in Trypanosoma brucei. ChemMedChem 4:2060–2069
131. Patterson S, Jones DC, Shanks EJ, Frearson JA, Gilbert IH, Wyatt PG, Fairlamb AH (2009) Synthesis and evaluation of 1-(1-(Benzo[b]thiophen-2-yl)cyclohexyl)piperidine (BTCP) analogues as inhibitors of trypanothione reductase. ChemMedChem 4:1341–1353
132. Blumenstiel K, Schoeneck R, Yardley V, Croft SL, Krauth-Siegel RL (1999) Nitrofuran drugs as common subversive substrates of Trypanosoma cruzi lipoamide dehydrogenase and trypanothione reductase. Biochem Pharmacol 58:1791–1799, PMID: 10571254
133. LaMontagne MP, Dagli D, Khan MS, Blumbergs P (1980) Analogs of 8-[[6-(diethylamino) hexyl]amino]-6-methoxy-4- methylquinoline as candidate antileishmanial agents. J Med Chem 23(9):981–985, PMID: 7411553
134. Yeates C (2002) Sitamaquine. Curr Opin Investig Drugs 3:1446–1452, PMID: 12431016
135. Berman JD, Lee LS (1983) Activity of 8-aminoquinolines against Leishmania tropica within human macrophages in vitro. Am J Trop Med Hyg 32(4):753–759, PMID: 6881421
136. Jha TK, Sundar S, Thakur CP, Felton JM, Sabin AJ, Horton J (2005) A phase II dose-ranging study of sitamaquine for the treatment of visceral leishmaniasis in India. Am J Trop Med Hyg 73(6):1005–1011, PMID: 16354802

137. Wasunna MK, Rashid JR, Mbui J, Kirigi G, Kinoti D, Lodenyo H, Felton JM, Sabin AJ, Albert MJ, Horton J, Albert MJ (2006) A phase II dose-increasing study of sitamaquine for the treatment of visceral leishmaniasis in Kenya. Am J Trop Med Hyg 73(5):871–876, PMID: 16282296
138. Fournet A, Gantier JC, Gautheret A, Leysalles L, Munos MH, Mayrargue J, Moskowitz H, Cavé A, Hocquemiller R (1994) The activity of 2-substituted quinoline alkaloids in BALB/c mice infected with Leishmania donovani. J Antimicrob Chemother 33:537–544
139. Fournet A, Barrios AA, Muñoz V, Hocquemiller R, Cavé A, Bruneton J (1993) 2-substituted quinoline alkaloids as potential antileishmanial drugs. Antimicrob Agents Chemother 37(4):859–863
140. Nakayama H, Loiseau PM, Bories C, Torres de Ortiz S, Schinini A, Serna E, Rojas de Arias A, Fakhfakh MA, Franck X, Figadère B, Hocquemiller R, Fournet A (2005) Efficacy of orally administered 2-substituted quinolines in experimental murine cutaneous and visceral leishmaniases. Antimicrob Agents Chemother 49(12):4950–4956
141. Urbina JA, Vivas J, Ramos H, Larralde G, Aguilar Z, Avilan L (1988) Alterations of lipid order profile and permeability of plasma membranes from Trypanosoma cruzi epimastigotes grown in the presence of ketoconazole. Mol Biochem Parasitol 30:185–196, PMID: 2845268
142. Urbina JA, Concepcion JL, Rangel S, Visbal G, Lira R (2002) Squalene synthase as a chemotherapeutic target in Trypanosoma cruzi and Leishmania mexicana. Mol Biochem Parasitol 125(1–2):35–45, PMID: 12467972
143. Lazardi K, Urbina JA, de Souza W (1990) Ultrastructural alterations induced by two ergosterol biosynthesis inhibitors, ketoconazole and terbinafine, on epimastigotes and amastigotes of Trypanosoma (Schizotrypanum) cruzi. Antimicrob Agents Chemother 34:2097–2105, PMID: 2073100
144. Maldonado RA, Molina J, Payares G, Urbina JA (1993) Experimental chemotherapy with combinations of ergosterol biosynthesis inhibitors in murine models of Chagas' disease. Antimicrob Agents Chemother 37:1353–1359, PMID: 8328786
145. McCabe R (1988) Failure of ketoconazole to cure chronic murine Chagas' disease. J Infect Dis 158:1408–1409, PMID: 3143768
146. Brener Z (1993) An experimental and clinical assay with ketoconazole in the treatment of Chagas disease. Mem Inst Oswaldo Cruz 88:149–153, PMID: 8246750
147. Molina J, Martins-Filho O, Brener Z, Romanha AJ, Loebenberg D, Urbina JA (2000) Activities of the triazole derivative SCH 56592 (posaconazole) against drug-resistant strains of the protozoan parasite Trypanosoma (Schizotrypanum) cruzi in immunocompetent and immunosuppressed murine hosts. Antimicrob Agents Chemother 44:150–155, PMID: 10602737
148. Bahia MT, Talvani A, Chang S, Ribeiro I (2009) Combination of benznidazole and nifurtimox plus posaconazole enhances activity against Trypanosoma cruzi in experimental Chagas disease. Am Soc Trop Med Hyg 58th Annual Meeting, Washington DC, #746
149. Olivieri BP, Molina JT, de Castro SL, Pereira MC, Calvet CM, Urbina JA, Araújo-Jorge TC (2010) A comparative study of posaconazole and benznidazole in the prevention of heart damage and promotion of trypanocidal immune response in a murine model of Chagas disease. Int J Antimicrob Agents 36(1):79–83, PMID: 20452188
150. Merck to Initiate Proof-of-Concept Study of Posaconazole for Chronic Chagas Disease, Recognized by WHO as One of the World's Neglected Tropical Diseases (2010) Merck website. http://www.merck.com/newsroom/news-release-archive/research-and-development/2010_0624.html. Accessed 19 July 2010
151. Szajnman SH, Montalvetti A, Wang Y, Docampo R, Rodriguez JB (2003) Bisphosphonates derived from fatty acids are potent inhibitors of Trypanosoma cruzi farnesyl pyrophosphate synthase. Bioorg Med Chem Lett 13(19):3231–3235, PMID: 12951099
152. Garzoni LR, Caldera A, Meirelles Mde N, de Castro SL, Docampo R, Meints GA, Oldfield E, Urbina JA (2004) Selective in vitro effects of the farnesyl pyrophosphate synthase inhibitor

risedronate on Trypanosoma cruzi. Int J Antimicrob Agents 23(3):273–285, PMID: 15164969
153. Garzoni LR, Waghabi MC, Baptista MM, de Castro SL, Meirelles Mde N, Britto CC, Docampo R, Oldfield E, Urbina JA (2004) Antiparasitic activity of risedronate in a murine model of acute Chagas' disease. Int J Antimicrob Agents 23(3):286–290, PMID: 15164970
154. Szajnman SH, García Liñares GE, Li ZH, Jiang C, Galizzi M, Bontempi EJ, Ferella M, Moreno SN, Docampo R, Rodriguez JB (2008) Synthesis and biological evaluation of 2-alkylaminoethyl-1,1-bisphosphonic acids against Trypanosoma cruzi and Toxoplasma gondii targeting farnesyl diphosphate synthase. Bioorg Med Chem 16(6):3283–3290, PMID: 18096393
155. Urbina JA (2010) Specific chemotherapy of Chagas disease: relevance, current limitations and new approaches. Acta Trop 115(1–2):55–68, PMID: 19900395
156. Azoles E1224 (Chagas) (2010) DNDi website. http://www.dndi.org/portfolio/azoles-e1224.html. Accessed 20 July 2010
157. Drug Combinations (Chagas) (2010) DNDi website. http://www.dndi.org/portfolio/drug-combination.html. Accessed 3 Aug 2010
158. Zeiman E, Greenblatt CL, Elgavish S, Khozin-Goldberg I, Golenser J (2008) Mode of action of fenarimol against Leishmania spp. J Parasitol 94(1):280–286
159. Keenan M, Best WM, Armstrong T, Thompson A, Charman S, White K, Don R, von Geldern TW (2009) Novel compounds for the treatment of Chagas disease. Am Soc Trop Med Hyg 58th Annual Meeting, Washington DC, #751
160. Keenan M (2010) New compounds for the treatment of Chagas disease. Internat'l Conf Parasitol (ICOPA), Melbourne, Australia, #244
161. Buckner FS, Griffin JH, Wilson AJ, Van Voorhis WC (2001) Potent anti-Trypanosoma cruzi activities of oxidosqualene cyclase inhibitors. Antimicrob Agents Chemother 45(4):1210–1215
162. Sealey-Cardona M, Cammerer S, Jones S, Ruiz-Pérez LM, Brun R, Gilbert IH, Urbina JA, González-Pacanowska D (2007) Kinetic characterization of squalene synthase from Trypanosoma cruzi: selective inhibition by quinuclidine derivatives. Antimicrob Agents Chemother 51(6):2123–2129, PMID: 17371809
163. Braga MV, Urbina JA, de Souza W (2004) Effects of squalene synthase inhibitors on the growth and ultrastructure of Trypanosoma cruzi. Int J Antimicrob Agents 24(1):72–78, PMID: 15225865
164. Orenes Lorente S, Gómez R, Jiménez C, Cammerer S, Yardley V, de Luca-Fradley K, Croft SL, Ruiz Perez LM, Urbina J, Gonzalez Pacanowska D, Gilbert IH (2005) Biphenylquinuclidines as inhibitors of squalene synthase and growth of parasitic protozoa. Bioorg Med Chem 13(10):3519–3529, PMID: 15848765
165. Urbina JA, Concepcion JL, Caldera A, Payares G, Sanoja C, Otomo T, Hiyoshi H (2004) In vitro and in vivo activities of E5700 and ER-119884, two novel orally active squalene synthase inhibitors, against Trypanosoma cruzi. Antimicrob Agents Chemother 48(7):2379–2387, PMID: 15215084
166. Hucke O, Gelb MH, Verlinde CL, Buckner FS (2005) The protein farnesyltransferase inhibitor Tipifarnib as a new lead for the development of drugs against Chagas disease. J Med Chem 48(17):5415–5418, PMID: 16107140
167. Chennamaneni NK, Arif J, Buckner FS, Gelb MH (2009) Isoquinoline-based analogs of the cancer drug clinical candidate tipifarnib as anti-Trypanosoma cruzi agents. Bioorg Med Chem Lett 19(23):6582–6584, PMID: 19875282
168. Kraus JM, Verlinde CL, Karimi M, Lepesheva GI, Gelb MH, Buckner FS (2009) Rational modification of a candidate cancer drug for use against Chagas disease. J Med Chem 52 (6):1639–1647, PMID: 19239254. Errata in: J Med Chem (2009) 52(14): 4549, J Med Chem (2009) 52(15): 4979
169. Price HP, Menon MR, Panethymitaki C, Goulding D, McKean PG, Smith DF (2003) Myristoyl-CoA: protein N-myristoyltransferase, an essential enzyme and potential drug target in kinetoplastid parasites. J Biol Chem 278:7206–7214

170. Frearson JA, Brand S, McElroy SP, Cleghorn LAT, Smid O, Stojanovski L, Price HP, Guther MLS, Torrie LS, Robinson DA, Hallyburton I, Mpamhanga CP, Brannigan JA, Wilkinosn AJ, Hodgkinson M, Hui R, Qiu W, Raimi OG, van Aalten DMF, Brenk R, Gilbert IH, Read KD, Fairlamb AH, Ferguson MAJ, Smith DF, Wyatt PG (2010) N-myristoyltransferase inhibitors as new leads to treat sleeping sickness. Nature 464:728–734
171. Robello C, Navarro P, Castanys S, Gamarro F (1997) A pteridine reductase gene ptr1 contiguous to a P-glycoprotein confers resistance to antifolates in Trypanosoma cruzi. Mol Biochem Parasitol 90:525–535
172. Bello AR, Nare B, Freedman D, Hardy LW, Beverley SM (1994) PTR1: a reductase mediating salvage of oxidized pteridines and methotrexate resistance in the protozoan parasite Leishmania major. Proc Natl Acad Sci USA 91:11442–11446
173. Nare B, Hardy LW, Beverley S (1997) The roles of pteridine reductase and dihydrofolate reductase-thymidylate synthase in pteridine metabolism in the protozoan parasite Leishmania major. J Biol Chem 272(21):13883–13891
174. Nare B, Luba J, Hardy LW, Beverley S (1997) Parasitology 114(Suppl):S101–S110
175. Cavazzuti A, Paglietti G, Hunter WN, Gamarro F, Piras S, Loriga M, Alleca S, Corona P, McLuskey K, Tulloch L, Gibellini F, Ferrari S, Costi MP (2008) Discovery of potent pteridine reductase inhibitors to guide antiparasite drug development. Proc Nat Acad Sci USA 105(5):1448–1453
176. Khabnadideh S, Pez D, Musso A, Brun R, Pérez LM, González-Pacanowska D, Gilbert IH (2005) Design, synthesis and evaluation of 2,4-diaminoquinazolines as inhibitors of trypanosomal and leishmanial dihydrofolate reductase. Bioorg Med Chem 13(7):2637–2649
177. Sirawaraporn W, Sertsrivanich R, Booth RG, Hansch C, Neal RA, Santi DV (1988) Selective inhibition of Leishmania dihydrofolate reductase and Leishmania growth by 5-benzyl-2,4-diaminopyrimidines. Mol Biochem Parasitol 31:79–86
178. Pez D, Leal I, Zuccotto F, Boussard C, Brun R, Croft SL, Yardley V, Ruiz Perez LM, Gonzalez Pacanowska D, Gilbert IH (2003) 2,4-Diaminopyrimidines as inhibitors of Leishmanial and Trypanosomal dihydrofolate reductase. Bioorg Med Chem 11(22):4693–4711
179. Berman JD, King M, Edwards N (1989) Antileishmanial activities of 2,4-diaminoquinazoline putative dihydrofolate reductase inhibitors. Antimicrob Agents Chemother 33 (11):1860–1863
180. Schormann N, Pal B, Senkovich O, Carson M, Howard A, Smith C, Delucas L, Chattopadhyay D (2005) Crystal structure of Trypanosoma cruzi pteridine reductase 2 in complex with a substrate and an inhibitor. J Struct Biol 152(1):64–75
181. Dawson A, Gibellini F, Sienkiewicz N, Tulloch LB, Fyfe PK, McLuskey K, Fairlamb AH, Hunter WN (2006) Structure and reactivity of Trypanosoma brucei pteridine reductase: inhibition by the archetypal antifolate methotrexate. Mol Microb 61(6):1457–1468
182. Schüttelkopf AW, Hardy LW, Beverley SM, Hunter WN (2005) Structures of Leishmania major pteridine reductase complexes reveal the active site features important for ligand binding and to guide inhibitor design. J Mol Biol 352(1):105–116
183. Tulloch LB, Martini VP, Iulek J, Huggan JK, Lee JH, Gibosn CL, Smith TK, Suckling CJ, Hunter WN (2010) Structure-based design of pteridine reductase inhibitors targeting African sleeping sickness and the leishmaniases. J Med Chem 53:221–229
184. Renslo AR, McKerrow JH (2006) Drug discovery and development for neglected parasitic diseases. Nat Chem Biol 2:701–710, PMID: 17108988
185. Mottram JC, Frame MJ, Brooks DR, Tetley L, Hutchison JE, Souza AE, Coombs GH (1997) The multiple Cpb cysteine proteinase genes of Leishmania mexicana encode isoenzymes that differ in their stage regulation and substrate preferences. J Biol Chem 272:14285–14293, PMID: 9162063
186. Robertson CD, Coombs GH (1994) Multiple high activity cysteine proteases of Leishmania mexicana are encoded by the lmcpb gene array. Microbiology 140:417–424, PMID: 8180705
187. McKerrow JH, Rosenthal PJ, Swenerton R, Doyle P (2008) Development of protease inhibitors for protozoan infections. Curr Opin Infect Dis 21:668–672, PMID: 18978536

188. McKerrow JH, Doyle PS, Engel JC, Podust LM, Robertson SA, Ferreira R, Saxton T, Arkin M, Kerr ID, Brinen LS, Craik CS (2009) Two approaches to discovering and developing new drugs for Chagas disease. Mem Inst Oswaldo Cruz 104(S1):263–269, PMID: 19753483
189. Jacobsen W, Christians U, Benet LZ (2000) In vitro evaluation of the disposition of a novel cysteine protease inhibitor. Drug Metab Dispos 28(11):1343–1351, PMID: 11038163
190. Engel JC, Doyle PS, Hsieh I, McKerrow JH (1998) Cysteine protease inhibitors cure an experimental Trypanosoma cruzi infection. J Exp Med 188:725, PMID: 9705954
191. Doyle PS, Zhou YM, Engel JC, McKerrow JH (2007) A cysteine protease inhibitor cures Chagas' disease in an immunodeficient-mouse model of infection. Antimicrob Agents Chemother 51(11):3932–3939, PMID: 17698625
192. Institute for One World Health (2010) http://www.oneworldhealth.org/chagas. Accessed 10 June 2010
193. Roush WR, Gwaltney SL II, Cheng J, Scheidt KA, McKerrow JH, Hansell E (1998) Vinyl sulfonate esters and vinyl sulfonamides: potent, irreversible inhibitors of cysteine proteases. J Am Chem Soc 120:10994–10995
194. Roush WR, Cheng J, Knapp-Reed B, Alvarez-Hernandez A, McKerrow JH, Hansell E, Engel JC (2001) Potent second generation vinyl sulfonamide inhibitors of the trypanosomal cysteine protease cruzain. Bioorg Med Chem Lett 11:2759–2762, PMID: 11591518
195. Du X, Guo C, Hansell E, Doyle PS, Caffrey CR, Holler TP, McKerrow JH, Cohen FE (2002) Synthesis and structure-activity relationship study of potent trypanocidal thio semicarbazone inhibitors of the trypanosomal cysteine protease cruzain. J Med Chem 45:2695–2707, PMID: 12061873
196. Brak K, Doyle PS, McKerrow JH, Ellman JA (2008) Identification of a new class of nonpeptidic inhibitors of cruzain. J Am Chem Soc 130(20):6404–6410, PMID: 18435536
197. Chen YT, Lira R, Hansell E, McKerrow JH, Roush WR (2008) Synthesis of macrocyclic trypanosomal cysteine protease inhibitors. Bioorg Med Chem Lett 18(22):5860–5863, PMID: 18585034
198. Ponte-Sucre A, Vicik R, Schultheis M, Schirmeister T, Moll H (2006) Aziridine-2,3-dicarboxylates, peptidomimetic cysteine protease inhibitors with antileishmanial activity. Antimicrob Agents Chemother 50:2439–2447, PMID: 16801424
199. St Hilaire PM, Alves LC, Herrera F, Renil M, Sanderson SJ, Mottram JC, Coombs GH, Juliano MA, Juliano L, Arevalo J, Meldal M (2002) Solid-phase library synthesis, screening, and selection of tight-binding reduced peptide bond inhibitors of a recombinant Leishmania mexicana cysteine protease B. J Med Chem 45:1971–1982, PMID: 11985465
200. Ascenzi P, Salvati L, Bolognesi M, Colasanti M, Polticelli F, Venturini G (2001) Inhibition of cysteine protease activity by NO-donors. Curr Protein Pept Sci 2:137–153, PMID: 12370021
201. Colasanti M, Salvati L, Venturini G, Ascenzi P, Gradoni L (2001) Cysteine protease as a target for nitric oxide in parasitic organisms. Trends Parasitol 17:575, PMID: 11756039
202. Vespa GNR, Cunha FQ, Silva JS (1994) Nitric oxide is involved in control of Trypanosoma cruzi-induced parasitemia and directly kills the parasite in vitro. Infect Immun 62:5177–5182, PMID: 7523307
203. Zeina B, Banfield C, al-Assad S (1997) Topical glyceryl trinitrate: a possible treatment for cutaneous leishmaniasis. Clin Exp Dermatol 22:244–245, PMID: 9536549
204. Lopez-Jaramillo P, Ruano C, Rivera J, Teran E, Salazar-Irigoyen R, Esplugues JV, Moncada S (1998) Treatment of cutaneous leishmaniasis with nitric-oxide donor. Lancet 351:1176–1177, PMID: 9643692
205. Davidson RN, Yardley V, Croft SL, Konecny P, Benjamin N (2000) A topical nitric oxide-generating therapy for cutaneous leishmaniasis. Trans R Soc Trop Med Hyg 94:319–322, PMID: 10975011
206. Salvati L, Mattu M, Colasanti M, Scalone S, Venturini G, Gradoni L, Ascenzi P (2001) NO donors inhibit Leishmania infantum cysteine proteinase activity. Biochim Biophys Acta 1545:357–366, PMID: 11342060

207. Ascenzi P, Bocedi A, Gentile M, Visca P, Gradoni L (2004) Inactivation of parasite cysteine proteinases by the NO-donor 4-(phenylsulfonyl)-3-((2-(dimethylamino)ethyl)thio)-furoxan oxalate. Biochim Biophys Acta 1703:69–77, PMID: 15588704
208. Buates S, Matlashewski GJ (1999) Treatment of experimental leishmaniasis with the immunomodulators imiquimod and S-28463: efficacy and mode of action. Infect Dis 179:1485–1494, PMID: 10228071
209. Arevalo I, Ward B, Miller R, Meng TC, Najar E, Alvarez E, Matlashewski G, Llanos-Cuentas A (2001) Successful treatment of drug-resistant cutaneous leishmaniasis in humans by use of imiquimod, an immunomodulator. Clin Infect Dis 33:1847–1851, PMID: 11692295
210. Miranda-Verástegui C, Llanos-Cuentas A, Arévalo I, Ward BJ, Matlashewski G (2005) Randomized, double-blind clinical trial of topical imiquimod 5% with parenteral meglumine antimoniate in the treatment of cutaneous leishmaniasis in Peru. Clin Infect Dis 40 (10):1395–1403, PMID: 15844060
211. Miranda-Verastegui C, Tulliano GF, Gyorkos TW, Calderon W, Rahme E, Ward B, Cruz M, Llanos-Cuentas A, Matlashewski G (2009) First-line therapy for human cutaneous leishmaniasis in Peru using the TLR7 agonist imiquimod in combination with pentavalent antimony. PLoS 3(7):e491, PMID: 19636365
212. Al-Mutairi N, Alshiltawy M, El Khalawany M, Joshi A, Eassa BI, Manchanda Y, Gomaa S, Darwish I, Rijhwani M (2009) Treatment of Old World cutaneous leishmaniasis with dapsone, itraconazole, cryotherapy, and imiquimod, alone and in combination. Int J Derm 48:862–869
213. Morgan RE, Werbovetz KA (2008) Selective lead compounds against kinetoplastid tubulin. Adv Exp Med Biol 625:33–47, PMID: 18365657
214. Ochola DO, Prichard RK, Lubega GW (2002) Classical ligands bind tubulin of trypanosomes and inhibit their growth in vitro. J Parasitol 88:600–604, PMID: 12099434
215. Darkin-Rattray SJ, Gurnett AM, Myers RW, Dulski PM, Crumley TM, Allocco JJ, Cannova C, Meinke PT, Colletti SL, Bednarek MA, Singh SB, Goetz MA, Dombrowski AW, Polishook JD, Schmatz DM (1996) Apicidin: a novel antiprotozoal agent that inhibits parasite histone deacetylase. Proc Natl Acad Sci USA 93:13143–13147, PMID: 8917558
216. Mai A, Cerbara I, Valente S, Massa S, Walker LA, Tekwani BL (2004) Antimalarial and antileishmanial activities of aroyl-pyrrolyl-hydroxyamides, a new class of histone deacetylase inhibitors. Antimicrob Agents Chemother 48:1435–1436, PMID: 15047563
217. Meinke PT, Colletti SL, Doss G, Myers RW, Gurnett AM, Dulski PM, Darkin-Rattray SJ, Allocco JJ, Galuska S, Schmatz DM, Wyvratt MJ, Fisher MH (2000) Synthesis of apicidin-derived quinolone derivatives: parasite-selective histone deacetylase inhibitors and antiproliferative agents. J Med Chem 43(25):4919–4922, PMID: 11124001
218. Colletti SL, Myers RW, Darkin-Rattray SJ, Gurnett AM, Dulski PM, Galuska S, Allocco JJ, Ayer MB, Li C, Lim J, Crumley TM, Cannova C, Schmatz DM, Wyvratt MJ, Fisher MH, Meinke PT (2001) Broad spectrum antiprotozoal agents that inhibit histone deacetylase: structure-activity relationships of apicidin. Part 1. Bioorg Med Chem Lett 11(2):107–111, PMID: 11206438
219. Colletti SL, Myers RW, Darkin-Rattray SJ, Gurnett AM, Dulski PM, Galuska S, Allocco JJ, Ayer MB, Li C, Lim J, Crumley TM, Cannova C, Schmatz DM, Wyvratt MJ, Fisher MH, Meinke PT (2001) Broad spectrum antiprotozoal agents that inhibit histone deacetylase: structure-activity relationships of apicidin. Part 2. Bioorg Med Chem Lett 11(2):113–117, PMID: 11206439
220. Shin BS, Chang HS, Park EH, Yoon CH, Kim HY, Kim J, Ryu JK, Zee OP, Lee KC, Cao D, Yoo SD (2006) Pharmacokinetics of a novel histone deacetylase inhibitor, apicidin, in rats. Biopharm Drug Dispos 27(2):69–75
221. Cavalcanti DP, Fragoso SP, Goldenberg S, de Souza W, Motta MC (2004) The effect of topoisomerase II inhibitors on the kinetoplast ultrastructure. Parasitol Res 94:439–448, PMID: 15517387

222. Das BB, Sengupta T, Ganguly A, Majumder HK (2006) Topoisomerases of kinetoplastid parasites: why so fascinating? Mol Microbiol 62(4):917–927, PMID: 17042788
223. Das BB, Sen N, Roy A, Dasgupta SB, Ganguly A, Mohanta BC, Dinda B, Majumder HK (2006) Differential induction of Leishmania donovani bi-subunit topoisomerase I-DNA cleavage complex by selected flavones and camptothecin: activity of flavones against camptothecin-resistant topoisomerase I. Nucleic Acids Res 34(4):1121–1132, PMID: 16488884
224. Sen N, Das BB, Ganguly A, Mukherjee T, Tripathi G, Bandyopadhyay S, Rakshit S, Sen T, Majumder HK (2004) Camptothecin induced mitochondrial dysfunction leading to programmed cell death in unicellular hemoflagellate Leishmania donovani. Cell Death Differ 8:924–936, PMID: 15118764
225. NECT – Nifurtimox-Eflornithine Co-Administration (HAT) (2010) DNDi website. http://www.dndi.org/portfolio/nect.html. Accessed 17 July 2010
226. Priotto G, Kasparian S, Ngouama D, Ghorashian S, Arnold U, Ghabri S, Karunakara U (2007) Nifurtimox-eflornithine combination therapy for second-stage Trypanosoma brucei gambiense sleeping sickness: a randomized clinical trial in Congo. Clin Infect Dis 45:1435–1442, PMID: 17990225
227. Priotto G, Kasparian S, Mutombo W, Ngouama D, Gharashian S, Arnold U, Ghabri S, Baudin E, Buard V, Kazadi-Kyanza S, Ilunga M, Mutangala W, Pohlig G, Schmid C, Karunakara U, Torreele E, Kande V (2009) Nifurtomox-eflornithine combination therapy for second-stage African Trypanosoma brucei gambiense trypanosomiasis: a multicentre, randomised, phase III, non-inferiority trial. Lancet 374:56–64, PMID: 19559476
228. Checchi F, Piola P, Ayikoru H, Thomas F, Legros D, Priotto G (2007) Nifurtimox plus eflornithine for late-stage sleeping sickness in Uganda: a case series. PLoS Negl Trop Dis 1: e64, PMID: 18060083
229. Baudin E (2008) Multicenter clinical trial of nifurtimox-eflornithine combination therapy for second-stage sleeping sickness. Am Soc Trop Med Hyg 57th Annual Meeting, New Orleans
230. NECT Added to WHO Essential Medicines List as Combination Treatment Against Sleeping Sickness (2010) DNDi website. http://www.dndi.org/press-releases/456-nect-added-to-who-essential-medicines-list-as-combination-treatment-against-sleeping-sickness.html. Accessed 16 July 2010
231. Torreele E (2009) Proposal for the inclusion of Nifurtimox-Eflornithine Combination as a treatment for Stage 2 Trypanosoma Brucei Gambiense Human African Trypanosomiasis (Sleeping Sickness) in the WHO Model List of Essential Medicines. 17th Expert Committee on the Selection and Use of Essential Medicines, Geneva
232. Yun O, Priotto G, Tong J, Flevaud L, Chappuis F (2009) NECT is next: implementing the new drug combination therapy for trypanosoma brucei gambiense sleeping sickness. PloS Negl Trop Dis 3:1–7
233. Chagas Disease Partnership Will Deliver Safe, Easy-to-Use Treatment for Children (2008) DNDi website. http://www.dndi.org/press-releases/2008/106-chagas-disease-partnership-will-deliver-safe-easy-to-use-treatment-for-children.html. Accessed 16 July 2010
234. Terlouw DJ, Alves FP, Sosa-Estani S, Freilij H, Altcheh J, Brutus L, Kiechel J-R, Ribeiro I (2009) A new pediatric tablet strength of benznidazole for the treatment of Chagas disease. Am Soc Trop Med Hyg 58th Annual Meeting, Washington DC, #834
235. Jaime Altcheh MD (study director) (2009) Population pharmacokinetics of benznidazole in children with Chagas disease. ClinicalTrials.gov identifier NCT00699387
236. Sundar S, Rai M (2005) Treatment of visceral Leishmaniasis. Expert Opin Pharmacother 6:2821–2829, PMID: 16318433
237. Berman J (2005) Clinical status of agents being developed for leishmaniasis. Expert Opin Investig Drugs 14:1337–1346, PMID: 16255674
238. van Griensven J, Boerlaert M (2009) Combination therapy for visceral leishmaniasis: why, what and where? WorldLeish4, 4th World Congress on Leishmaniasis, Lucknow, India

239. Combination Therapy (VL in Asia) (2010) DNDi website. http://www.dndi.org/portfolio/combination-therapy-asia.html. Accessed 2 Aug 2010
240. Combination Therapy (VL in Africa) (2010) DNDi website. http://www.dndi.org/portfolio/combination-therapy-africa.html. Accessed 2 Aug 2010
241. Combination Therapy (VL in Latin America) (2010) DNDi website. http://www.dndi.org/portfolio/combination-therapy-latin-america.html. Accessed 2 Aug 2010
242. Mudawi AM (2009) New treatments for visceral leishmaniasis in East Africa: the story so far. WorldLeish4, 4th World Congress on Leishmaniasis, Lucknow, India
243. Hailu A, Musa A, Wasunna M, Balasegaram M, Yifru S, Mengistu G, Hurissa Z, Hailu W, Weldegebreal T, Tesfaye S, Makonnen E, Khalil E, Ahmed O, Fadlall A, El-Hassan A, Raheem M, Mueller M, Koummuki Y, Rashid J, Mbui J, Mucee G, Njoroge S, Manduku V, Musibi A, Mutuma G, Kirui F, Lodenyo H, Mutea D, Kirigi G, Edward T, Smith P, Muthami L, Royce C, Ellis S, Alobo M, Omollo R, Kesusu J, Owiti R, Kinuthia J (2010) PLoS Negl Trop Dis 4:e709
244. Musai AM, Younis B, Fadlalla A, Royce C, Balasegaram M, Wasunna M, Hailu A, Edward T, Omollo R, Mudawi M, Kokwaro G, El-Hassan A, Khalil E (2010) Paromomycin for the treatment of visceral leishmaniasis in Sudan: a randomized, open-label, dose-finding study. PLoS Negl Trop Dis 4:e855
245. van Griensven J, Boelaert M (2011) Combination therapy for visceral leishmaniasis. Lancet 377(9764):443–444
246. Sundar S, Sinha PK, Rai M et al (2011) Comparison of short-course multidrug treatment with standard therapy for visceral leishmaniasis in India: an open-label, non-inferiority, randomised controlled trial. Lancet 377(9764):477–486
247. Torreele E (2008) Fexinidazole: a rediscovered compound progresses as a potential candidate for sleeping sickness. Am Soc Trop Med Hyg 57th Annual Meeting, New Orleans. http://www.youtube.com/watch?v=8QCrq99yzUc&feature=player_embedded
248. Best WM, Sims CG, Scaffidi A, Giuseppe L, Thompson RCA, Armstrong T. Antiparasitic compounds. WO2010/009508
249. Bray MA, Torreele E, Mazué G, Tweats D, Sassella D (2009) Fexinidazole investigator's brochure. DNDi
250. Jennings FW, Urquhart GM (1983) The use of the 2 substituted 5-nitroimidazole, Fexinidazole (Hoe 239) in the treatment of chronic T. brucei infections in mice. Z Parasitenkd 69(5):577–581
251. Raether W, Seidenath H (1983) The activity of fexinidazole (HOE 239) against experimental infections with Trypanosoma cruzi, trichomonads and Entamoeba histolytica. Ann Trop Med Parasitol 77(1):13–26
252. CEREP Study No. 14929 (2008) In vitro pharmacology – study of fexinidazole, fexinidazole sulfone and fexinidazole sulfoxide
253. Torreele E, Bourdin B, Tweats D, Kaiser M, Brun R, Mazué G, Bray MA, Pécoul B (2010) Fexinidazole – a new oral nitroimidazole drug candidate entering clinical development for the treatment of sleeping sickness. PLoS Negl Trop Dis 4:e923
254. Best WM, Sims CG, Thompson RCA, Reid SA, Armson A, Reynoldson JA. Antiparasitic compounds. WO2006/108224
255. Wylie S, Sokolova A, Patterson S, Fairlamb AH (2010) Cross-resistance to nitro-drugs and implications for the treatment of human African trypanosomiasis. Internat'l Conf Parasitol (ICOPA), Melbourne, Australia, #197
256. Bourdin B, Jędrysiak R, Tweats D, Brun R, Kaiser M, Suwiński J, Torreele E (2011) 1-Aryl-2202 4-nitroimidazoles, a new promising series for the treatment of Human African Trypanosomiasis. Eur J Med Chem, 46(5):1524–1535
257. Freeman JC, Perales JB, Woodland A, Bacchi CJ, Bowling TS, Cleghorn L, Gamon CG, Grimaldi R, Hauser D, Mercer LT, Nare B, Nguyen TM, Noe RA, Rewerts CE, Wring SA, Wyatt PG, Yarlett NR, Jacobs RT, Don R (2009) Hit-to-lead and lead optimization of novel small molecules for the treatment of human african trypanosomiasis. Amer Chem Soc 238th National Meeting, Washington DC, #MEDI 346
258. Brenk R, Schipani A, James D et al (2008) Lessons learnt from assembling screening libraries for drug discovery for neglected diseases. ChemMedChem 3(3):435–444

259. Martin D (2010) Identification and development of new chemical entities to treat visceral leishmaniasis: a bumpy road. Internat'l Conf Parasitol (ICOPA), Melbourne, Australia, #243
260. Rock FL, Mao W, Yaremchuk A, Tukalo M, Crepin T, Zhou H, Zhang YK, Hernandez V, Akama T, Baker SJ, Plattner JJ, Shapiro L, Martinis SA, Benkovic SJ, Cusack S, Alley MRK (2007) An antifungal agent inhibits an aminoacyl-tRNA synthetase by trapping tRNA in the editing site. Science 316:1759–1761
261. Akama T, Baker SJ, Zhang YK, Hernandez V, Zhou H, Sanders V, Freund Y, Kimura R, Maples KR, Plattner JJ (2009) Discovery and structure-activity study of a novel benzoxaborole anti-inflammatory agent (AN2728) for the potential topical treatment of psoriasis and atopic dermatitis. Bioorg Med Chem Lett 19:2129–2132
262. Ding D, Zhao Y, Meng Q, Xie D, Nare B, Chen D, Bacchi CJ, Yarlett N, Zhang Y-K, Hernandez V, Xia Y, Freund Y, Abdulla M, Ang K-H, Ratnam J, McKerrow JH, Jacobs RT, Zhou H, Plattner JJ (2010) Discovery of novel benzoxaborole-based potent antitrypanosomal agents. ACS Med Chem Lett 1(4):165–169
263. Xie D, Zhang YK, Hernandez V, Ding D, Xia A, Nare B, Jacobs RT, Freund Y, Don R, McKerrow J, Zhou H, Plattner J (2008) A new class of benzoxaborole-based potent anti-trypanosomal agents: probing the effect of different linkage groups on trypanosoma brucei growth inhibition. Am Soc Trop Med Hyg 57th Annual Meeting, New Orleans, #162
264. Jacobs RT, Nare B, Wring SA, Orr MD, Chen D, Sligar JM, Jenks MX, Noe RA, Bowling TS, Mercer LT, Rewerts C, Gaukel E, Owens J, Parham R, Randolph R, Beaudet B, Bacchi CJ, Yarlett N, Plattner JJ, Freund Y, Ding C, Akama T, Zhang YK, Brun R, Kaiser M, Scandale I, Don R (2011) SCYX-7158, an Orally-Active Benzoxaborole for the Treatment of Stage 2 Human African Trypanosomiasis. PLoS Negl Trop Dis. 5(6)e1151
265. Jacobs R, Ding C, Freund Y, Jarnagin K, Plattner J, Bacchi C, Yarlett N, Orr M, Nare B, Rewerts C, Chen D, Noe A, Sligar J, Jenks M, Wring S, Don R (2009) Lead optimization of novel boron-containing drug candidates for the treatment of human African trypanosomiasis. Am Soc Trop Med Hyg 58th Annual Meeting, Washington DC, #829
266. Nare B, Mercer L, Bowling T, Orr M, Chen D, Sligar J, Jenks M, Noe A, Wring S, Bacchi C, Yarlett N, Freund Y, Plattner J, Jarnagin K, Don R, Jacobs R (2009) In vitro pharmacodynamics and mechanism of action studies of oxaborole 6-carboxamides: a new class of compounds for the treatment of African trypanosomiasis. Am Soc Trop Med Hyg 58th Annual Meeting, Washington DC, #132
267. Wring S, Bacchi C, Beaudet B, Bowling T, Chen D, Don R, Freund Y, Gaukel E, Jarnagin K, Jenks M, Mercer L, Nare B, Noe A, Orr M, Parham R, Plattner J, Rewerts C, Sligar J, Yarlett N, Jacobs R (2009) SCYX-6759, an orally bioavailable oxaborole 6-carboxamide, achieves therapeutically relevant exposure in brain and CSF leading to 100% cures in a mouse model of CNS-stage human African trypanosomiasis. Am Soc Trop Med Hyg 58th Annual Meeting, Washington DC, #748
268. Nare B, Wring S, Bacchi C, Beaudet B, Bowling T, Brun R, Chen D, Ding C, Freund Y, Gaukel E, Hussain A, Jarnagin K, Jenks M, Kaiser M, Mercer L, Mejia E, Noe A, Orr M, Parham R, Plattner J, Randolph R, Rattendi D, Rewerts C, Sligar J, Yarlett N, Don R, Jacobs R (2010) Discovery of novel orally bioavailable oxaborole 6-carboxamides that demonstrate cure in a murine model of late-stage central nervous system African trypanosomiasis. Antimicrob Agents Chemother 54(10):4379–4388
269. Oxaboroles for HAT (2010) DNDi web site. http://www.dndi.org/portfolio/oxaborole.html. Accessed 16 July 2010
270. Nare B, Wring S, Bacchi C et al (2010) Discovery of novel orally bioavailable oxaborole 6-carboxamides that demonstrate cure in a murine model of late-stage central nervous system african trypanosomiasis. Antimicrob Agents Chemother 54(10):4379–4388
271. Freund Y (2010) AN4169, a novel boron-containing small molecule with in vivo efficacy against T. cruzi in mice. Internat'l Conf Parasitol (ICOPA), Melbourne, Australia, #373
272. Kerfoot M (2010) Three simple, rapid in vitro assays for determining stage specificity and the cidal or static activity of anti-Typanosoma cruzi compounds. Internat'l Conf Parasitol (ICOPA), Melbourne, Australia, #508

Top Med Chem 7: 243–276
DOI: 10.1007/7355_2011_16

Published online: 27 May 2011

Dengue Drug Discovery

Pei-Yong Shi, Zheng Yin, Shahul Nilar, and Thomas H. Keller

Abstract Dengue is the most common viral disease in the tropical and subtropical regions of the world. At present no vaccine or antiviral drugs are available to treat a dengue infection, which leaves supportive care in hospitals as the only available treatment option. Since dengue epidemics often put great strain on healthcare systems, there is an urgent need for novel strategies to combat this disease. The development of vaccines for dengue is complicated by the presence of four different serotypes, which has increased the importance of antiviral drug discovery. In the past 10 years, dengue has changed from a disease that was basically unknown to drug discovery scientists into a vibrant field of research for both biologists and chemists. In this chapter, we will review the different drug targets that have been identified for dengue and critically assess the progress that has been made in antiviral drug discovery. While these efforts have not yet resulted in a clinical development candidate, the progress has been impressive, considering the limited resources. Nevertheless, a concerted effort is required to identify drug candidates for the most promising targets, dengue NS3 protease and dengue NS5 polymerase.

Keywords Dengue, E-protein, Flavivirus, Glucosidase, NS3 protease, NS5 polymerase

---

P.-Y. Shi and S. Nilar
Novartis Institute for Tropical Diseases, 10 Biopolis Road, #05-01, Chromos, Singapore 138670, Singapore

Z. Yin
College of Pharmacy and The State Key Laboratory of Elemento-Organic Chemistry, Nankai University, Tanjin 300071, China

T.H. Keller (✉)
Experimental Therapeutics Center, 31 Biopolis Way, #03-01, Nanos, Singapore 138669, Singapore
e-mail: thkeller@etc.a-star.edu.sg

Contents

1 Introduction

The interest in drug discovery for dengue has increased remarkably in the last 10 years as can be seen in Fig. 1, which shows the results from a search for the term "drug for dengue" in SciFinder. A number of factors have contributed to the increasing awareness in the research community for this neglected viral disease. First of all, dengue continues to be a major health problem for countries in the tropical regions of the world, where every year between 50 and 100 million people are infected with the virus. Second, there has been a renewed interest from both funding agencies and parts of the pharmaceutical industry, which has provided increased funding for basic and applied research. Finally, the economic development in Southeast Asia and

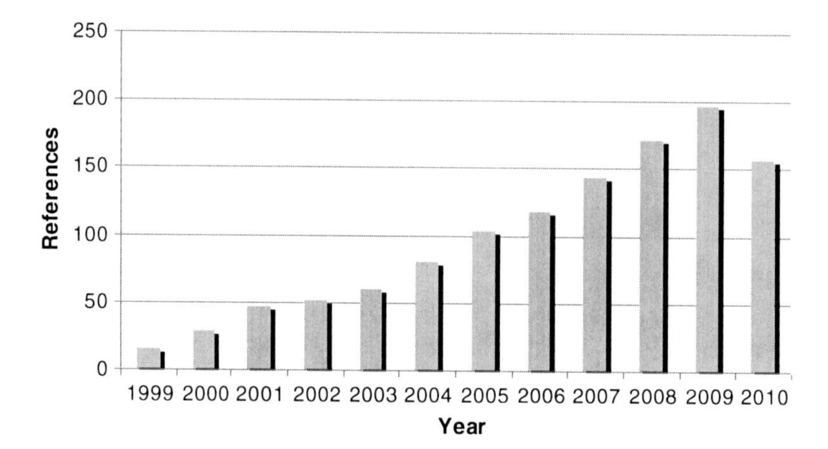

Fig. 1 Number of references in response to a Scifinder query for the term "drug for dengue" (all references with the exact term and dengue and drug in close context were used for the graph). The search was performed on 12 November 2010

South America, two regions severely affected by dengue, has spurred local interest in finding solutions to contain the impact of epidemics on the health care system.

A number of aspects of dengue antiviral research have recently been reviewed [1, 2]. In this chapter, we provide an overview of the most promising areas of dengue drug discovery, with a special emphasis on a realistic evaluation of the different approaches from the view of industrial medicinal chemistry. In each section, we will discuss the available chemical starting points and provide an overview of the issues that need to be addressed to make further progress toward a therapeutic for dengue infections.

2 Dengue Disease Complex

Dengue is a disease caused by the infection of the dengue virus (DENV). The four serotypes of DENV (DENV-1 to -4) are mainly transmitted by the mosquito *Aedes aegypti*. People bitten by an infected mosquito may develop a mild form of the disease, characterized by sudden onset of high fever, headaches, retro-orbital pain, and muscle/joint/bone pain. Some infected individuals progress into severe, life-threatening disease forms, dengue hemorrhagic fever (DHF) and dengue shock syndrome (DSS), characterized by plasma leakage, thrombocytopenia, bleeding, low pulse and tension, and hypovolemic shock [3].

Because no anti-DENV therapy is currently available, only supportive care can be offered to the patients. The challenge faced by doctors is the lack of methods to predict which patients with dengue fever (DF) will develop into DHF and DSS. As a result, two types of care are practiced in the clinical setting: (1) patients with DF are treated with supportive care, in the hope that complications may be prevented; (2) once patients develop DHF and DSS, careful fluid therapy is used to effectively resuscitate plasma volume. For more details on the management of dengue patients, the reader is referred to the extensive review by Wills [4]. Although the mechanism of vascular leakage remains poorly understood, dengue patient management has improved considerably over the years. The WHO-sponsored treatment guidelines, the increased availability of fluid treatment in hospitals, and the education of health care personnel have led to a significant reduction in mortality in Southeast Asia. Similar efforts are needed to improve the dengue patient care in the Americas and the Pacific regions.

DENV belongs to genus *Flavivirus* in the family *Flaviviridae*. The genus *Flavivirus* consists of more than 70 viruses, many of which are arthropod-borne human pathogens, including DENV, yellow fever virus (YFV), West Nile virus (WNV), Japanese encephalitis virus (JEV), and tick-borne encephalitis virus (TBEV) [3]. In the past century, DENV has increased in geographic distribution and disease severity, becoming the most common mosquito-transmitted viral pathogen in the tropical and subtropical regions of the world (Fig. 2). The WHO estimates that DENV is endemic in more than 100 countries and poses a public health threat to 2.5 billion people. The virus causes 50–100 million human infections annually, leading to 500,000 patients

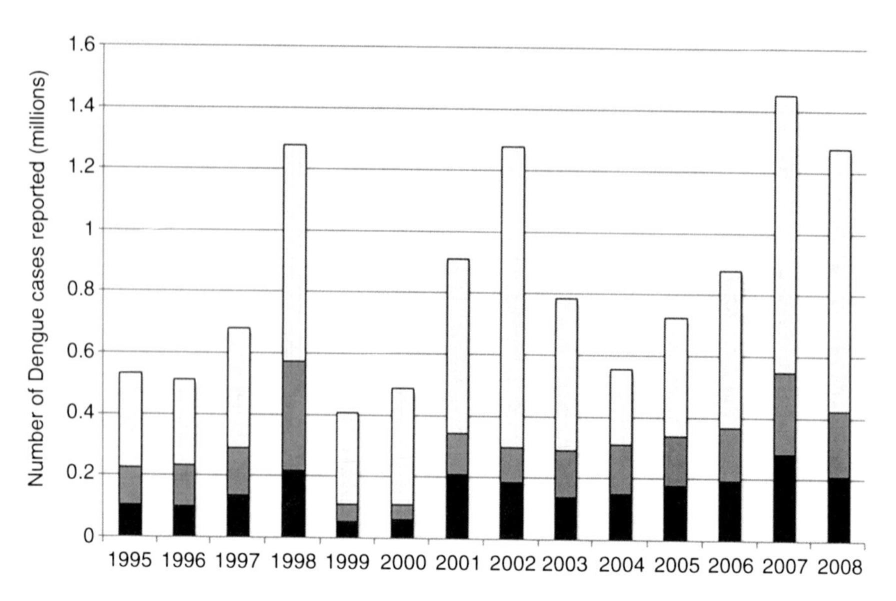

Fig. 2 Number of cases of dengue reported to the WHO; *black*, South East Asia; *dark gray*, Western Pacific; *light gray*, The Americas
Source: WHO

with DHF/DSS requiring hospitalization, of whom 2.5% die from disease complications [3]. No effective antiviral therapy is currently available for flavivirus infections. Although human vaccines are available for YFV, JEV, and TBEV, development of a DENV vaccine has been challenging. When a person is infected by one serotype of DENV, the antibodies that the body produces can only neutralize the homologous serotype of DENV. Unfortunately it has been shown that these antibodies can enhance the uptake of DENV of a heterologous serotype through an Fc-receptor mediated endocytosis, leading to an increased infection [5]. Therefore, a successful dengue vaccine needs to simultaneously immunize and induce a long-lasting protection against all four serotypes of DENV. An incompletely immunized individual may be sensitized to DHF/DSS. This challenge has underlined the importance of finding an antiviral therapy for DENV [2].

2.1 Medicinal Chemistry Perspective

When considering the chances of success in dengue drug discovery, a look at the available treatments for human immunodeficiency virus (HIV) is very helpful [6]. In the last 20 years, a large number of HIV drugs have been discovered, some targeting the entry of the virus into the host cell, while others inhibit the viral proteins that are necessary for replication. This success has been spawned by advances in the understanding of the virus life cycle and detailed information on the tertiary structure of the viral proteins. In addition, the attractiveness of the area for the pharmaceutical industry ensured that a large number of drug discovery researchers were focused on finding a treatment.

While DENV is a neglected disease and therefore the resources for drug discovery are limited, there is no reason why the strategies that lead to success in HIV cannot be applied for DENV. From a medicinal chemistry perspective, the availability of protein targets with structural information is very important, since it allows drug discovery teams to design inhibitors with the desired physicochemical, pharmacokinetic, and toxicological properties. The target product profile [7] for a dengue drug is rather demanding, since one of the primary patient populations are children under the age of 15. As a consequence, dengue drugs will need to be carefully optimized to avoid side effects that could limit the use of the drug.

3 Dengue Genome and Replication

DENV is a small enveloped virus of about 50 nm in diameter (Fig. 3a). The virion contains an outer shell that is made of a bilipid membrane together with the viral envelope (E) and membrane (prM/M) proteins. Inside the envelope shell is the nucleocapsid that is formed by the viral capsid (C) protein and genomic RNA. Like other flaviviruses, the genome of DENV is a single positive sense RNA of approximately 11 kb in length (Fig. 3b). The genomic RNA contains a single open reading

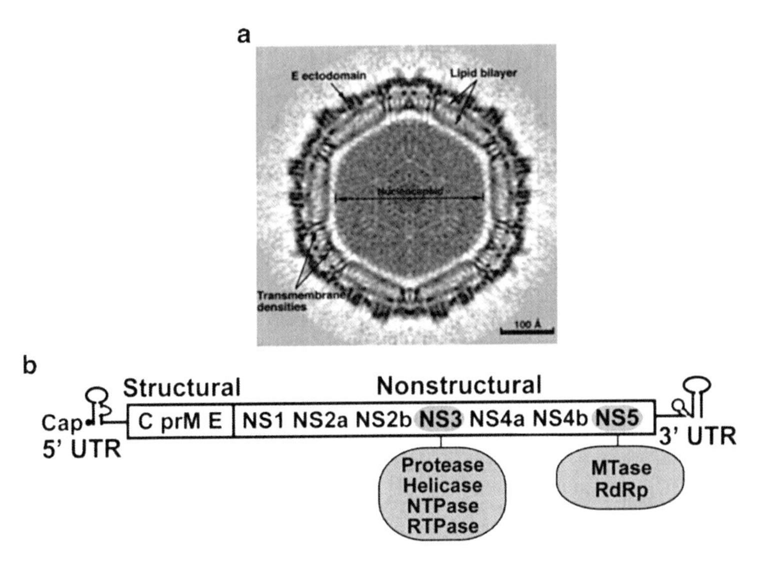

Fig. 3 DENV virion and genome. (**a**) A cross-section of DENV cryo-electronic microscopy image. Viral components are labeled. The image is reproduced from [8]. (**b**) DENV genome structure. The genomic RNA contains a 5′ UTR, a single open reading frame, and a 3′ UTR. The open reading frame encodes three structural and seven nonstructural proteins. The enzymatic activities of NS3 and NS5 are indicated

frame that is flanked with a 5′and a 3′ untranslated region (UTR). The single open reading frame encodes a long polyprotein that is processed into ten mature proteins by a combination of host and viral proteases. The N-terminal region of the open reading frame encodes three structural proteins (C, prM, and E) and seven non-structural proteins (NS1, NS2A, NS2B, NS3, NS4A, NS4B, and NS5). The structural proteins form the viral particle. The nonstructural proteins participate in the replication of the RNA genome, virion assembly [9, 10], and evasion of innate immune response [11, 12].

Among the ten flaviviral proteins, only NS3 and NS5 have known enzymatic activities. The N-terminal domain of NS3, together with NS2B, contains a serine protease activity; the C-terminal domain functions as an RNA helicase, an RNA triphosphatase, and an NTPase [13–15]. The N-terminal domain of NS5 contains a methyltransferase (MTase) activity; the C-terminal domain serves as an RNA-dependent RNA polymerase (RdRp) [16–18]. Other nonstructural proteins are required for RNA replication, among which NS2A, NS2B, NS4A, and NS4B are transmembrane proteins that form the scaffold for the viral replication complex [19–21]. Besides viral proteins, host proteins also participate in virus replication [22, 23]. Although the exact topology and components of the replication complex remain to be determined, the nonstructural proteins with or without known enzymatic activity are valid antiviral targets. Over the past decade, crystal or NMR structures have been reported for the DENV C [24], prM [25], E [26], NS2B/NS3protease [27], helicase [28], MTase [17], and RdRp [29].

4 Dengue Drug Targets

4.1 Envelope Protein

The infectious entry of DENV into a cell is still not well understood, but single particle tracking of living cells has provided valuable insights [30]. The virus first binds to cellular receptors and then moves along the cell surface toward a clathrin-coated pit. It is then internalized by clathrin-mediated endocytosis followed by membrane fusion between the virion and the endosome [30]. The dengue E-protein mediates both the initial binding of the virus to receptors on the cell surface and the fusion process that eventually leads to the release of the genetic material into the cell [2]. Both of these events are attractive targets for antiviral therapy, as has been demonstrated in HIV drug discovery [6].

4.1.1 Inhibition of Cellular Attachment

Unfortunately the cellular attachment process of DENV is rather complex, and a number of different receptors appear to be involved depending on the cell type.

For example, the C-type lectins DC-SIGN and L-SIGN [31, 109], the high-affinity laminin receptor [32], the mannose receptor [33], GRP78 [34], and glycosaminoglycan heparan sulfate [35] have all been shown to mediate the binding of the virus to the cell. Accordingly, only a few highly charged molecules have been shown to block the attachment of DENV to cell membranes. While polyanionic compounds such as suramine, heparane sulfate [35], sulfated galactomannans [36], and carrageenans (sulfated polysaccharides) [37] have antiviral activity, they are poor starting points for an oral drug, suggesting that cell attachment is unlikely to be a fruitful avenue for dengue drug discovery.

4.1.2 Inhibition of the Fusion Process

Once the dengue virion is internalized to an endosome, the acidic pH leads to several important molecular events. First the E-protein, which exists as a homodimer in the mature virus, dissociates into monomers (Fig. 4a) and then changes conformation. The conformational change facilitates the formation of E-protein trimers (Fig. 4b), which then induce the fusion of the viral and cell membranes and eventually the injection of the viral capsid into the cytoplasm of the cell. Since this change in conformation of the E-protein seems to be essential for the entry of the virus into the host cell, it has been suggested that inhibition of this structural transition would be an attractive antiviral target [38].

One avenue to interfere in the fusion process would be to block the final stages of the conformational change with peptides that bind to domains II of the E-protein (Fig. 4a) [38]. Michael and coworkers [39] used this strategy which has successfully been used for the peptidic HIV entry inhibitor Enfuvirtide [6]. A number of peptides were synthesized, and the best results were obtained with a 33-mer which matches a sequence of dengue E-protein domain II. The peptide inhibited DENV infectivity with an IC_{50} in the range of 10 μM and was also active against WNV.

The second way to block the conformational rearrangement of the E-protein would be to target a ligand-binding pocket in the hinge region [26]. Such a pocket was observed when the dengue E-protein was crystallized in the presence of a detergent, and one of the *n*-octyl-β-D-glucoside (β-OG) detergent molecules was found to co-crystallize with the E-protein (Fig. 4c). Since it has been suggested that inhibitors targeting this site could have antiviral activity, a number of research groups have used computational approaches to identify lead molecules. Unfortunately there is no high-throughput screening (HTS) assay, which can measure the direct, functional effect of such a hinge pocket binder, so that computational screening is the method of choice for the identification of such inhibitors.

A research group from Purdue University used GOLD to dock the NCI library into the β-OG pocket of dengue E-protein, leading to the identification of 23 compounds that were synthesized and tested in a yellow fever (YFV) replication assay [40], where the reduction in viral growth was determined by measuring the luciferase activity in YFV-infected BHK cells. Two compounds were identified as starting points for an optimization program.

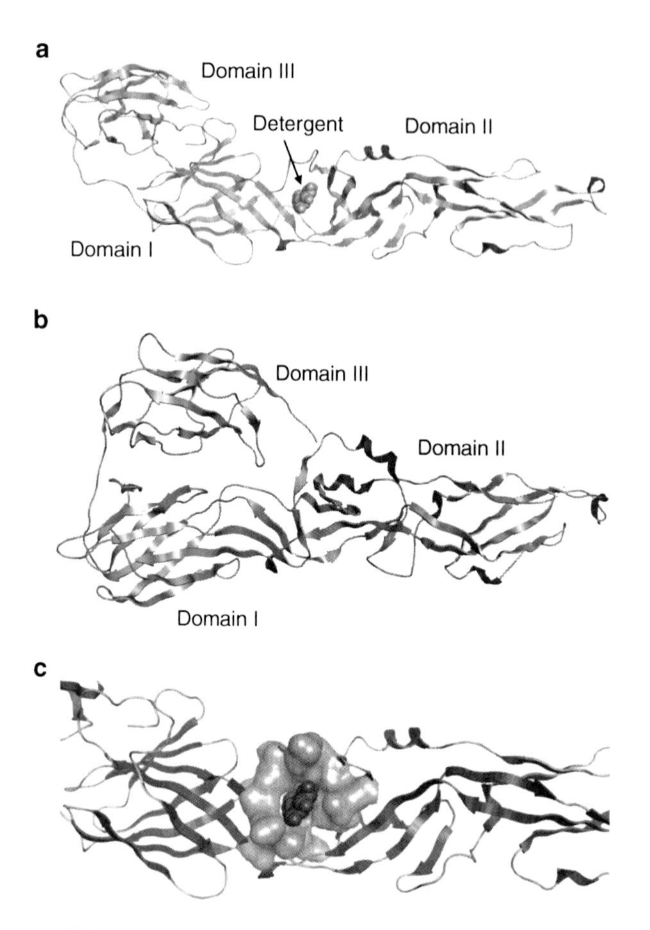

Fig. 4 Structure of dengue E-protein [38]. (**a**) Pre-fusion conformation; (**b**) post-fusion conformation; (**c**) close-up of the detergent-binding site

O
N
H
S
N
N
N
Cl
Cl

1
YFV EC_{50} = 0.9 μM
CC_{50} = 153 μM

O
MeO
S
Br
N
Br
Cl

2
YFV EC_{50} = 5 μM
CC_{50} = 186 μM

H_2N H N N NH O N H N H N N H NH NH_2 NH

3

Analyses of the ligand–protein interactions lead to the design of compounds with a more favorable fit into the hydrophobic pocket. In addition, problematic functional groups were removed from the molecules to reduce cellular cytotoxicity. The most promising derivatives from this optimization effort are **1** and **2,** both having improved potency and an improved therapeutic window compared to the original hits.

Zhou [41] identified ligand candidates from 142,000 virtual molecules (a subset of the NCI library) using a computational HTS protocol targeting again the β-OG pocket of the dengue E-protein. The resulting compounds were ranked by interaction energy (solvation contributions were included), subjected to thermal relaxation and visually inspected for quality of fit. Following this docking procedure, the hits were tested in the same YFV cellular assay as the one used by Li and coworkers [40]. The most promising inhibitor **3** was examined in an YFV replicon assay, since a replicon lacks structural proteins like the E-protein. Unfortunately **3** exhibited the same level of antiviral activity in the replicon, suggesting at the very least that this amindine-containing compound (**3**) interacts with multiple targets in the YFV cellular assay (Table 1). Nevertheless, the authors demonstrated that **3** binds to the whole dengue virus using saturation transfer difference NMR. Compound **3** also binds to purified dengue E-protein, and competition experiments with β-OG provided an estimate of a dissociation constant of 1 μM. Overall, the evidence suggests that the virtual screening was successful and identified a ligand for the β-OG pocket of DENV E-protein. Unfortunately the replicon result is equally clear in showing that **3** interacts with at least one other target besides the E-protein. A comparison of **3** with the compounds shown in Table 5 suggests that another target for the compound may very well be NS-3 protease.

Table 1 Biological activity for compound **3**[a]

Compound	IC_{50} (μM) Inhibition of viral growth	IC_{50} (μM) Inhibition of replicon	CC_{50} (μM) Host cell cytotoxicity
3	13	17	371

[a]Data from [41]

Young and coworkers [42] have used GOLD to dock a 3D database of 135,000 compounds from the Maybridge chemical database into the crystal structure of the DENV E-protein. After evaluating the hits from the virtual screening for synthetic tractability, predicted solubility, and drug-like properties, five compounds were purchased and evaluated in DENV, WNV, YFV, and respiratory scyncytial virus (RSV) proliferation assays (plaque reduction). Compound **4** inhibited all flavivirus assays at low micromolar concentrations without any sign of cytotoxicity (Table 2).

Table 2 Antiviral activity of compound **4**[a]

Compound	IC_{50} (μM)				CC_{50} (μM)
	DENV	WNV	YFV	RSV	Cytotoxicity
4	1.2	3.8	1.6	>100	>100

[a]Data from [42]

As would be expected, the compound was not active on RSV (Family: *Paramyxoviridae*). Since the virtual screening was based on the hypothesis of Modis [38] that blocking the β-OG pocket (Fig. 4c) would inhibit the conformational rearrangement of the E-protein, **4** was tested in a cell-based fusion assay. When infected cells were exposed to low pH media in the absence or presence of compound **4** at 5 μM, an approximately 60% reduction in fusion was observed. This result provides the first conclusive evidence that inhibition of E-protein-mediated fusion may indeed be a viable drug discovery approach.

Table 3 Antiviral activity of compound **6**[a]

Compound	EC_{50} (μM)						
	DENV1	DENV2	DEN3	DEN4	WNV	YFV	JEV
6	0.11	0.07	0.45	0.33	0.56	0.47	1.42

[a]Data from [43]

Virtual screening of the Novartis corporate library led to the identification of a compound class with interesting activity in a DENV cellular assay (detection of E-protein production in BHK21 cells) [43]. Optimization of the chemical starting point **5** produced quinazoline **6** with excellent activity on all four serotypes of DENV, YFV, WNV, and JEV (Table 3). A number of experiments were performed to determine the mode of action of **6**. Time addition studies suggested that the compound works in the early stages of virus infection, and binding of the compound with E-protein was demonstrated using a biotinylated derivative. The presented data [43] strongly suggest that **6** interacts with the E-protein and interferes with the virus entry process. Nevertheless, despite its overall attractive biological profile, the compound has poor physical properties and could therefore not be further developed as an antiviral. All attempts to improve this compound during lead optimization failed.

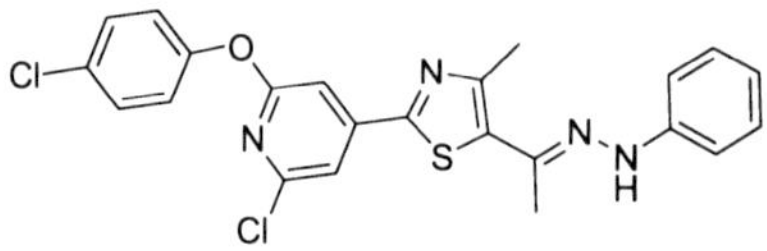

Cl O N Cl N S N N H 4

H N NH N N S O 5

N H N N N S Cl 6

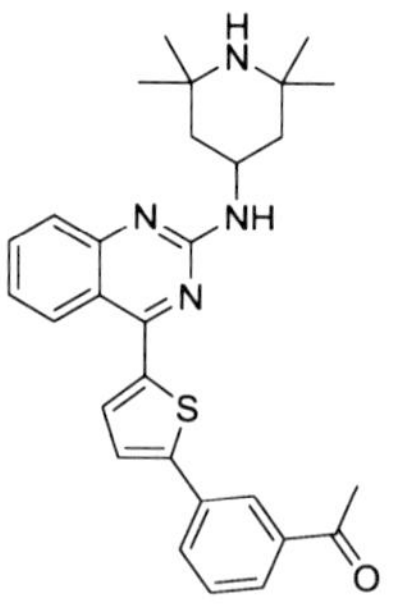

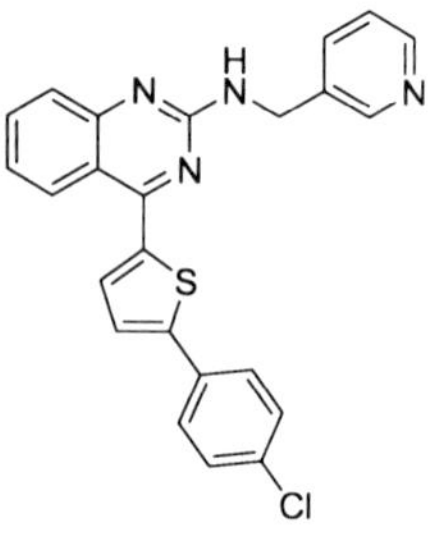

4.1.3 Medicinal Chemistry Perspective

The available data suggest that inhibiting the entry of a DENV virion into a cell is more challenging than in the case of HIV. The cell attachment seems to be mediated by a plethora of cellular receptors through carbohydrate–protein interactions which are very likely not druggable. In contrast, the binding pocket in the hinge region of the DENV E-protein seems to be a viable target for drug discovery. A few compounds have been identified that inhibit the fusion process of DENV; however, their potency and physicochemical properties will need to be improved. A lead optimization of, for example, compound **4** would be facilitated by a direct binding assay and biophysical techniques like NMR, since it is unlikely that both the potency and the physical properties can be optimized on a cellular assay. All of these compounds will have to face the fact that dengue is a self-limiting disease, and therefore an entry inhibitor will only be effective if patients can be dosed in the very early stages of the disease.

4.2 NS3 Helicase

The C-terminal domain of the NS3 protein contains an RNA helicase/NTPase that belongs to superfamily 2. Both the helicase domain [44] and the full NS3 protein [28] have been crystallized; however, the exact role of NS3 in the replication of DENV is still not fully established [45]. The two catalytic domains (protease and helicase) are connected by a flexible linker, which allows the protein to adopt a number of different conformations. This domain movement has been inferred from X-ray crystal structural studies on DENV [46], and similar results have been reported for the Murray Valley encephalitis virus [45]. Based on site-directed mutagenesis studies and the structural data, it has been proposed that the flexibility may be important for the modulation of RNA processing by the helicase/NTPase domain [46].

4.2.1 Medicinal Chemistry Perspective

While it has been shown that a functional helicase domain is important for the replication of the DENV, there are few indications from the available data that small molecule inhibitors with acceptable physicochemical and pharmacokinetic properties can be developed. The most obvious way to produce a functional inhibitor for DENV helicase is the NTP domain, especially since the energy gained from the hydrolysis of ATP drives the RNA processing [28]. Unfortunately the ATP pocket is solvent exposed and the nucleoside interacts only weakly with the protein, suggesting that the identification of potent, drug-like inhibitors will be very challenging (Fig. 5). Other potential ways to block the RNA processing, like inhibiting the conformational flexibility of the protein or blocking RNA binding, look daunting from a medicinal chemistry perspective.

HCV helicase has been a very challenging target for drug discovery. While a few weak inhibitors have been reported [2], so far none of these molecules have entered the HCV clinical development pipeline.

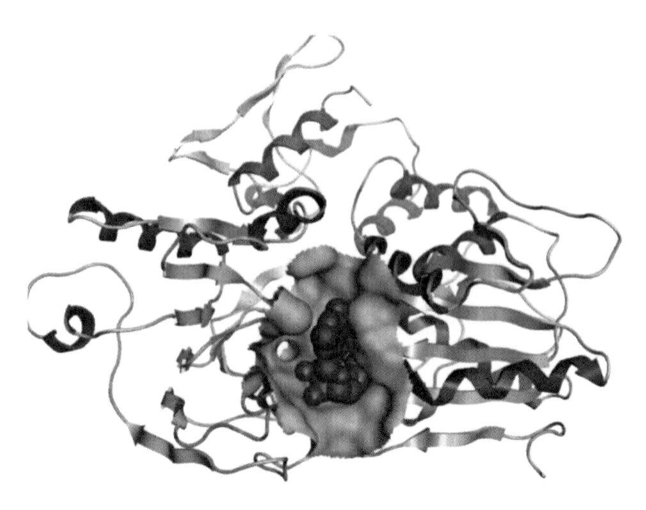

Fig. 5 Dengue helicase NTP-binding pocket with adenylyl imidodiphosphate, a non-hydrolyzable form of ATP [28]

4.3 *NS3 Protease*

Virus-encoded proteases have been shown to be essential for the replication of many viruses and have a successful history as primary molecular targets for antiviral drug discovery [47]. Up to now, nine protease inhibitors are in clinical use for HIV [48] and two in late Phase III trials for HCV [49]. The DENV genome encodes a trypsin-like serine protease in the N-terminal 180 residues of the NS3 protein [50], which requires association with NS2B for activity [51]. Similar to other viruses, dengue protease is responsible for posttranslational processing of the polyprotein, which is essential for viral replication and virion assembly. Therefore, it serves as an attractive target for drug discovery.

4.3.1 Inhibitors of Dengue Protease

Dengue NS2b/NS3 protease catalyzes the proteolytic cleavage of the polyprotein between NS2A-NS2B, NS2B-NS3, NS3-NS4A, and NS4B-NS5 [52]. The viral protease prefers two basic amino acid residues at P1 and P2 position (Arg or Lys) and a short side-chain amino acid (Gly, Ala, or Ser) at the P1′ position of the cleavage sites [53, 54].

Based on the cleavage sites, a hexapeptide (Ac-RTSKKR-$CONH_2$) representing non-prime site amino acids was shown to be a competitive inhibitor of dengue protease with a K_i of 12 μM, while prime site analogs appeared to be inactive at concentrations up to 1 mM [55]. These findings suggest that, as is common for serine proteases, the non-prime site is mostly responsible for high-affinity binding of the inhibitors.

Table 4 Importance of C-terminal functional group for inhibitor activity[a]

	Bz-Nle-Lys-Arg-Arg-R					
	R=OH	R=NH_2	R=H	R=CF_3	R=$CONH_2$	R=$B(OH)_2$
K_i (μM)	>500	128	5.8	0.85	ns[b]	0.04

[a]Inhibition of dengue NS3 protease [57]
[b]Not stable

Substrate screening of a tetrapeptide library resulted in the identification of Bz-Nle-Lys-Arg-Arg-ACMC as optimal substrate, showing >150-fold more sensitivity than other published substrates [56]. Modification of this tetrapeptide sequence with various electrophilic functional groups such as aldehydes, trifluoromethyl ketones, and boronic acids led to a series of potent non-prime site substrate-based inhibitors (Table 4) [57]. A structure–activity relationship (SAR) study of peptide aldehydes revealed the relative importance of the various side chains in the respective binding pockets and suggested that a P2 arginine residue is more important for protease interaction than P1 arginine [58].

Standard serine protease inhibitors are rarely active against DENV protease with the exception of aprotinin which is a nanomolar inhibitor. Aprotinin is a monomeric (single-chain) globular polypeptide derived from bovine lung tissue. It inhibits the protease activity by engulfing the protein and blocking substrate from accessing of active site [59], and is therefore not interesting from the standpoint of drug discovery.

A number of research groups have attempted to identify non-peptidic inhibitors for DENV NS3 protease. Using both computational methods and biochemical screening, several active inhibitors have been identified (Table 5). A series of strongly positively charged inhibitors containing guanine functionalities were designed to mimic the signature bifurcated arginine side chain in the P1 region. Weak activities against dengue protease were observed with both mono- and bis-guanidine compounds **7** and **8** [60]. Similar, weak inhibitors have been identified by screening of the Novartis corporate archive (e.g., **9**) [61]. HTS resulted in the identification of **10**, which inhibits both DENV and WNV proteases selectively as compared to the host serine proteases trypsin or Factor Xa [62]. Natural products (e.g., **11**) isolated from *Boesenbergia rotunda* also have been reported as active inhibitors for DENV protease [63]. Virtual screening using a EUDOC docking program successfully identified protease inhibitors, which can inhibit viral replication in cell culture experiments (compounds **12** and **13**) [64].

4.3.2 Structural Studies of NS3 Protease

The 3D structures of proteins, especially those complexed with inhibitors, provide insight at the molecular level and significantly benefit the discovery of inhibitors. Since the amino acid homology of the protease domain is >50% among various members of the flavivirus family, the structural information for all flavivirus NS3 proteases is relevant for the design of DENV protease inhibitors. So far, only the

Table 5 Non-peptidic inhibitors of DENV NS3 protease

	Compound	K_i (μM)	Reference
7	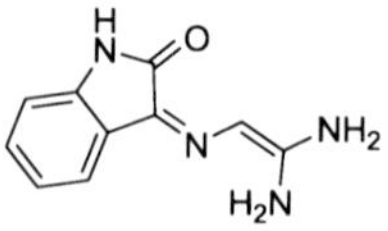	23	[60]
8	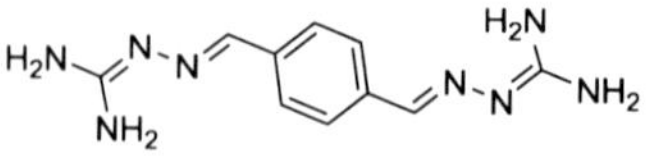	44	[60]
9	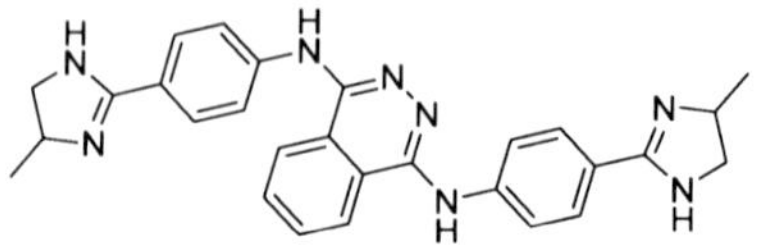	36	[61]
10	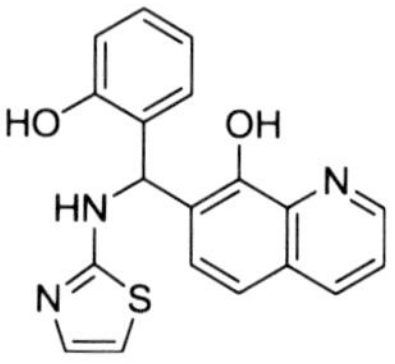	30	[62]
11	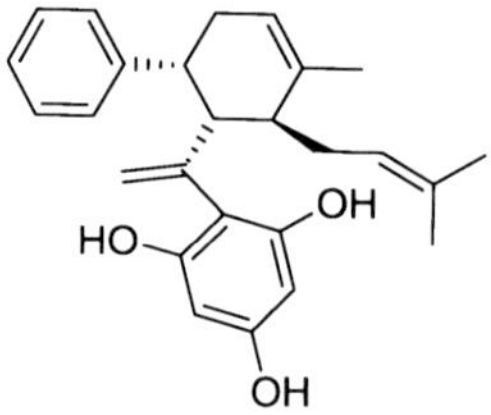	21	[63]
12	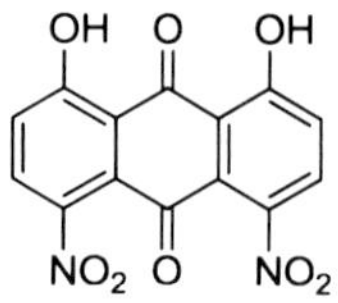	NA[a]	[64]
13	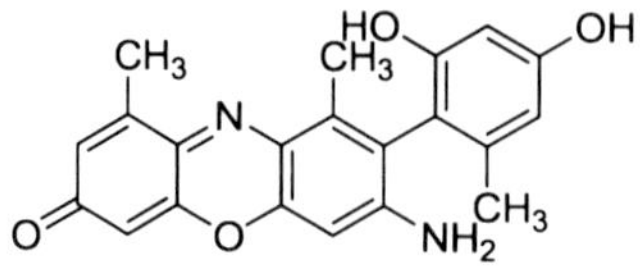	NA[a]	[64]

[a]*NA* not available; compounds measured in a cellular replication assay

structures of DENV and WNV protease domains have been reported (Table 6), while the structure of the complete NS3 protein allows intriguing insight into the interaction between the DENV protease and the helicase domains [28, 46]. Unfortunately only WNV protease so far has yielded ligand–inhibitor complexes that are truly relevant for drug design (Table 6).

The proteolytic activity of the NS3 protein is strongly dependent on NS2B as a cofactor. Structural studies have shown that the N-terminus of NS2B is an important structural component of the active site, where the N-terminal residues 49–66 of NS2B form a β-strand which interacts with the N-terminus of the protease β-barrel covering hydrophobic residues and providing stabilization for the protease [27] (Fig. 6).

The availability of structures with a bound inhibitor is critical for structure-based drug design. Unfortunately, such data are currently not available for dengue

Table 6 X-ray crystal structures available in RCSB[a] for flavivirus viral proteases

Entry	Flavivirus	PDB	Resolution (Å)	Comment	Reference
1	DENV2	2FOM	1.50	NS2B/NS3pro	[27]
2[b]	WNV	2FP7	1.68	NS2B/NS3pro complex with Bz-NKRR-H	[27]
3	WNV	2GGV	1.80	NS2B/NS3pro	[59]
4[b]	WNV	2IJO	2.30	NS2B/NS3pro complex with aprotinin/ BPTI	[59]
5	DENV4	2VBC	3.15	NS2B/NS3FL	[28]
6[b]	WNV	3E90	2.45	NS2B/NS3pro complex with Naph-KKR-H	[65]
7	DENV1	3L6P	2.20	NS2B/NS3pro	[66]
8	DENV1	3LKW	2.00	NS2B/NS3pro (active site mutant)	[66]
9	DENV4	2WHX	2.80	NS2B/NS3FL	[46]
10	DENV1	2WZQ	2.20	NS2B/NS3FL	[46]

[a]Protein data bank (PDB)
[b]Structures with bound inhibitors

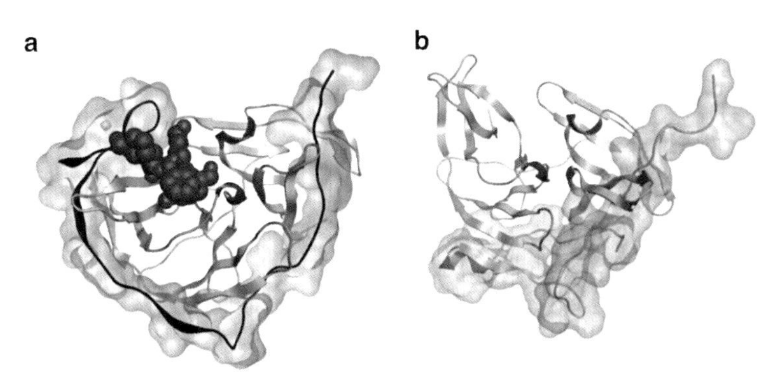

Fig. 6 Schematic representation of the NS2B-NS3 protease; (**a**) complex of WNV NS2B/NS3pro with the Bz-NKRR-H (cpk) [27]; (**b**) DENV2 NS2B/NS3pro; NS2B is shown as solvent accessible surface in *light gray*, while NS3pro is shown in a *ribbon diagram*

NS2B/NS3. In structures of the WNV NS2B/NS3 protease complexed with peptide analogs, the conformation of C-terminal part of NS2B cofactor is significantly different between inhibitor-free and inhibitor-bound proteases. The C-terminal residues 66–88 of NS2B are rearranged to interact with the inhibitor, which results in the formation of a belt around the second chymotrypsin-like β-barrel. The central hydrophilic portion of NS2B rearranges to form an important component of the protease catalytic site, and the C-terminus of NS2B contributes crucially the formation of S2 and S3 pocket (Fig. 6).

The observed SAR for tetrapeptide aldehydes has been rationalized with modeling studies using the crystal structure of the West Nile NS3 protease. It is evident that the key peptide recognition sites are the S1 and S2 pockets. The study suggests that the crystal structure is relevant for computer-based drug design of West Nile protease [67].

4.3.3 Medicinal Chemistry Perspective

Considerable effort has been made to identify inhibitors of the dengue protease. However, all of the compounds that have been reported in the literature are either very weak or peptidic in nature and therefore do not have the profile required for an oral drug candidate.

The druggability of a target is an extremely important consideration in any drug discovery project. While serine proteases are generally considered to be suitable for drug discovery, the shallow, solvent-exposed active site of flavivirus proteases constitutes an exceptional challenge [68]. This makes it unlikely that small molecular weight compounds with sufficient affinity will be identified, as illustrated by the compounds in Table 5, which all have low affinity/potency with low ligand efficiency and therefore look unattractive from a drug discovery standpoint.

Up to now, peptide-based inhibitors are still at the very early stage, serving mostly as tools to study the topology of the DENV protease active site. Nevertheless, peptidic inhibitors are interesting starting points, since warheads can be used to increase the affinity of peptidomimetic inhibitors [49]. The preferred substrate sequence of flavivirus proteases is strongly biased toward amino acids with positively charged side chains such as lysine and arginine at P1 and P2. As far as we know, two or three positively charged amino acids are required in all known peptidic inhibitors to achieve satisfactory enzyme inhibition. Since the target product profile for a dengue drug requires an orally active drug, poly-cationic inhibitors are not acceptable. This means that in addition to the challenge of changing a peptidic inhibitor into an orally bioavailable peptidomimetic, most of the charged side chain need to be replaced with neutral moieties in a development compound. As a consequence, dengue NS3 protease can be considered a very challenging target that will need a large chemistry effort to achieve success.

Fortunately the experience in HCV suggests that despite the challenges such an undertaking can be successful [49]. The HCV protease inhibitors currently in clinical trials are primarily derived from the natural peptide substrates, which were considered extremely difficult starting points from a medicinal chemistry perspective. Similar to dengue protease, the active site is solvent exposed and the major interactions in the binding sites featured charged groups and a thiol.

4.4 NS5 Polymerase

DENV contains a single-strand positive sense RNA, which is injected into the cell during infection. The RNA-dependent RNA polymerase (RdRp) domain of NS5, whose enzymatic activity is essential for virus replication, uses the plus-strand RNA as a template and synthesizes complementary minus-strand RNA, which again serves as a template for the production of new plus-strand genomic RNA [69]. The understanding of the mechanistic details of the enzyme has been facilitated by the X-ray crystal structure of the dengue NS5 RdRp domain [29]. While this structure does not contain any nucleic acid, a combination of mechanistic studies [70] and a comparison with the HCV RdRp [69] have shed light on the template-dependent RNA synthesis.

Figure 7 shows the dengue RdRp domain with the classic right-hand shape common to many polymerases. It is thought that the single-stranded RNA enters through the tunnel between the finger and the palm domains resulting in a binary complex, which binds nucleoside triphosphate (NTP). Then the RNA synthesis is initiated followed by elongation, termination, and dissociation of the nucleic acid from the enzyme. The proposed mechanism for the enzyme catalysis is shown in Fig. 7b, suggesting that Mg^{2+} ions play a central role in the activation of the 3′-OH group as well as the stabilization of the reaction complex during the transition state [71].

Flavivirus RdRps are thought to be excellent targets for drug discovery, since their enzymatic activity is essential for viral replication and the host cells do not contain comparable polymerases [70]. Furthermore, polymerase inhibitors are the

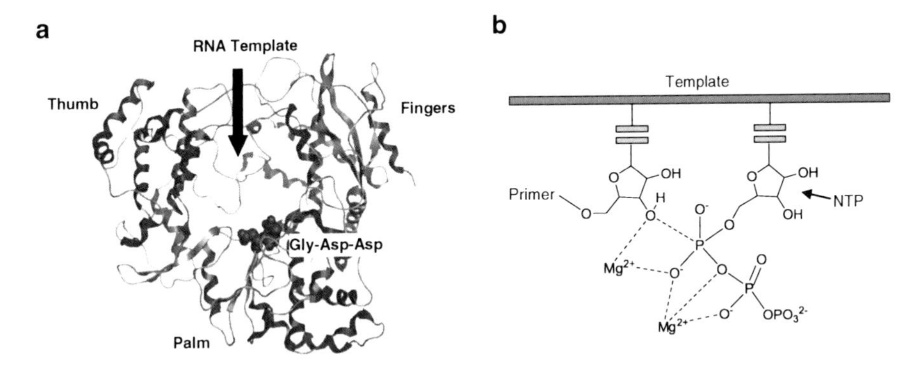

Fig. 7 (**a**) Structure of dengue RdRp domain with Gly-Asp-Asp (cpk) denoting the active site [29]. (**b**) Mechanism for the polymerase reaction [71]

mainstay of antiviral therapy for HIV, hepatitis B, and the herpes virus, and a large effort is currently underway to develop such drug candidates for HCV [72]. In principle, there are two ways to inhibit the RNA synthesis of polymerases. One can either interfere directly with the binding of the substrate at the active site (Fig. 7), or perturb the three dimensional structure of the protein so that one of the essential steps in the RNA synthesis is blocked.

4.4.1 Chain Termination

One of the most straightforward strategies to interfere with the function of the polymerase is to provide it with NTP molecules that either compete with the binding of the natural nucleotides or terminate the synthesis of the nucleic acid when they are incorporated into the growing RNA strand. Since such NTPs are highly charged and therefore not suitable as drugs, medicinal chemists have taken advantage of the cellular phosphorylation machinery to transform nucleosides into the molecules (triphosphates) that are required for the inhibition of the polymerases [73].

The first nucleoside with anti-dengue activity was reported by researchers at Merck [74]. Compound **14a** had micromolar activity on African green monkey kidney cells infected with DENV2. Unfortunately this compound is a substrate for cellular adenosine deaminase, and therefore exhibits poor oral pharmacokinetics [75]. Nevertheless, further studies clearly demonstrated that **14a** is a chain-terminating viral polymerase inhibitor and therefore a possible chemical starting point for drug discovery. A limited SAR around this compound has recently been reported [76], suggesting that changes in the structure of **14a** are not well tolerated (Table 7).

The absence of clear SAR is quite common in the area of nucleoside drug discovery. Since the nucleosides are transformed to the active triphosphate by cellular kinases, the EC_{50} is a composite of the molecular recognition events of the kinases and the DENV RdRp. The picture is further complicated by the presence

Table 7 Nucleoside analogs as dengue inhibitors

Compound	R	EC_{50} (μM)[a]	CC_{50} (μM)[b]
14a[c]	Me	4	18
14a[d]	Me	1.12	>50
14b[d]	Et	>50	>50
14c[d]	$CH=CH_2$	>50	>50
14d[d]	C≡CH	1.41	40
14e[d]	C≡CMe	>50	>50
15[c]	na[e]	13.6	>50

[a]Dengue 2
[b]Cytotoxicity
[c]Migliaccio et al. [74]
[d]Chen et al. [76]
[e]Not applicable

of nucleoside transporters in certain cell lines and the general poor passive diffusion of the polar nucleosides [77].

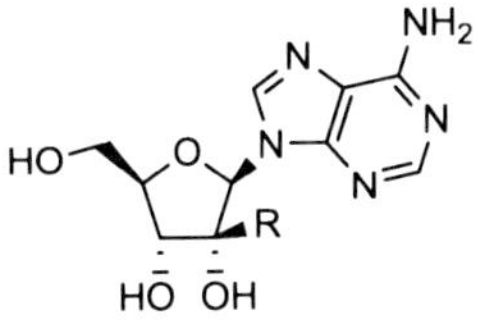

14a-e 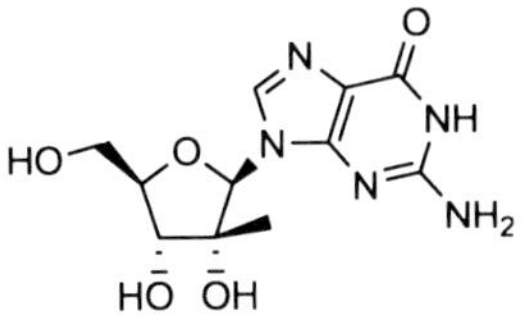 **15**

Faced with all these challenges, the medicinal chemists at Merck have concentrated on purine modifications to come up with derivatives that exhibit improved stability in vivo [75, 78]. Simply replacing the nitrogen atom at the 9-position of the heterocycle with carbon produced a nucleoside (**16**) with potent activity against HCV both in vitro [78] and in vivo [79].

While the preclinical work on **16** is clearly aimed at HCV, and there is very little information available about its activity in Dengue, the Novartis Institute for Tropical Diseases (NITD) has comprehensively profiled a related nucleoside **17** [76, 80]. 9-Deaza-2′C-ethynyl-adenosine (**17**) potently inhibits DENV, other flaviviruses, and HCV (Table 8), but was inactive when tested on vesicular stomatitis virus (Family: *Rhabdoviridae*) and Western equine encephalitis virus (Family: *Togaviridae*).

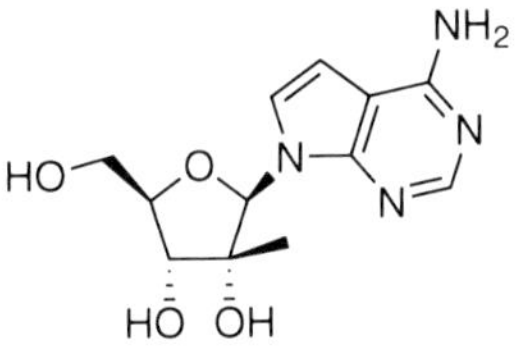

16 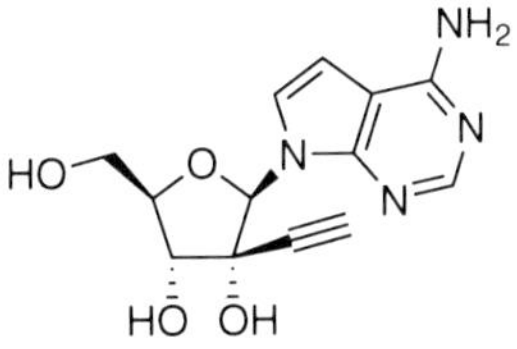 **17**

Table 8 Spectrum of antiviral activity of compound **17**[a]

Virus	Cell type	Assay[b]	EC_{50} (μM)
DENV1	BHK	CFI	0.16
DENV2	BHK	CFI	0.65
DENV3	BHK	CFI	0.46
DENV4	BHK	CFI	0.22
DENV2	BHK	Replicon	0.23
HCV	Huh7	Replicon	0.11
DENV2	Huh7	CPE	1.25
JEV	BHK	CPE	1.25
WNV	BHK	CPE	3.75
YFV	A549	CPE	0.85

[a]See Yin et al. [80, 81]
[b]*CFI* cell-based flavivirus immunodetection assay; *CPE* see [2]

It was shown that **17** is a substrate for adenosine kinase, and that the triphosphate is a potent inhibitor of DENV RdRp. Furthermore, in vivo pharmacokinetic studies with ^{14}C-labeled **17** demonstrated that the triphosphate is formed in vivo. Taken together, these results strongly suggest that **17** is transformed in vivo to its triphosphate which functions as a chain terminator in the DENV RNA synthesis [76]. Both single and repeated dosing of **17** reduced the viremia in dengue-infected AG129 mice, and the same mouse strain was protected from death when infected with the mouse-adapted dengue strain with a single dose of 75 mg/kg [76, 80].

Overall, the preclinical profile of **17** was very encouraging; however, preclinical development was halted when a "no observed adverse effect level" (NOAEL) could not be achieved after daily administration of 50 mg/kg for 2 week by the oral route [80].

A search for RdRp inhibitors of HCV at Roche produced 4′-azidocytidine (**18**) as a promising lead compound [82]. Because of its polarity, **18** is not an acceptable drug candidate. To circumvent this problem, the physical properties were modified by the attachment of easily cleavable ester groups. The resulting prodrug, Balapiravir (**19**), was successfully tested in combination with pegylated α-interferon in treatment-naïve patients with HCV infections (Phase IIa) [83]. Despite the encouraging results in viral load reduction, the development of **19** was discontinued for HCV in late 2008 due to unspecified safety findings [72, 84]. Since there is considerable overlap in the RdRp activity between DENV and HCV, it would not be surprising if **19** is an inhibitor of DENV. While there is no published data for flaviviruses, the clinical trials database of the US National Institute of Health suggests that the safety, tolerability, and efficacy of balapiravir are currently being studied in patients with confirmed DENV infection (www.clinicaltrials.gov/ct2/show/NCT01096576; accessed 8 December 2010).

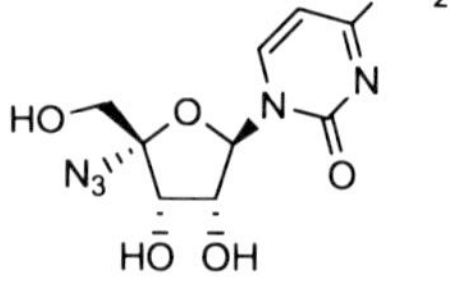

18 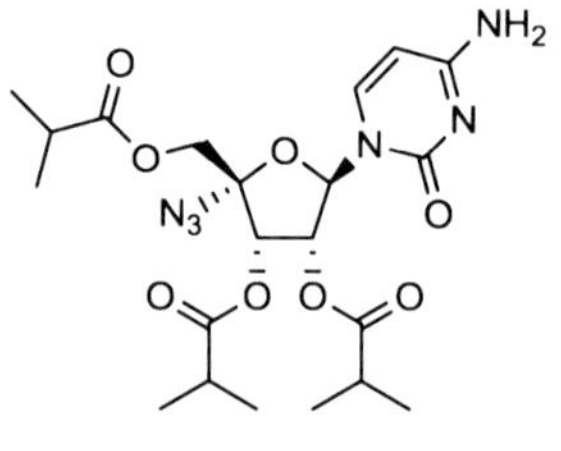 **19**

4.4.2 Allosteric Inhibition

Nucleoside active site inhibitors have the advantage of being broadly active against all serotypes of dengue and other *flaviviridae* (see Table 8). However, the well-known toxicology hurdles and the limited structural space that is available for derivatization have spawned the search for alternative approaches. Since the enzymatic activity of the RdRp is dependent on its tertiary structure, any compound that locks the protein in an inactive conformation or inhibits necessary domain movement will be a functional inhibitor of the polymerase reaction. In HCV polymerase, at least four allosteric binding sites have been identified (Fig. 8) [72]. The structural diversity of the inhibitors is much greater than for the nucleosides, and at least for HCV, a much greater proportion of allosteric inhibitors have progressed to clinical studies [72].

It has been suggested that the computational prediction of cavities in the protein may lead to the identification of allosteric binding sites for DENV RdRp inhibitors [70]. Unfortunately our understanding of the enzyme mechanism is not detailed enough to design functional inhibitors with confidence. Instead HTS still remains the most promising avenue for the discovery of novel allosteric inhibitors, as demonstrated by scientists from NITD who identified compound **20** as a promising lead for DENV RdRp inhibition. The compound was optimized to the submicromolar inhibitor **21**, which exhibits excellent selectivity against HCV, WNV polymerase, and human DNA polymerase α and β [81]. Photoaffinity labeling of a derivative of **21** showed that the compound likely binds at the entrance of the RNA tunnel, and biochemical experiments demonstrated that the order of addition of the RNA template and **21** affected compound potency [85]. These results confirm that the compound competes with RNA for its binding site (Fig. 9). While **21** is an interesting tool compound for biological studies, it proved not to be a viable starting point for further optimization [81].

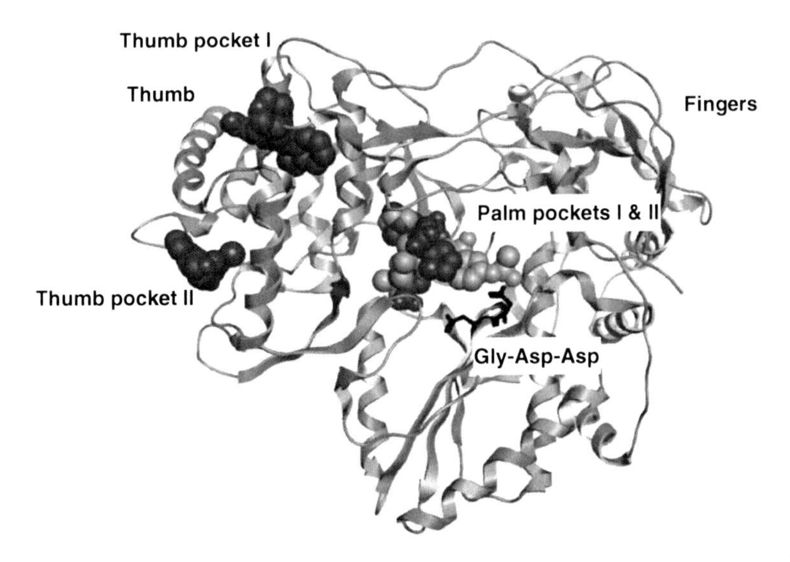

Fig. 8 Allosteric binding sites in HCV polymerase (inhibitors in cpk) [72]

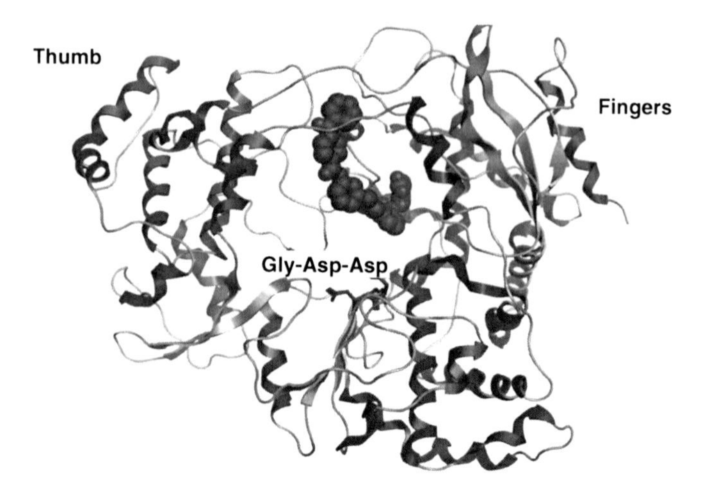

Fig. 9 Dengue RdRp domain with allosteric inhibitor **22**. The compound was modeled into the RdRp crystal structure using information from the affinity labeling experiment [80, 81]

20

21

21 with extension for photoaffinity labelling **(22)**

4.4.3 Medicinal Chemistry Perspective

There is little question that the RdRp domain is the most attractive target for dengue drug discovery. Even a relatively small effort (when compared to HCV) has identified two to three nucleosides with interesting anti-dengue activities. While the hurdles in toxicology are higher than for other classes of compounds, there is no reason to believe that a safe and effective nucleoside cannot be identified for

DENV. The only concern in this area is the very cramped intellectual property space, a common problem for antiviral nucleosides. Unfortunately a combination of a very limited structure–activity space (for all viral RdRps) and the large amount of work that has been done for other viral diseases creates a tough challenge for any medicinal chemist in this area.

At present, very little work has been done on allosteric inhibitors of DENV RdRp. In order to make progress in this area, a concerted effort both in HTS and the biophysical validation of inhibitors is absolutely crucial. HTS is probably the best way to identify new allosteric inhibitors, since it is not clear which binding pockets will be best for functional inhibition. In our view, screening of the full length NS5 protein would be preferable for the identification of promising hits. Once an inhibitor is identified, structural information is necessary for rapid progress toward a drug candidate.

4.5 *NS5 Methyltransferase*

The 5′ end of the flavivirus genome contains a type 1 cap structure (m7GpppAm). The flavivirus methyltransferase (MTase) methylates the guanine N-7 and ribose 2′-O positions of the viral RNA cap in a sequential manner (i.e., GpppA-RNA → m7GpppA-RNA → m7GpppAm-RNA) [17, 86]. The methylation reaction uses *S*-adenosyl-L-methionine (SAM) as a methyl donor and generates *S*-adenosyl-L-homocysteine (SAH) as a by-product. In contrast to cellular RNA cap methyltransferases, flavivirus MTase requires a specific viral RNA sequence and structure for its methylation reactions [87–89]. Structure-based mutagenesis studies suggest that flavivirus MTase catalyzes the two methylation events through an RNA substrate repositioning model [90]. In DENV and WNV, a mutation that abolished both N-7 and 2′-O methylation activities is lethal for viral replication, whereas a mutation that knocks out only the 2′-O-methylation activity is competent in replication. These results indicate that the N-7 methylation is essential for flavivirus replication and should be targeted for antiviral development [88, 91].

4.5.1 Medicinal Chemistry Perspective

Up to now, DENV MTase has not been targeted for drug discovery, mainly due to difficulties encountered for this target class in cancer research.

One challenge for the development of DENV MTase inhibitor is selectivity. Many host MTases play important roles in normal physiology through methylating DNA, RNA, and proteins [92]. To eliminate potential toxicity, the inhibitor will need to specifically suppress viral MTase without affecting host MTases. A recent structural study revealed a hydrophobic pocket located next to the SAH-binding site in the WNV MTase; importantly, the newly identified pocket is conserved among different flavivirus MTases [93] and therefore may serve as a starting point for fragment-based screening or de novo design.

5 Host Targets

Host proteins that are important for DENV cell entry and replication are attractive targets for drug discovery. Nevertheless, most research so far has concentrated on viral proteins, mainly because of the challenges associated with on- and off-target toxicities. For DENV, the problem is more pronounced compared to diseases such as HIV or HCV, since dengue fever in most cases is not a life-threatening disease and an important part of the patient population is children. It is therefore likely that the tolerance for side effects will be very low, which suggests that the design of pharmaceutically useful inhibitors will be challenging.

5.1 Glucosidase

It is well established that N-linked oligosaccharide processing in the endoplasmic reticulum (ER) is important for the proper folding of viral proteins and the secretion of enveloped viruses [94]. The biosynthesis of glycosylated proteins begins with the transfer of the oligosaccharide $Glc_3Man_9GlcNAc_2$ to arginine residues on the nascent protein, followed by stepwise removal of glucose residues by glucosidase I and II [95]. Inhibitors of these glucosidases have been used since the 1970s to study the mechanism of glycoprotein processing [96], and more recently their potential as treatments for viral infections has been explored [97]. The dengue virus contains three glycoproteins (prM, E, and NS1), and glycosidase inhibitors have been shown to have antiviral effects in vitro [98] and in vivo [99].

The mechanism of glycosidase catalysis proceeds through an oxocarbenium ion-like transition state **A**, which is mimicked by the two prototypical glucosidase inhibitors deoxynojirimycin (**23**) and castanospermine (**24**) [100].

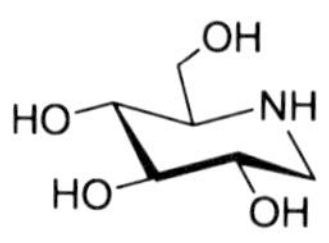

Deoxynojirimycin (**23**)

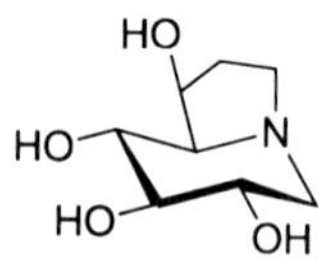

Castanospermine (**24**)

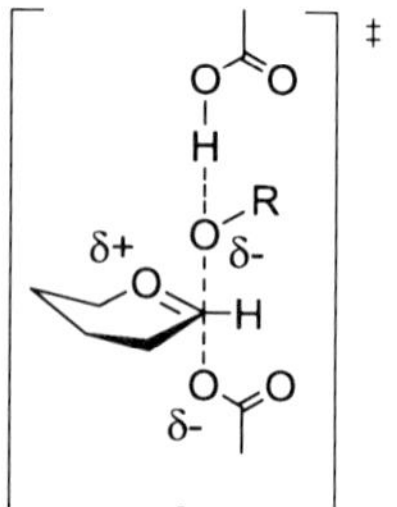

A

Compound **23** which is thought to exist as the conjugate acid at neutral pH [100] inhibits α-glucosidase with an IC_{50} in the low micromolar region; however, cellular effects of this highly polar natural product can only be seen at millimolar concentrations [101]. This has led to an intense effort to synthesize derivatives with improved properties, mainly by derivatization of the secondary nitrogen [102].

Table 9 Summary of in vitro activity of deoxynojirimycin derivatives

Compound	α-Glucosidase, IC_{50} (μM)	Ceramide glucosyltransferase, IC_{50} (μM)	MTS proliferation, CC_{50} (μM)
23[a]	3.5	NA	>1,000
25[b]	0.65	34.4	>1,000
26[b]	1.33	7.4	118.9
27[b]	1.95	4	36.6

NA not available
[a]Data from Taylor et al. (1994)
[b]Data from [102]

When comparing compounds with n-alkyl chains between 4 and 18 carbons, cellular uptake was generally proportional to chain length, while α-glucosidase activity was unaffected (Table 9). Increasing the hydrophobicity of *N*-alkylated deoxynorjirimycin derivatives, however, increased the cytotoxicity of the compounds and led to higher promiscuity (Table 9), for example, inhibition of ceramide glucosyltransferase [102].

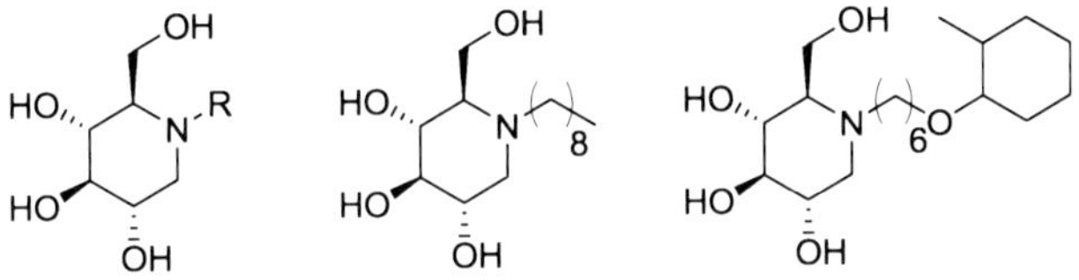

23 R=H
25 R=butyl
27 R=octadecenyl

26

28

Compound **26** was shown to have antiviral effects against DENV2 by interfering with the posttranslational modification of viral proteins and their interaction with chaperons in the ER [103]. This imino-sugar derivative also significantly reduced viremia in DENV2-infected AG129 mice, when administered orally twice a day for 3 days at 75 mg/kg [99]. While **26** has a dramatically improved antiviral activity compared to the original natural product, the narrow therapeutic window precluded clinical development as a treatment for viral infections [104]. The structure activity of this series of compounds suggested that the imino-sugar head group is responsible for the α-glucosidase activity, while the tail could be used to modulate permeability and cytotoxicity. Further modifications of the side chain lead to the identification of **28** which had a more than tenfold improvement in anti-dengue activity compared to **26** with an improved cytotoxicity window [105] (Table 10).

Table 10 Activity of deoxynojirimycin derivatives against DENV2[a]

Compound	Activity against DEN2, EC_{50} (μM)	MTT cytotoxicity assay, CC_{50} (μM)
26	1.1	>40
28	0.075	65

[a]Data from [105]

Castanospermine (**24**) is an indolizidine alkaloid that as early as 1983 was shown to inhibit the processing of the oligosaccharide portion of influenza viral hemagglutinin [96]. While **24** is a very potent inhibitor of α-glycosidase I, it shows poor activity in cellular assays due to its highly polar nature (Table 11). For example, its antiviral activity in DENV2-infected Huh-7 cells was only moderate, exhibiting an IC_{50} of 87 μM [106].

Table 11 Activity of castanospermine and prodrug against glucosidase and syncytium reduction in HIV-I[a]

Compound	Syncytium reduction, IC_{50} (μM)	α-Glucosidase I inhibition, IC_{50} (μM)
24	39	0.12
29	0.89	1.27

[a]Data from Taylor et al. (1994)

This problem could be solved through a prodrug approach. 6-*O*-butanoyl castanospermine (**29**) showed 40-fold greater uptake by H9 cells chronically infected with HIV, which resulted in a much improved cellular activity compared to **24** (Table 11; [94]).

Interestingly, especially when contrasted with the deoxynojirimycin derivatives discussed previously (e.g., compound **25** and **26**), both castanospermine (**24**) and its prodrug **29** showed minimal cytotoxicity against human hepatocytes [107].

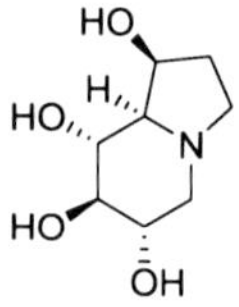

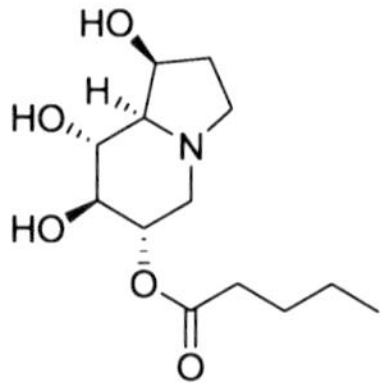

Prodrug **29** is rapidly converted to the parent compound **24** when taken up into a cell in vitro, or when given to mice by the oral or the intravenous route [107]. Pharmacokinetics in rats and dogs was excellent with 94% of **29** being absorbed and 92% excreted as **24** in the urine within 24 h.

With the excellent pharmacokinetic parameters, it was not surprising that compound **29** showed dose-dependent activity in a mouse model of dengue fever viremia [99]. While this animal model of dengue viremia is not a realistic simulation of dengue fever in human patients, the overall profile of 6-*O*-butanoyl canstanospermine (**29**) suggests that it may be a promising drug candidate for DENV.

Unfortunately the clinical track record of **29** has been disappointing [108]. Initial Phase I and Phase II trials against HIV were performed in the 1990s. While the drug was generally well tolerated (main adverse events included loose stool, diarrhea, and flatulence), the aniviral effects were marginal and the development for HIV was abandoned. It has been noted that the lack of efficacy could have been due to the rather low plasma concentration that were achieved [108].

A number of Phase II clinical studies have recently been reported which tested the potential of **29** as an anti-HCV treatment. Again mild-to-moderate gastrointestinal symptoms were the most frequently observed adverse events; however, these side effects of the drug could be managed with anti-diarrheal agents and low sucrose/starch and high glucose diet. Unfortunately the compound was again not efficacious as monotherapy [108].

5.2 *Medicinal Chemistry Perspective*

The two currently available classes of glucosidase inhibitors have been used as tools since the early 1990s. It is clear that compounds based on deoxynojirimycin have a small cytoxicity window, and the recent modifications have not been able to convincingly solve this issue.

Despite the similarity between the structures, castanospermine does not seem to suffer the same problem and is therefore a superior compound. The ester prodrug 6-*O*-butanoyl castanospermine has an excellent preclinical profile. Unfortunately the lack of success of this compound in HIV and HCV clinical trials suggests that a clinical trial for dengue would be at high risk. Further work on glucosidase inhibitors should focus on new scaffolds that do not have the well-known drawbacks of the two classical natural products.

6 Conclusions

Dengue drug discovery has made impressive progress in the last 10 years. Crystal structures of all the major drug targets are available, and biological assays and animal models are of sufficient quality to quickly progress a potential development compound. In our view, the progress toward a drug for dengue is mostly hampered by the limited resources that are available to identify novel chemical starting points and then progress these compounds to a development candidate. Since resources are constrained for this neglected disease, it may be advisable to focus on the most promising targets and bundle the available resources.

The current state of knowledge in DENV and HCV drug discovery suggests that NS3 protease, NS5 polymerase, and to a lesser extent the envelope protein are the most promising targets. While one can speculate about the biological attractiveness of NS3 helicase, NS5 MTase, and host proteins, it is clear that the hurdles for successful drug discovery are very high in all of these areas.

NS3 protease is a tractable but difficult target, since small molecular weight compounds are unlikely to deliver drug candidates and peptidomimetics approaches are resource intensive. Depeptidization of a NS3 protease hit and optimization of a compound for oral bioavailability is a multiyear undertaking with large demands on medicinal chemistry and structural biology. It is likely that a high quality, oral inhibitor of DENV NS3 protease can be identified, but this can only be done with an industrial-sized team (20–30 FTEs) in a multiyear project.

NS5 polymerase is probably the most tractable target for DENV, since it should be possible to identify a nucleoside with a toxicity window that is sufficient for clinical development. The identification of non-nucleoside, allosteric inhibitors should also be achievable, but again this needs a sizable investment in HTS, biochemistry, structural biology, and medicinal chemistry. Since the best pockets for allosteric inhibitors cannot be predicted and the emergence of resistance could be different depending on the mechanism of inhibition, it may be important to optimize several scaffolds in parallel. Again this needs a large project team and a longer term time horizon.

After the protease and the polymerase, the envelope protein is probably the next attractive drug target for DENV. The current batch of compounds has validated the inhibition of E-protein-mediated fusion as a viable approach. However, since all of these hits have a number of flaws (functional groups with toxicological liabilities, high lipophilicity), it may be best to set up an HTS or fragment-based screening campaign to find new, more tractable chemical starting points.

Overall, the prospect for a dengue drug look promising, although it may take some time until a compound can be progressed to clinical development.

References

1. Stevens AJ, Gahan ME et al (2009) The medicinal chemistry of dengue. J Med Chem 52:7911–7926
2. Noble CG, Chen YL et al (2010) Strategies for the development of dengue virus inhibitors. Antivir Res 85:450–462
3. Gubler D, Kuno G et al (2007) Flaviviruses. In: Knipe DM, Howley PM (eds) Fields virology, vol 1, 5th edn. Lippincott William & Wilkins, Philadelphia
4. Wills B (2008) Management of dengue. In: Halstead SB (ed) Dengue. Imperial College Press, London
5. Halstead SB (2003) Neutralization and antibody-dependent enhancement of dengue viruses. Adv Virus Res 60:421–67
6. Esté JA, Telenti A (2007) HIV entry inhibitors. Lancet 370:81–88
7. Keller TH, Chen YL et al (2006) Finding new medicines for flaviviral diseases. Novartis Found Symp 277:102–114
8. Zhang W, Chipman PR et al (2003) Visualization of membrane protein domains by cryo-electron microscopy of dengue virus. Nat Struct Biol 10:907–12
9. Kummerer BM, Rice CM (2002) Mutations in the yellow fever virus nonstructural protein NS2A selectively block production of infectious particles. J Virol 76:4773–4784
10. Liu WJ, Chen HB et al (2003) Molecular and functional analyses of Kunjin virus infectious cDNA clones demonstrate the essential roles for NS2A in virus assembly and for a nonconservative residue in NS3 in RNA replication. J Virol 77:7804–13
11. Liu W, Wang X et al (2005) Inhibition of interferon signaling by the New York 99 strain and Kunjin subtype of West Nile virus involves blockage of STAT1 and STAT2 activation by nonstructural proteins. J Virol 79:1934–42

12. Munoz-Jordan JL, Sanchez-Burgos GG et al (2003) Inhibition of interferon signaling by dengue virus. Proc Natl Acad Sci USA 100:14333–8
13. Falgout B, Miller RH et al (1993) Deletion analysis of dengue virus type 4 nonstructural protein NS2B: identification of a domain required for NS2B-NS3 protease activity. J Virol 67:2034–42
14. Wengler G, Wengler G (1991) The carboxy-terminal part of the NS 3 protein of the West Nile flavivirus can be isolated as a soluble protein after proteolytic cleavage and represents an RNA-stimulated NTPase. Virology 184:707–15
15. Wengler G, Wengler G (1993) The NS3 nonstructural protein of flaviviruses contains an RNA triphosphatase activity. Virology 197:265–273
16. Ackermann M, Padmanabhan R (2001) De novo synthesis of RNA by the dengue virus RNA-dependent RNA polymerase exhibits temperature dependence at the initiation but not elongation phase. J Biol Chem 276:39926–39937
17. Egloff MP, Benarroch D et al (2002) An RNA cap (nucleoside-2'-O-)-methyltransferase in the flavivirus RNA polymerase NS5: crystal structure and functional characterization. EMBO J 21:2757–68
18. Tan BH, Fu J (1996) Recombinant dengue type 1 virus NS5 protein expressed in Escherichia coli exhibits RNA-dependent RNA polymerase activity. Virology 216:317–25
19. Lindenbach B, Rice C (1997) trans-Complementation of yellow fever virus NS1 reveals a role in early RNA replication. J Virol 71:9608–9617
20. Miller S, Sparacio S (2006) Subcellular localization and membrane topology of the dengue virus type 2 non-structural protein 4B. J Biol Chem 281:8854–63
21. Miller S, Kastner S et al (2007) The non-structural protein 4A of dengue virus is an integral membrane protein inducing membrane alterations in a 2K-regulated manner. J Biol Chem 282:8873–82
22. Blackwell JL, Brinton MA (1997) Translation elongation factor-1 alpha interacts with the 3' stem-loop region of West Nile virus genomic RNA. J Virol 71:6433–44
23. Davis W, Blackwell J et al (2007) Interaction between the cellular protein eEF1A and the 3' terminal stem loop of the West Nile virus genomic RNA facilitates viral RNA minus strand synthesis. J Virol 81:10172–10187
24. Ma L, Jones CT (2004) Solution structure of dengue virus capsid protein reveals another fold. Proc Natl Acad Sci USA 101:3414–9
25. Li L, Lok SM et al (2008) The flavivirus precursor membrane-envelope protein complex: structure and maturation. Science 319:1830–4
26. Modis Y, Ogata S et al (2003) A ligand-binding pocket in the dengue virus envelope glycoprotein. Proc Natl Acad Sci USA 100:6986–6991
27. Erbel P, Schiering N et al (2006) Structural basis for the activation of flaviviral NS3 proteases from dengue and West Nile virus. Nat Struct Mol Biol 13:372–273
28. Luo D, Xu T et al (2008) Crystal structure of the NS3 protease-helicase from dengue virus. J Virol 82:173–183
29. Yap TL, Xu T et al (2007) Crystal structure of the dengue virus RNA-dependent RNA polymerase catalytic domain at 1.85-angstrom resolution. J Virol 81:4753–4765
30. van der Schaar HM, Rust MJ et al (2008) Dissecting the cell entry pathway of dengue virus by single-particle tracking in living cells. PLoS Pathog 4:e1000244
31. Navarro-Sanchez E, Altmeyer R et al (2003) Dendritic-cell-specific ICAM3-grabbing non-integrin is essential for the productive infection of human dendritic cells by mosquito-cell-derived dengue viruses. EMBO Rep 4:723–728
32. Thepparit C, Smith DR (2004) Serotype-specific entry of dengue virus into liver cells: identification of the 37-kilodalton/67-kilodalton high-affinity laminin receptor as a dengue virus serotype 1 receptor. J Virol 78:12647–12656
33. Miller JL, deWet BJM et al (2008) The mannose receptor mediates dengue virus infection of macrophages. PLoS Pathog 4:e17
34. Jindadamrongwech S, Thepparit S (2004) Identification of GRP 78 (BiP) as a liver cell expressed receptor element for dengue virus serotype 2. Arch Virol 149:915–927

35. Marks RM, Lu H et al (2001) Probing the interaction of dengue virus envelope protein with heparin: assessment of glycosaminoglycan-derived inhibitors. J Med Chem 44:2178–2187
36. Ono L, Wollinger W (2003) In vitro and in vivo antiviral properties of sulfated galactomannans against yellow fever virus (BeH111 strain) and dengue 1 virus (Hawaii strain). Antivir Res 60:201–208
37. Talarico LB, Damonte EB (2007) Interference in dengue virus adsorption and uncoating by carrageenans. Virology 363:473–485
38. Modis Y, Ogata S et al (2004) Structure of the dengue virus envelope protein after membrane fusion. Nature 427:313–319
39. Hrobowski YM, Garrey RF et al (2005) Peptide inhibitors of dengue virus and West Nile virus infectivity. Virol J 2:49–59
40. Li Z, Khaliq M et al (2008) Design, synthesis, and biological evaluation of antiviral agents targeting flavivirus envelope proteins. J Med Chem 51:4660–4670
41. Zhou Z, Kalique M et al (2008) Antiviral compounds discovered by virtual screening of small molecule libraries against dengue virus E protein. ACS Chem Biol 3:765–775
42. Kampmann T, Yennamalli R et al (2009) In silico screening of small molecule libraries using the dengue virus envelope E protein has identified compounds with antiviral activity against multiple flaviviruses. Antivir Res 84:234–241
43. Wang QY, Patel SJ et al (2009) A small-molecule dengue virus inhibitor. Antimicrob Agents Chemother 53:1823–1831
44. Xu T, Sampath A et al (2005) Structure of the dengue virus helicase/nucleoside triphosphatase catalytic domain at a resolution of 2.4 Å. J Virol 79:10278–10288
45. Assenberg R, Mastrangelo E et al (2009) Crystal structure of a novel conformational state of the flavivirus NS3 protein: implications for polyprotein processing and viral replication. J Virol 83:12895–12906
46. Luo D, Wei N et al (2010) Flexibility between the protease and helicase domains of the dengue virus NS3 protein conferred by the linker region and its functional implications. J Biol Chem 285:18817–18827
47. Hsu JT, Wang HC (2006) Antiviral drug discovery targeting to viral proteases. Curr Pharm Des 12:1301–1314
48. Menendez-Arias L (2010) Molecular basis of human immunodeficiency virus drug resistance: an update. Antivir Res 85:210–231
49. Thompson AJV, McHutchison JG (2009) Investigational agents for chronic hepatitis C. Aliment Pharmacol Ther 29:689–705
50. Bazan JF, Fletterick RJ et al (1989) Detection of a trypsin-like serine protease domain in flaviviruses and pestiviruses. Virology 171:637–639
51. Falgout B, Pethel M et al (1991) Both nonstructural proteins NS2B and NS3 are required for the proteolytic processing of dengue virus nonstructural proteins. J Virol 65:2467–2475
52. Murthy HMK, Clum S et al (1999) Dengue virus NS3 serine protease. J Biol Chem 274:5573–5580
53. Chambers TJ, Weir RC et al (1990) Evidence that the N-terminal domain of nonstructural protein NS3 from yellow fever virus is a serine protease responsible for site-specific cleavages in the viral polyprotein. Proc Nat Acad Sci USA 87:8898–8902
54. Preugschat F, Yao CW et al (1990) In vitro processing of dengue virus type 2 nonstructural proteins NS2A, NS2B and NS3. J Virol 64:4364–4374
55. Chanparpaph S, Saparpakorn P et al (2005) Competitive inhibition of the dengue virus NS3 serine protease by synthetic peptides representing polyprotein cleavage sites. Biochem Biophys Res Commun 330:1237–1246
56. Li J, Lim SP et al (2005) Functional profiling of recombinant of NS3 proteases from all four serotypes of dengue virus using tetrapeptide and octapeptide substrate libraries. J Biol Chem 280:48535–49542
57. Yin Z, Patel SJ et al (2006) Peptide inhibitors of dengue virus NS3 protease. Part 1: warhead. Bioorg Med Chem Lett 16:36–39

58. Yin Z, Pate SJ et al (2006) Peptide inhibitors of dengue virus NS3 protease. Part 2: SAR study of tetrapeptide aldehyde inhibitors. Bioorg Med Chem Lett 16:40–43
59. Aleshin A, Shiryaev S et al (2007) Structural evidence for regulation and specificity of flaviviral proteases and evolution of the *Flaviviridae* fold. Protein Sci 16:795–806
60. Ganesh VK, Muller N et al (2005) Identification and characterization of nonsubstrate based inhibitors of the essential dengue and West Nile virus proteases. Biorg Med Chem 13:257–264
61. Bodenreider C, Beer D et al (2009) A fluorescence quenching assay to discriminate between specific and nonspecific inhibitors of dengue virus protease. Anal Biochem 395:195–204
62. Mueller NH, Pattabiraman N et al (2008) Identification and biochemical characterization of small-molecule inhibitors of West Nile virus serine protease by a high-throughput screen. Antimicrob Agents Chemother 52:3385–3393
63. Kiat TS, Pippen R et al (2006) Inhibitory activity of cyclohexenyl chalcone derivatives and flavonoids of fingerroot, Boesenbergia rotunda (L.), towards dengue-2 virus NS3 protease. Int J Biol Sci 16:3337–3340
64. Tomlinson SM, Malmstrom RD et al (2009) Structure-based discovery of dengue virus protease inhibitors. Antivir Res 82:110–114
65. Robin G, Chappel K et al (2009) Structure of west Nile virus NS3 protease: ligand stabilization of the catalytic conformation. J Mol Biol 385:1568–1577
66. Chandramouli S, Joseph J et al (2010) Serotype-specific structural differences in the protease-cofactor complexes of the dengue virus family. J Virol 84:3059–3067
67. Knox JE, Ma NL et al (2006) Peptide inhibitors of West Nile NS3 protease: SAR study of tetrapeptide aldehyde inhibitors. J Med Chem 49:6585–6590
68. Keller TH, Pichota A et al (2006) A practical view of druggability. Curr Opin Chem Biol 10:357–361
69. Liu Y, Jiang WW et al (2006) Mechanistic study of HCV polymerase inhibitors at individual steps of the polymerization reaction. Biochemistry 45:11312–11323
70. Malet H, Massé N et al (2008) The flavivirus polymerase as a target for drug discovery. Antivir Res 80:23–35
71. Steitz TA, Smerdon SJ et al (1994) A unified polymerase mechanism for nonhomologous DNA and RNA polymerases. Science 266:2022–2025
72. Li H, Shi ST (2010) Non-nucleoside inhibitors of hepatitis C virus polymerase: current progress and future challenges. Future Med Chem 2:121–141
73. Perrone P, Daveria F et al (2007) First example of phosphoramidate approach applied to a 4'-substituted purine nucleoside (4'-azidoadenosine): conversion of an inactive nucleoside to a submicromolar compound versus hepatitis C virus. J Med Chem 50:5463–5470
74. Migliaccio G, Tomassini JE et al (2003) Characterization of resistance to non-obligate chain-terminating ribonucleoside analogs that inhibit hepatitis C virus replication in vitro. J Biol Chem 278:49164–49170
75. Eldrup AB, Prhavc M et al (2004) Structure-activity relationship of heterobase-modified 2′-C-methyl ribonucleosides as inhibitors of hepatitis C virus RNA replication. J Med Chem 47:5284–5297
76. Chen YL, Yin Z et al (2010) Inhibition of dengue virus RNA synthesis by an adenosine nucleoside. Antimicrob Agents Chemother 54:2932–2939
77. Pastor-Anglada M, Cano-Soldado P et al (2005) Cell entry and export of nucleoside analogues. Virus Res 107:151–164
78. Butora G, Olsen DB et al (2007) Synthesis and HCV inhibitory properties of 9-deaza- and 7,9-dideaza-7-oxa-20-C-methyladenosine. Bioorg Med Chem 15:5219–5229
79. Carroll SS, Ludmerer S et al (2009) Robust antiviral efficacy upon administration of a nucleoside analog to hepatitis C virus-infected chimpanzees. Antimicrob Agents Chemother 53:926–934
80. Yin Z, Chen YL et al (2009) An adenosine nucleoside inhibitor of dengue virus. Proc Natl Acad Sci USA 106:20435–20439

81. Yin Z, Chen YL et al (2009) N-sulfonylanthranilic acid derivatives as allosteric inhibitors of dengue viral RNA-dependent RNA polymerase. J Med Chem 52:7934–7937
82. Smith DB, Martin JA et al (2007) Design, synthesis, and antiviral properties of 40-substituted ribonucleosides as inhibitors of hepatitis C virus replication: the discovery of R1479. Bioorg Med Chem Lett 17:2570–2576
83. Pockros PJ, Nelson D et al (2008) R1626 Plus peginterferon alfa-2a provides potent suppression of hepatitis C virus RNA and significant antiviral synergy in combination with ribavirin. Hepatology 48:395–397
84. Li H, Tatlock J et al (2009) Discovery of (*R*)-6-cyclopentyl-6-(2-(2,6-diethylpyridin-4-yl)ethyl)-3-((5,7-dimethyl-[1,2,4]triazolo[1,5-*a*]pyrimidin-2-yl)methyl)-4-hydroxy-5,6-dihydropyran-2-one(PF-00868554) as a potent and orally available hepatitis C virus polymerase inhibitor. J Med Chem 52:1255–1258
85. Niyomrattanakit P, Chen YL et al (2010) Inhibition of dengue virus polymerase by blocking of the RNA tunnel. J Virol 84:5678–5686
86. Ray D, Shah A et al (2006) West Nile virus 5'-cap structure is formed by sequential guanine N-7 and ribose 2'-O methylations by nonstructural protein 5. J Virol 80:8362–70
87. Chung KY, Dong H et al (2010) Higher catalytic efficiency of N-7-methylation is responsible for processive N-7 and 2'-O methyltransferase activity in dengue virus. Virology 402:52–60
88. Dong H, Chang D et al (2010) Biochemical and genetic characterization of dengue virus methyltransferase. Virology 405:568–578
89. Dong H, Ray D et al (2007) Distinct RNA elements confer specificity to flavivirus RNA cap methylation events. J Virol 81:4412–4421
90. Dong H, Zhang B et al (2008) Flavivirus methyltransferase: a novel antiviral target. Antivir Res 80:1–10
91. Zhou Y, Ray D et al (2007) Structure and function of flavivirus NS5 methyltransferase. J Virol 81:3891–3903
92. Copeland RA, Solomon ME et al (2009) Protein methyltransferases as a target class for drug discovery. Nat Rev Drug Discov 8:724–32
93. Dong H, Liu L et al (2010) Structural and functional analyses of a conserved hydrophobic pocket of flavivirus methyltransferase. J Biol Chem 285:32586–32595
94. Jacoby GS (1995) Glycosylation inhibitors in biology and medicine. Curr Opin Struct Biol 5:605–611
95. Elbein AD (1991) Glycosidase inhibitors: inhibitors of N-linked oligosaccharide processing. FASEB J 5:3055–3063
96. Pan YT, Hori H et al (1983) Castanospermine inhibits the processing of the oligosaccharide portion of the influenza hemagglutinin. Biochemistry 22:3975–3983
97. Block TM, Lu X et al (1998) Treatment of chronic hepadnavirus infection in woodchuck animal model with an inhibitor of protein folding and trafficking. Nat Med 4:610–614
98. Courageot MP, Frenkiel MP (2000) α-Glucosidase inhibitors reduce dengue virus production by affecting the initial steps of virion morphogenesis in the endoplasmic reticulum. J Virol 74:564–572
99. Schul W, Wei L et al (2007) A dengue fever viremia model in mice show reduction in viral replication and suppression of the inflammatory response after treatment with antiviral drugs. J Infect Dis 195:665–674
100. Gloster TM, Madsen R et al (2006) Dissection of conformationally restricted inhibitors binding to a β-glucosidase. Chembiochem 7:738–742
101. Tan A, van den Broek L et al (1991) Chemical modifications of the glucosidase inhibitor 1-deoxynojirimycin. J Biol Chem 22:14504–14510
102. Mellor HR, Nolan J (2002) Preparation, biochemical characterization and biological properties of radiolabelled N-alkylated deoxynojirimycins. Biochem J 366:225–233
103. Wu SF, Lee CJ et al (2002) Antiviral effects of an iminosugar derivative on flavivirus infections. J Virol 76:3595–3604
104. Gu B, Mason P (2007) Antiviral profiles of novel iminocyclitol compounds against bovine viral diarrhea virus, West Nile virus, dengue virus and hepatitis B virus. Antivir Chem Chemother 18:49–59

105. Chang J, Wang L (2009) Novel imino sugar derivatives demonstrate potent antiviral activity against flaviviruses. Anitmicrob Agents Chemother 53:1501–1508
106. Whitby K, Pierson TC et al (2005) Castanospermine, a potent inhibitor of dengue virus infection in vitro and in vivo. J Virol 79:8698–8706
107. Sorbera LA, Castaner J et al (2005) Celgosivir. Drug Future 30:545–552
108. Durantel D (2009) Celgosivir, an α-glucosidase I inhibitor for the potential treatment of HCV infection. Curr Opin Investig D 10:860–870
109. Tassaneetrithep B, Burgess TH et al (2003) DC-SIGN (CD209) mediates dengue virus infection of human dendritic cells. J Exp Med 7:823–829

Top Med Chem 7: 277–300
DOI: 10.1007/7355_2011_15

Published online: 19 April 2011

Recent Advances in Discovery and Development of Medicines for the Treatment of Secretory Diarrhea in the Developing World

Dan Marquess

Abstract Secretory diarrhea remains a global health challenge in the developing and developed countries. Secretory or infectious diarrhea is caused by infection of the gastrointestinal tract with a bacterial, viral, or parasitic pathogen. Acute diarrheal diseases have a major impact on morbidity and mortality across the world, with as many as four billion cases occurring annually. Diarrheal disease is responsible for the approximately 1.7 million deaths in children under 5 years old annually. Preventative healthcare measures include safe and effective vaccines and improved water, sanitation, hygiene systems, and nutritional practices. Treatment options in developing countries are usually supportive, replacing intestinal fluid losses with oral rehydration solution (ORS). Oral rehydration therapy (ORT) is well accepted as the most effective treatment for rehydration of children with acute diarrhea and is recommended by the World Health Organization for the prevention and management of dehydration. Its use has resulted in a significant reduction in diarrhea-related mortality. ORT has been recently complemented by zinc supplementation therapy. Rehydration has little effect on stool volume or frequency; therefore, the World Health Organization has recommended that anti-secretory drug treatment be added to rehydration therapy as long as the treatment has proven safe and efficacious in the pediatric population. In the developing countries, Hydrasec™ (racecadotril) is the most widely used antisecretory antidiarrheal, particularly in children. This review focuses on recent advances in the discovery and development of medicines and pharmacological mechanisms that modulate the secretory component of diarrhea that could be efficacious in the treatment of infectious diarrhea in the developing countries, particularly in the pediatric population.

Keywords μ-Opioid, Crofelemer, Cystic fibrosis transmembrane conductance regulator (CFTR), Enteric disease, Enterocyte, Cholera, Hydrasec™, Loperamide, Neprilysin, Norovirus, Oral rehydration therapy, Racecadotril, Rotavirus, Secretory diarrhea, Tiorfan™ thiorphan

---

D. Marquess
Theravance Inc., 901 Gateway Blvd., South San Francisco, CA 94080, USA
e-mail: dmarquess@theravance.com

Contents

1 Introduction and Scope of This Review

1.1 Prevalence and Epidemiology

Enteric and diarrheal diseases include secretory (or infectious) diarrhea and non-diarrheal enteric diseases such as typhoid (*Salmonella typhi*, *Salmonella paratyphi*), hepatitis A and E, geohelminths, and a host of other viral, bacterial, and parasitic pathogens (www.gatesfoundation.org/diarrhea).

Secretory diarrhea remains a global health challenge in the developing and developed countries [1]. Secretory or infectious diarrhea is caused by infection of the gastrointestinal (GI) tract with an infectious pathogen. These pathogens can be viral e.g., norovirus [2] or rotavirus [3], they can be bacterial pathogens e.g., the gram-negative bacillus, *Vibrio cholerae*, the causative agent of cholera or *Escherichia coli*, or they can be parasitic such as *Giardia lamblia*. To use the well-studied example of bacterial pathogen *V. cholerae*, bacterial enterotoxins produced by this pathogen bind to the intestinal enterocyte. Following endocytosis of the toxin, it causes constitutive activation of adenylate cyclase (AC) that elevates levels of intracellular cyclic adenosine monophosphate (cAMP) causing increased fluid secretion and secretory diarrhea [4].

Acute diarrheal diseases have a major impact on morbidity and mortality across the world. Each year in developing countries, approximately four billion episodes of acute diarrhea occur in children under 5 years of age [5]. In the most economically deprived areas of the world such as the Indian subcontinent, Africa and Latin America, children may experience between three and ten episodes per year. These repeated bouts of diarrhea and persistent diarrheal disease radically impair gut function, which is the single greatest contributor to childhood malnutrition and growth retardation. Diarrheal diseases are also a major cause of childhood hospitalization, primarily to dehydration. Morbidity is related to intestinal inflammation, bacteremia, and extraintestinal complications.

Diarrheal disease is responsible for the approximately 1.7 million deaths [6] in children under 5 years old annually. Most of the deaths from acute infectious diarrhea are a result of excessive fluid and electrolyte loss, which in turn results in dehydration and acidosis [7].

Preventative healthcare measures include safe and effective vaccines and improved water, sanitation, hygiene systems, and nutritional practices.

1.2 Current Treatment Options for Secretory Diarrhea

Treatment options in developing countries are usually supportive, replacing intestinal fluid losses with oral rehydration solution (ORS). Oral rehydration therapy (ORT) is well accepted as the most effective treatment for rehydration of children with acute diarrhea and is recommended by the World Health Organization for the prevention and management of dehydration. Its use has resulted in significant reductions in diarrhea-related mortality with an estimated 50 million lives saved [8]. This has recently been complemented by zinc supplementation therapy, which has been shown to significantly reduce deaths when used as part of the ORT regimen [9]. Zinc considerably reduces the duration and severity of diarrhea episodes, decreases stool output, lessens the need for hospitalization, and may also prevent future diarrhea for up to 3 months [10].

Rehydration has little effect on stool volume or frequency; therefore, the World Health Organization has recommended that antisecretory drug treatment be added to rehydration therapy as long as the treatment has proven safety and efficacious in the pediatric population. Thus, more therapeutic options are needed to further reduce mortality and morbidity from secretory diarrhea. In high-volume watery diarrhea such as cholera, replacing fluid losses still pose a major challenge for caregiver and healthcare professionals. In this context, antisecretory approaches for diarrhea are attractive in that they (a) will likely work across multiple pathogens;

(b) focus on host targets so are not susceptible to resistance issues; and (c) can be used as adjuvants to drive ORS therapy uptake, which despite having been proven safe and effective are still woefully underutilized in the developing world. In the developing countries, Hydrasec™ (racecadotril) is the most widely used antisecretory agent, particularly in children. This dual prodrug of thiorphan, a neprilysin (NEP) inhibitor, has shown clinical efficacy in some types of secretory diarrhea and is safe and well tolerated in adults and in the pediatric population [11].

Currently, two rotavirus vaccines, Rotarix™ and Rotateq™, are approved in many countries [3]. Human rotavirus vaccine significantly reduces the incidence of severe rotavirus gastroenteritis among African infants during the first year of life [12]. Additionally, introduction of rotavirus vaccine in Mexico has resulted in significant decline in diarrhea-related deaths among Mexican children [13]. For vaccination against typhoid, administration of the Vi polysaccharide domain in more than 37,000 residents in India showed an overall protective effectiveness of 61% and 80% for children between the ages of 2 and 5 years [14].

1.3 Scope of This Review

This review focuses on recent advances in the discovery and development of medicines and pharmacological mechanisms that modulate the secretory component of diarrhea and that could be efficacious in the treatment of infectious diarrhea in the developing countries, particularly in the pediatric population. Such approaches would complement treatment with ORS in that they may reduce stool volume and frequency of diarrhea.

2 Physiology of Increased GI Secretion and Pharmacological Mechanisms for Treatment

2.1 Pharmacological Mechanisms of Secretion, Pathophysiology of Secretion and Morbidity/Mortality

Fluid secretion plays a key role in intestinal physiology. The small intestine secretes fluid and electrolytes under basal conditions and in response to the ingestion of food. Secretion occurs from the small intestinal crypts. Under normal conditions, the intestine carries out the reabsorption of luminal fluid, electrolytes, and nutrients. A small amount of basal secretion facilitates hydration of the intestinal mucosa and mixing of intestinal content. Fluid secretion results from the active secretion of chloride and bicarbonate ions.

Active chloride secretion from the apical membrane of the enterocyte is carried out by the cystic fibrosis transmembrane regulator (CFTR). On the basolateral side of the membrane, there are three other components of this chloride secretory

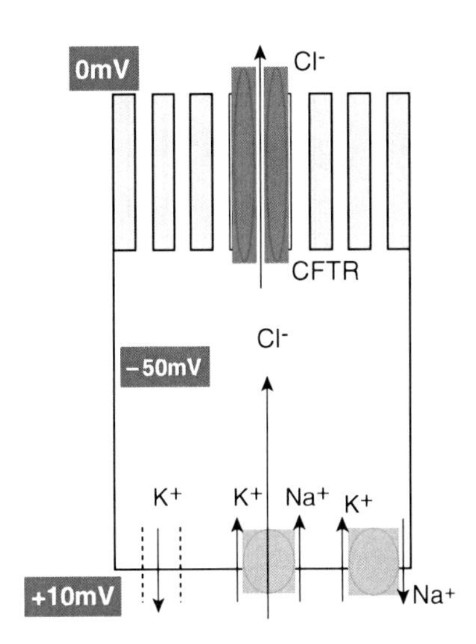

Fig. 1 Four-component model of chloride ion secretion that includes the cAMP-activated chloride channel; with the cystic fibrosis transmembrane conductance regulator (CFTR) on the apical membrane. The sodium ion-potassium ion-chloride cotransporter, the sodium ion-potassium ion ATPase and the potassium ion channel are shown on the basolateral membrane ([15], reproduced with permission)

system, the sodium ion-potassium ion-chloride cotransporter, the sodium ion-potassium ion ATPase and the potassium ion channel, that each operate to maintain electroneutrality of the enterocyte (Fig. 1).

The mechanism of chloride transport in health and under pathophysiological conditions has been studied extensively and is elegantly discussed in two recent articles [2, 4]. Phosphorylation of the CFTR by cAMP-mediated protein kinase A (PKA) activation opens the chloride channel, and since the interior of the enterocyte is electronegative chloride ion flows out of the cell.

A recent review details the changes in the GI tract during bacterial pathogenesis [16]. Under pathophysiological conditions such as intestinal infection with enterotoxigenic bacteria, secretion exceeds absorption such that the colon is not able to reabsorb the significant secretion of fluid and electrolytes, resulting in secretory diarrhea. In this situation, the chloride secretory process is activated directly by the enterotoxins. Cholera enterotoxin, released from *V. cholerae*, consists of an A subunit coupled to a B-subunit that is composed of five identical peptides assembled into a pentameric ring (Fig. 2). Cholera toxin binds to the enterocyte through an interaction of the B-subunit with a GM1 ganglioside receptor. Following endocytosis, the A-subunit leading to constitutive activation of AC driving elevated intracellular levels of cAMP causing active secretion of chloride and fluid via the activated CFTR. Prevention of diarrhea can be achieved via (a) elimination of the *V. cholerae* through host immunity or antibiotics, (b) inhibition of cholera enterotoxin binding, (c) inhibition of increased cAMP levels, such as by the action of enkephalins on opioid receptors and subsequent activation of phosphodiesterase (PDEs) [17], or (d) direct inhibition of the CFTR channel.

Most of the deaths from acute infectious diarrhea result from excessive fluid and electrolyte losses and associated dehydration and acidosis. The majority of the deaths may be avoided by replacing fluid and electrolyte losses with ORS. Recent

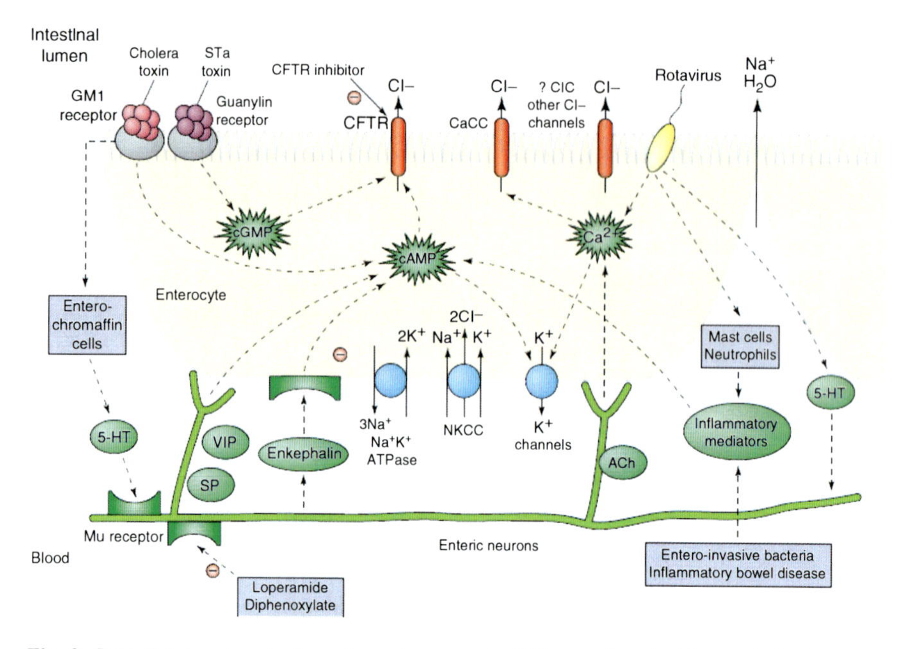

Fig. 2 Potential drug targets for antisecretory agents ([17] reproduced with permission)

research has provided further information on the modulation of the pharmacological mechanisms of these secretory processes [4]. It is hoped that this will lead to development of novel antisecretory medicines to complement rehydration therapy and further safely reduce the incidence and prevalence of diarrheal disease.

2.2 Clinically Validated Pharmacological Mechanisms for the Treatment of Secretory Diarrhea

This review focuses on recent developments that relate to current medicines or clinical candidates for treating secretory (or infectious) diarrhea, which operate via the following pharmacological mechanisms (see Fig. 3 for examples of molecules that are approved or are in late stage clinical trials for the treatment of diarrhea):

- *Inhibition of the enkephalinase enzyme, NEP,* to elevate levels of endogenous enkephalins that activate the δ-opioid receptor resulting in reduced cAMP levels in the enterocyte and reduce secretion e.g., Hydrasec™ (racecadotril) (**1**) and Tiorphan™ (thiorphan) (**2**) [11]
- *Inhibition of cystic fibrosis transmembrane regulator (CFTR)* to reduce secretion e.g., crofelemer (**3**) Phase 3 trials for different types of diarrhea [18]

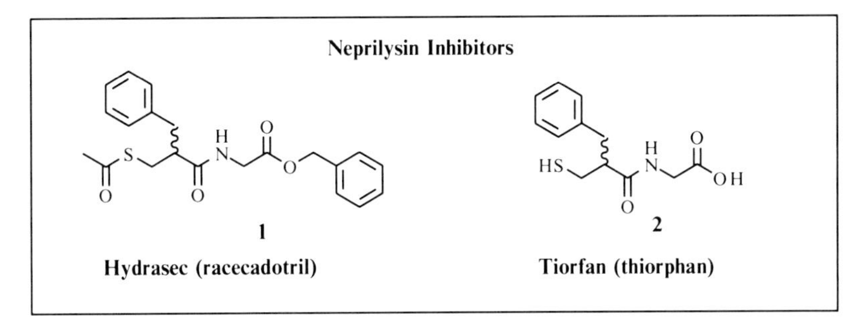

CFTR Inhibitor

3 (R=H or OH)
crofelemer (n = 5-11)

DNA polymerase Inhibitor

4
Xifaxan (rifaximin)

Peripheral Opioid Antagonists

5
Immodium (loperamide)

6
Lomotil (diphenoxylate)

Fig. 3 Selected medicines that are approved for the treatment of acute diarrhea or are in late-stage clinical trials

- *Antibacterials to reduce bacterial overgrowth* in the GI tract e.g., Zifaxan™ (rifaximin) (**4**), an antibiotic inhibitor of bacterial DNA polymerase for which distribution is limited to the GI lumen and is not systemically absorbed [19]
- *Agonism of peripheral opioid receptors* to reduce GI motility e.g., Imodium™ (loperamide) (**5**) [20] and Lomotil™ (diphenoxylate) (**6**)

For each of these different mechanisms, this review will outline in greater detail advances in drug discovery and development:

- Introduction to modulating secretion via this specific pharmacological mechanism
- Preclinical evidence validating the role of this pharmacological mechanism in modulating GI secretion

- Clinical efficacy and safety via this mechanism for the treatment of secretory diarrhea
- Recent advances in medicinal chemistry of new molecules in this class
- Future outlook for this class in the treatment of secretory diarrhea

3 Advances in Discovery and Development of NEP Inhibitors

3.1 Role of NEP in Modulating Enkephalin Metabolism and Altering GI Function via Opioid Receptor Signaling

NEP (EC.3.4.24.11, neutral endopeptidase, enkephalinase) is the endogenous endopeptidase that is involved in the metabolism of important regulatory peptides in both the CNS and in the periphery. NEP is implicated in the degradation of the enkephalins, endogenous opioid peptides Met-enkephalin (**7**) and Leu-enkephalins (**8**) (Fig. 4) and peptides that modulate blood pressure, such as the vasodilator atrial natriuretic peptide (ANP) [21].

NEP is the prototype of the type II integral membrane Zn^{2+}-dependent endopeptidases and is composed of approximately 750 residues with a short N-terminal cytoplasmic domain followed by a 23-residue sequence. NEP is located at the cell surface with the majority of the protein facing the extracellular space. It is widely distributed in mammalian tissues and is involved in the inactivation of a variety of signaling peptides.

There are significant levels of this enzyme in the GI tract, where it is responsible for the cleavage of the endogenous enkephalin peptides that are agonists predominantly at the δ-opioid receptor. Inhibition of this enzyme increases levels of enkephalins that, in the GI tract, act as agonists at the δ-opioid G-protein coupled receptor [22]. These enkephalin peptides (**6** and **7**) can act directly on the enterocyte and reduce the levels of cAMP via PDE activation leading to reduced secretion of chloride via the CFTR. They can also activate the δ-opioid receptor on the secretomotor neuron in the submucosal plexus, which leads to reduced intestinal secretion [23, 24].

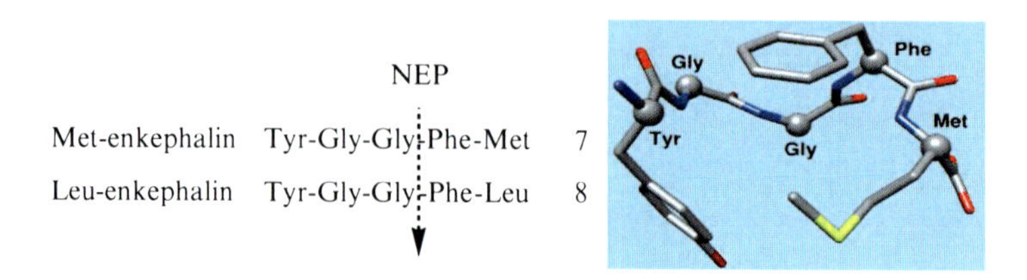

Fig. 4 Site of enzymatic cleavage of enkephalins by NEP

3.2 Recent Preclinical Studies with NEP Inhibitors

Potent inhibitors of NEP have been designed and studied that produce a range of pharmacological responses by increasing opioid peptides or vasoactive peptide such as ANP levels indicating the breadth of their therapeutic potential as GI, pain or cardiovascular medicines [25].

With regard to GI function, preclinical studies in rodent models of pharmacology demonstrated that acetorphan reduced castor oil-induced diarrhea in a dose-dependent manner via a peripheral, not a central mechanism. Treatments were administered 15 min before a 1 ml dose of castor oil and total cumulative stool weight was calculated at 20 min. Intravenous acetorphan (5, 10 and 20mg/kg) significantly reduced stool weight by 61, 79 and 100%, respectively, compared with rats dosed with vehicle) [26].

Five crystal structures of the NEP enzyme have been reported in the Cambridge Crystallographic Database. Analyses of these X-ray crystal structures have provided useful information concerning the environment close to the Zn^{2+} cation defining the catalytic domain. The analysis of several inhibitors revealed the presence of a large hydrophobic pocket at the S1' subsite level. The Zn^{2+} cation is coordinated by three ligands from the protein His583, His587, and Glu646.

The crystal structure of the soluble extracellular domain (residues Asp57-Trp749) of human NEP complexed with the metalloproteinase inhibitor phosphoramidon (**9**) at 2.1A resolution has been reported (Fig. 5) [27]. The structure reveals two multiply connected folding domains, which create a large central cavity containing the active site. The Zn^{2+} cation is coordinated with a tetrahedral geometry. Its binding involves a single oxygen atom of the tetrahedral N-phosphoryl of phosphoramidon. The sidechain of R717 forms a hydrogen bond with the carbonyl oxygen atom of the L-leucyl residue in the P1' position. The rhamnose moiety of the P1 residue is mainly exposed to solvent (Fig. 5).

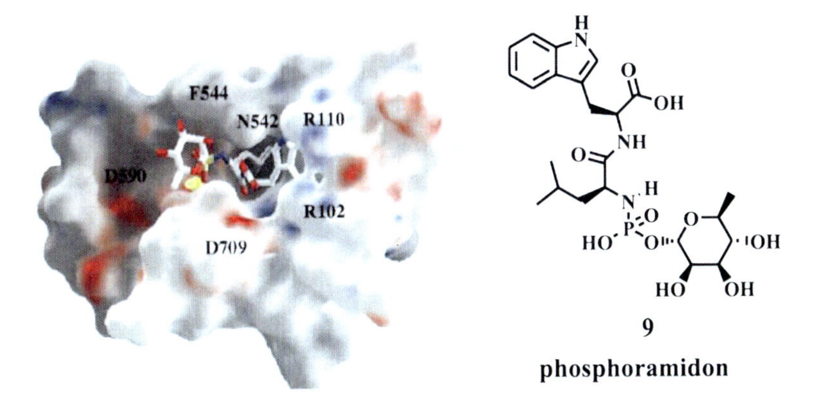

Fig. 5 PBD representation of NEP co-crystallized with the NEP inhibitor phosphoramidon (**9**) (1dmt) ([27] Reproduced with permission)

Subsequent X-ray structures of NEP with thiol-containing inhibitors demonstrated that these inhibitors bind in a similar manner [28]. As before, the lipophilic P1 group points into the large, hydrophobic S1' subsite and the central amide bond is involved in a network of H-bonds with the protein that includes Asn542 and Arg747 residues. The thiolate group binds to the Zn^{2+} cation. Carboxylate groups that are oriented in the S2' subsite region of the protein interact with the Arg102 residue.

3.3 Recent Clinical Safety and Efficacy Studies with NEP Inhibitors

As already introduced in this review, Hydrasec (**1**, racecadotril) and Tiorfan (**2**, thiorphan) are NEP inhibitors that are approved for the treatment of acute diarrhea [25]. The molecule racecadotril (**1**) contains a benzyl ester and thioacetate and is a dual prodrug of thiorphan (**2**) which is a potent nanomolar inhibitor of human NEP enzyme. This benzyl ester thioacetate, upon oral dosing, is converted to the active NEP inhibitor thiorphan.

Data from clinical studies provide evidence that this pharmacological mechanism of NEP inhibition provides safe and efficacious antidiarrheal treatment. Hydrasec (racecadotril) is approved for use in South East Asia, Central and South America. Tiorfan (thiorphan) is approved in most of Europe, Russia and some African countries.

A significant number of well-controlled clinical studies have been conducted with racecadotril in healthy volunteers and in both adult and children patients with acute secretory diarrhea of varied etiology.

Two different studies have been reported in healthy volunteers. Baumer et al. demonstrated that racecadotril reduced the number and weight of stools in healthy volunteers with castor-induced diarrhea without constipation-like side effects. A single dose of racecadotril (11.1 mg/kg) significantly reduced diarrhea during the 24 h period by 37% and stool number by 49% compared with placebo [29]. Bergmann et al. demonstrated that racecadotril did not affect GI transit times in healthy volunteers [30].

In 193 adult patients with acute diarrhea, Baumer et al. demonstrated that racecadotril decreased the incidence of diarrhea by 30% ($p < 0.01$) when compared with placebo [29].

A randomized, placebo-controlled study of racecadotril in the treatment of acute watery diarrhea in boys [31] demonstrated a significant reduction in 48-h stool output, total stool output before recovery, total intake of ORS and duration of diarrhea.

A randomized, placebo controlled study of racecadotril (1.5 mg/kg, t.i.d) in 172 infants aged 3 months to 4 years demonstrated significant reduction in stool output by approximately 50% after 24 h of treatment ($p = 0.004$) compared with placebo in the first 48 h of treatment [32]. This study confirmed the efficacy and tolerability of racecadotril as an adjuvant therapy to ORS.

A systematic review [33] of these three randomized controlled trials on the treatment of acute diarrhea in children concluded that racecadotril reduced stool output and duration of diarrhea. It also opined that more data in out-patients were needed and that safety and cost-effectiveness needed to be explored before routine therapy with racecadotril was recommended. None of the studies reported any significant adverse events resulting from the administration of racecadotril.

3.4 Recent Advances in Medicinal Chemistry of NEP Inhibitors

Recent developments in medicinal chemistry of NEP inhibitors have been to create potent, nanomolar inhibitors that contain a range of Zn^{2+}-chelating moieties. This section will focus on the discovery of NEP inhibitors designed in have different physiochemical properties to create molecules with different pharmacokinetic, pharmacodynamic and stability properties.

Racemic thiorphan (**2**) is the prototypical, selective NEP inhibitor that has been most studied in clinical and nonclinical GI pharmacology) [25]. (2*R*)-Thiorphan (**10**) belongs to the thiol class of NEP inhibitors, where it is the thiol functional group that binds to the Zn^{2+} cation in the NEP active site (Fig. 6). Generally, the lipophilic benzyl P1' substituent of the ligand occupies the large hydrophobic S1' subsite. The central amide bond is involved in a network of H-bonds with the protein that includes Asn542 and Arg747 residues and the thiolate group binds to the Zn^{2+} cation. Carboxylate groups distal to the zinc-chelating thiol are oriented in the S2' subsite region of the protein and interact with the Arg102 residue. Other members of this class of thiol-containing NEP inhibitors are the retrothiorphan (**11**),

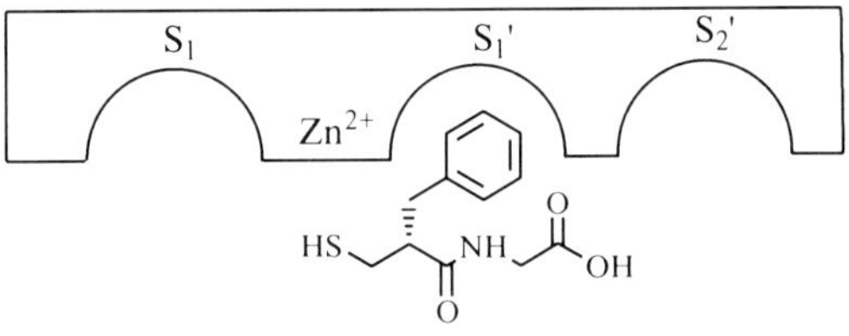

S_1 S_1' S_2' Zn^{2+}

10

(2*R*)-thiorphan

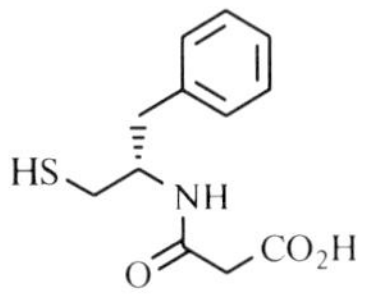

11

(2*R*)-retrothiorphan

Fig. 6 Representative thiol NEP inhibitors

which contains a reverse amide [28] but utilizes the same binding interaction with the NEP protein surface to provide nanomolar inhibition.

Current advances in synthesis of the single enantiomers of thiorphan and corresponding prodrugs have been extensively reviewed [34] and are beyond the scope of this review.

Significant medicinal chemistry efforts have also been directed toward the discovery of more potent compounds that have an alternative Zn^{2+}-chelating group to the potent thiol chelator such as phosphonic acids, hydroxamic acids, and carboxylic acids. While chelation of the Zn^{2+} cation with thiol provides very potent NEP inhibitors, there is interest in leveraging a Zn^{2+} cation chelating group that has different stability over the thiol-containing NEP inhibitors. The rationale for finding an alternate Zn^{2+} chelator is to improve chemical stability to facilitate chemistry, manufacturing and control considerations for development and that does not interact with other thiol-containing molecules in in vitro or in vivo biological studies. This is particularly important since these differing functional groups can dramatically alter the pharmacokinetic profile of the candidate compound.

In the carboxylate series of NEP inhibitors (Fig. 7), candoxatrilat, UK-73,967 (**12**) is a well-studied member of this class and is a potent, nanomolar inhibitor of NEP. Candoxatril, UK-79,300 (**13**) is the orally active indanyl ester prodrug of candoxatrilat (**12**) and was extensively studied in the treatment of chronic heart failure [35]. Recently, the carboxylate chelating NEP inhibitor LBQ-6577 (**14**) has been studied in hypertension as a fixed combination of its ethyl ester prodrug AHU-377 (**15**) with

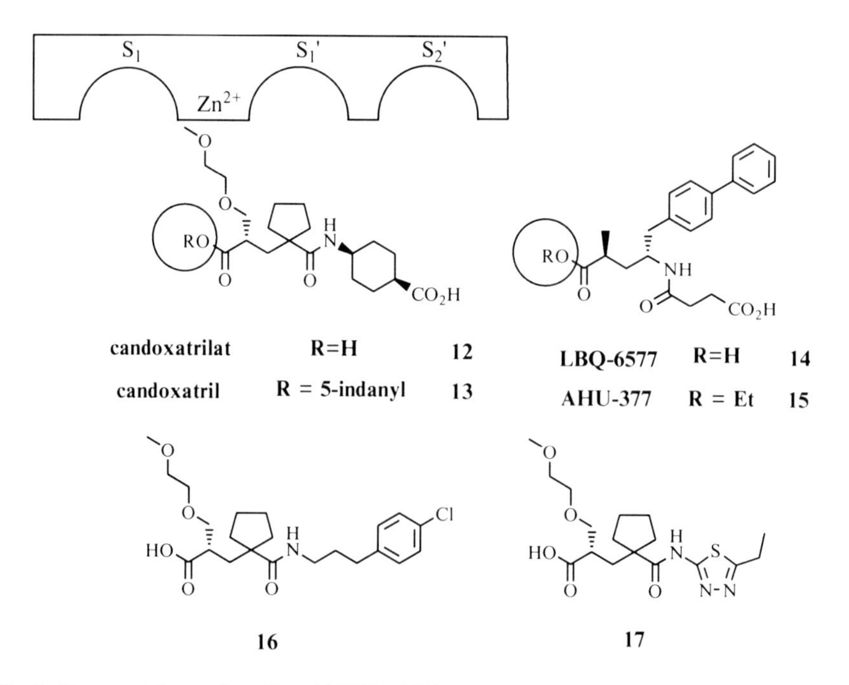

Fig. 7 Representative carboxylic acid NEP inhibitors

valsartan to provide the complex LCZ-696 [36]. This fixed combination is in currently in Phase 3 clinical trials for chronic heart failure. Discovery of carboxylate containing NEP inhibitors (**16**) and (**17**) have also been reported and are being developed as medicines for treatment of female sexual arousal disorder [37, 38]. Their discovery was based on candoxatrilat as a start point. These advanced-stage clinical candidates may have potential utility as antisecretory agents for the treatment of acute diarrhea.

In the phosphonic acid class (Fig. 8), the prototypical metalloprotease inhibitor is phosphoramidon (**9**). The inhibitor CGS-24592 (**18**) is a potent inhibitor of NEP and its diphenyl phosphonate prodrug (**19**) demonstrated high potency, long duration of action and oral bioavailability [39]. As with previous NEP inhibitors, the therapeutic focus on this compound was in cardiovascular medicine through its ability to increase circulating levels of ANP. This prodrug (**19**) induced potent inhibition of NEP ex vivo for at least 8 h following oral administration to rats (30 mg/kg po). Its antihypertensive activity was demonstrated in DOCA-salt rats.

Hydroxamic acids (Fig. 9) are also very potent inhibitors of Zn^{2+}-containing proteases because of their high affinity to chelate the Zn^{2+} cation. Recently, N-formyl hydroxylamine-containing NEP inhibitors (**20**, **21**) that are similar in structure to omapatrilat were described [40, 41].

Lead optimization efforts have also explored removing the central amide of NEP inhibitors (Fig. 10). A structure-based design approach [42] provided a novel series

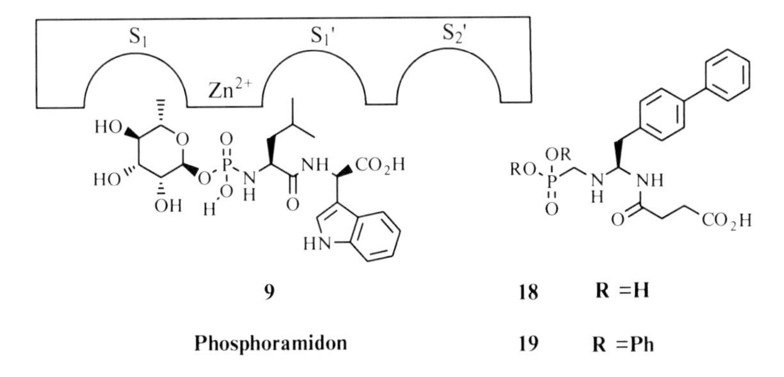

Fig. 8 Representative phosphonic acid NEP inhibitors

Fig. 9 Representative hydroxamic acid NEP inhibitors

Fig. 10 Representative nonpeptidic thiol NEP inhibitors

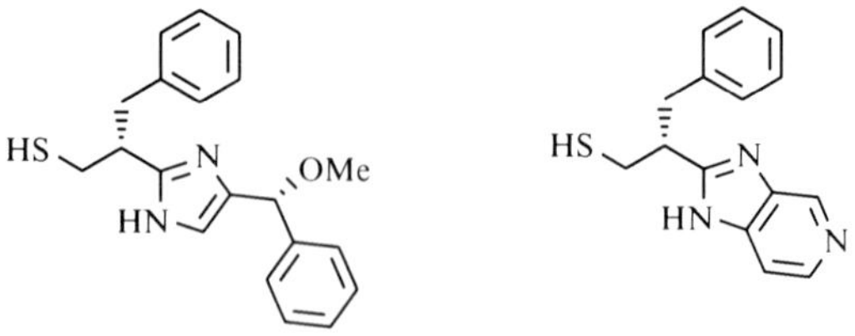

of imidazoles exemplified by **22** novel series of imidazo[4,5-c]pyridine NEP inhibitors exemplified by the thiol **23** (NEP IC_{50} 0.20 μM). Computational analyses indicated that the imidazole was a good amide isostere undergoing H-bonding to both Asn542 and Arg717.

Additionally, NEP inhibitors that have designed in a second pharmacology to provide additional clinical efficacy are also worthy of further discussion. In the cardiovascular therapeutic area, there was significant interest in studying the safety and efficacy by design of dual pharmacology ACE-NEP inhibitors. One of the most advanced compounds in development for hypertension and congestive heart failure was omapatrilat [36, 43], this was discontinued in late stage development because of reports of severe angioedema due to ACE inhibition interfering with bradykinin metabolism.

3.5 Conclusions and Outlook for the Discovery and Development of NEP Inhibitors for the Treatment of Secretory Diarrhea

In summary, Hydrasec (racecadotril) and Tiorfan (thiorphan) are approved for the treatment of the secretory diarrhea in certain regions of the world. Importantly, both have demonstrated that they are efficacious and tolerated in both the adult and pediatric populations. These medicines represent an important advance for the treatment of secretory diarrhea in the developing world via inhibition of NEP. As yet, clinical use of these medicines would be appear to be modest. Tormo et al. recommend that racecadotril should be considered as an antisecretory medicine in addition to ORT [11] to reduce the amount and frequency of stool output.

Given the recent advances in the medicinal chemistry of NEP inhibitors, there is an increased understanding on how to design more potent, nonthiol NEP inhibitors. Of particular note is the ability to potently inhibit NEP using carboxylates, hydroxamates, and phosphonates as the Zn^{2+}-chelating group. These different Zn^{2+} cation chelating groups will have very different physicochemical properties from each other and this may translate into different pharmacokinetics, stability, and efficacy.

Currently, there is both the opportunity and the need to design novel NEP inhibitors with different distribution profiles and evaluate how that alters efficacy.

There is also opportunity to design a compound that would be dosed less frequently or that would have a different stability profile to the currently approved NEP inhibitors. At this point, further studies are required to determine whether an NEP inhibitor that acts topically, in the GI lumen, on the surface of the enterocyte would have similar efficacy to racecadotril, which is orally bioavailable.

4 Advances in Discovery and Development of Cystic Fibrosis Transmembrane Conductance Regulator Inhibitors

4.1 Role of CFTR in Modulating GI Secretion

The cystic fibrosis transmembrane conductance regulator (CFTR) gene encodes a cAMP-regulated chloride channel. CFTR is a 1,480 amino acid protein that is a member of the ATP-binding cassette (ABC) transporter family. This channel is the genetic basis of the lethal hereditary disease cystic fibrosis. It is expressed in epithelial cells of the lungs, intestine and other fluid transporting tissues. CFTR is a transmembrane protein composed of two six helix membrane spanning domains each followed by a nucleotide binding domain (Fig. 11) [44].

In the GI tract, the CFTR is expressed on the apical membrane of the enterocyte. It is the primary pathway for chloride and fluid secretion into the intestinal lumen. CFTR chloride flux is the final rate-limiting step for intestinal chloride secretion and for fluid secretion in infectious diarrhea.

There is good evidence for the central role of CFTR-mediated chloride ion secretion in enterotoxin-mediated chloride secretion in infectious diarrheas including cholera and traveler's diarrhea. Toxin-mediated CFTR activation leads to dehydration from fluid loss from electrolyte transport across the epithelial cells lining the GI tract. These enterotoxins produced by enterotoxigenic *E. coli* (ETEC) and *V. cholerae* bind to the receptors on the luminal surface of enterocytes and generate intracellular second messengers. These then activate the CFTR and secretion of chloride that creates the driving force for secretion of sodium ions and water into the lumen. CFTR is an attractive target for the development of inhibitors with anti-diarrheal efficacy in cholera and other disorders of intestinal fluid secretion [16].

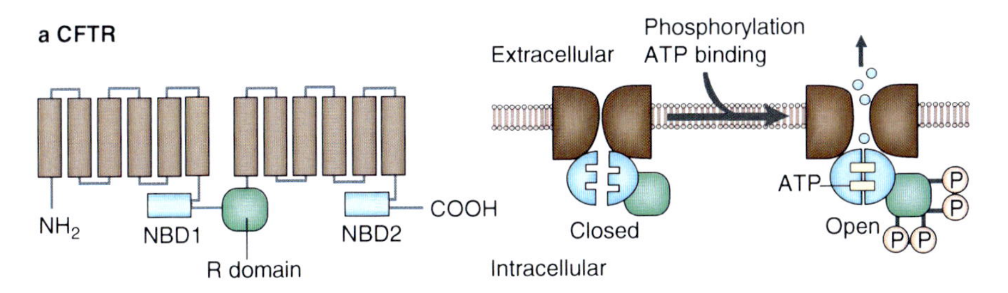

Fig. 11 Structure of CFTR channel ([44], reproduced with permission)

4.2 Recent Preclinical Studies with CFTR Inhibitors

The mouse closed-loop model is a well-established in vitro intestinal-tissue model to evaluate the potency and efficacy of CFTR inhibitors to reverse toxin-induced secretion. In this assay, following in vivo or intra-lumenal administration of test compound, mid jejunal loops are removed, physically closed off (often into two loops for A control/treatment pair), and injected with either saline or saline/cholera toxin. Secretion is then evaluated by studying the weight to length ratio of the mid-jejunal loop, with "secretion" leading to an increase in intestinal loop volume and weight due to increased fluid transport across the intestinal membrane into the loop. In this model, fluid transport was unaffected by CFTR inhibitor administration [e.g., CFTRinh-172 (**31**)]. Injection of cholera toxin into the loop produced fluid secretion (i.e., increased loop weight) over 6 h after a slow onset. Intraperitoneal injection of CFTRinh-172 (**31**) reduced fluid accumulation by approximately 90%, with an IC_{50} of 5 μm [17].

4.3 Recent Clinical Safety and Efficacy Studies with CFTR Inhibitors

Crofelemer (**3**) (SP-303, Provir) [18] is the most extensively studied CFTR inhibitor in diarrheal disease. It has been studied clinically for the treatment of several types of secretory diarrhea including AIDS-associated diarrhea, traveler's diarrhea, cholera and diarrhea predominant irritable bowel syndrome (D-IBS).

Crofelemer (**3**) is a mixture of polyphenolic oligomer isolated from the bark latex of the Amazonian tree *Croton lechleri* (family Euphorbiaceae). This naturally occurring extract is an acid labile purified proanthocyanidin oligomer of varying chains lengths. It has been characterized by NMR and MS and the proposed structural representation is depicted in **3** (Fig. 12). The monomers that comprise this oligomer include (+)-catechin (**24**), (+)-gallocatechin (**25**), (−)-eipcatechin (**26**),

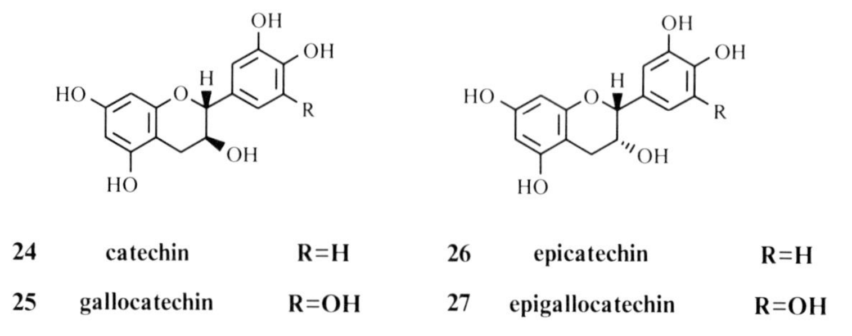

Fig. 12 Representative structures of the component monomers that comprise the crofelemer (**3**) oligomer

and (−)-epigallocatechin (**27**) (Fig. 12). The oligomer is composed of 5–11 linearly linked monomers to provide crofelemer with an average molecular weight of 2,200 Da. The large chemical structure of this oligomer prevents it oral absorption and it is proposed that its likely site of action is on the apical side of the lumen of the GI tract on the enterocyte.

Crofelemer's (**3**) mechanism of action is still under investigation and is proposed to be a novel mechanism of action acting through dual inhibition of the CFTR channel and the calcium activated chloride (CaCC) ion channel in the intestinal lumen to attenuate the flow of chloride ions into the GI tract, thus reducing secretion [45]. Crofelemer has been demonstrated to have no effect on GI motility.

A preliminary clinical study [46] from Bangladesh investigated the use of crofelemer for the treatment of acute, severely dehydrating watery diarrhea due to cholera. This was a placebo-controlled trial in 100 adults studying a single dose of 125 mg and 250 mg of crofelemer. The primary objective of the study was to evaluate safety and efficacy of crofelemer in reducing watery stool output normalized to body weight in the first 24 h. Results showed that those treated with crofelemer appeared to have a greater reduction of watery stool output compared to the placebo group. Crofelemer is currently being studied clinically in infectious diarrhea, traveler's diarrhea, and HIV-related diarrhea (www.clinicaltrials.gov).

4.4 *Recent Advances in the Medicinal Chemistry of CFTR Inhibitors*

This section will outline the recent advances in the lead optimization of CFTR inhibitors that have resulted in the discovery of more potent and more selective CFTR inhibitors. Importantly, this has been achieved while also providing novel classes of inhibitors that have different biological properties in electrophysiology assays.

Small molecule inhibitors of CFTR chloride conductance with submicromolar inhibitory potency have been reported. Early CFTR inhibitors such as diphenylamine-2-carboxylate (**28**) and 3-nitro-2-(3-phenylpropyl-amino)benzoate (**29**) were originally discovered as inhibitors of kidney tubule conductance (Fig. 13). They are not selective for this chloride channel and have micromolar potency chloride channels in the kidney. These inhibitors are nonspecific chloride channel inhibitors that electrophysiology and structural biology studies have indicated likely result from pore occlusion [17]. These small molecules have been shown to be effective in preventing chloride and fluid secretion induced by cholera toxin in human intestinal cell and in rodent models.

The sulfonylurea class of inhibitors includes glibenclamide (**30**) (Fig. 13). Electrophysiology studies indicate that these inhibit CFTR function via binding to the open conformation of the chloride channel. Both the arylaminobenzoates and sulfonylureas require micromolar concentrations to inhibit the CFTR and will likely inhibit other anion transporters at these high concentrations [17].

Fig. 13 Representative prototypical chloride channel inhibitors

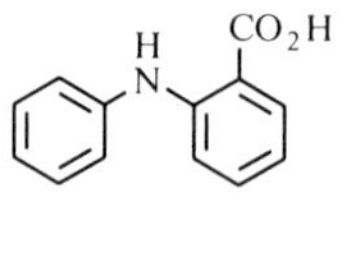

28

DPC

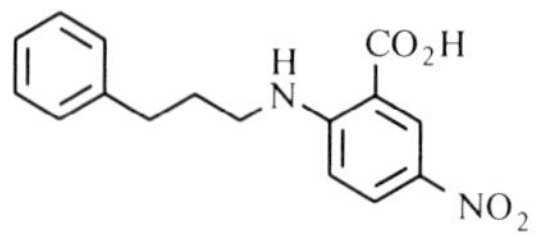

29

NPPB

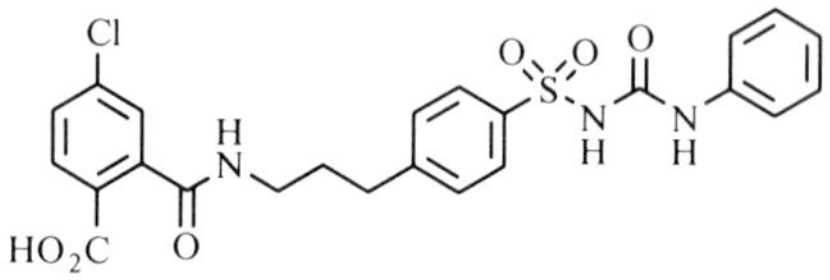

30

Glibenclamide

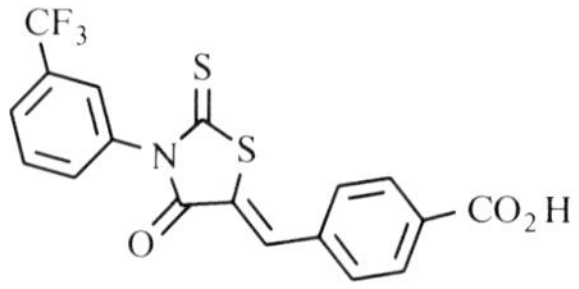

31

CFTRinh-172

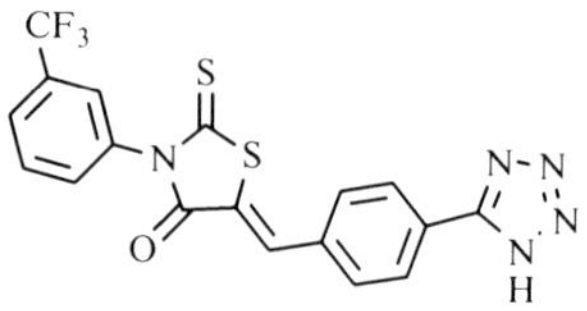

32

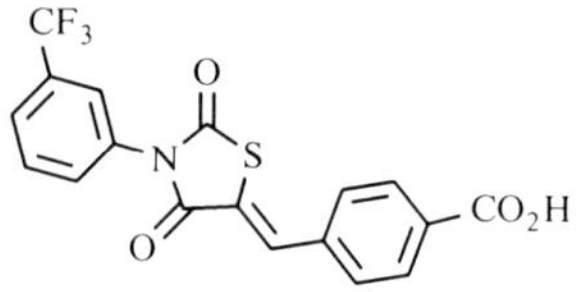

33

Fig. 14 Thiazolidinone series of CFTR inhibitors

Screening endeavors have identified three new, more potent classes of CFTR inhibitors that have now been studied in a range of in vitro and in vivo pharmacology studies.

The first class is the thiazolidinone series of CFTR inhibitors (Fig. 14). The thiazolidinone $CFTR_{inh}$-172 (**31**) inhibits CFTR chloride conductance (IC_{50} 0.3–3.0 μM) and binds to the cytoplasmic side of the plasma membrane. Electrophysiology studies demonstrated voltage-independent channel inhibition with $CFTR_{inh}$-172 binding to the closed state of the channel. CFTR mutagenesis studies support this inhibitor making a key interaction with Arg347 located near the cytoplasmic entrance of the pore. In vivo pharmacology studies have examined the pharmacokinetics and safety of CFTR$_{inh}$-172. Lead optimization has provided more soluble analogs of CFTR$_{inh}$-172 that include the tetrazole (**32**) and the oxo analog (**33**) [47].

A second series are the glycine hydrazides (Fig. 15) exemplified by GlyH-101 (**34**) that inhibits the CFTR channel (IC$_{50}$ 5 μM) via occlusion of the external pore [48]. Further discovery efforts have provided nonabsorbed derivatives of glycine hydrazides that include polyethylene glycol (PEG) (**35**) and lectin adducts [49]. These conjugates inhibited CFTR when applied to the mucosal surface. This is important since, in a similar fashion to crofelemer, this indicates that these inhibitors could act directly on the enterocyte at the epithelial layer. This is an important feature since it provides the potential for efficacy without oral absorption and may remove the potential safety or tolerability findings from systemic exposure of molecule.

Verkman et al. have recently described the identification of novel, third class of CFTR inhibitors (Fig. 16): nanomolar pyrimido-pyrrolo-quinoxazolinedione CFTR inhibitors [50]. PPQ-102 (**36**) inhibited the CFTR chloride channel (IC$_{50}$ 90 μM) and that did not demonstrate the voltage-dependent inhibition exhibited by the

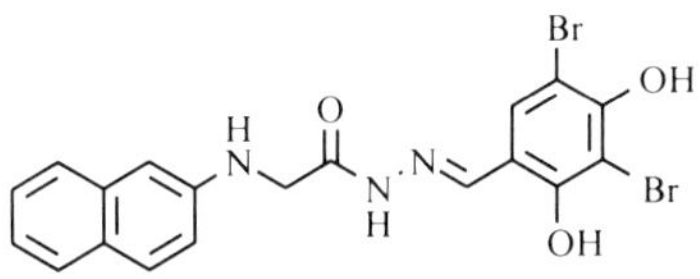

34
GlyH-101

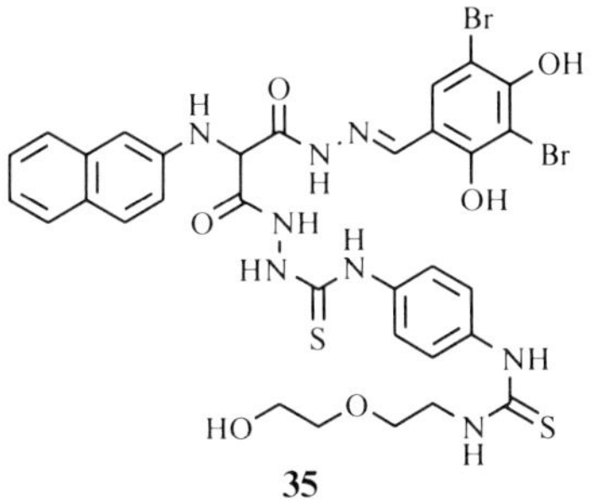

35
MalH-(PEG)$_1$

Fig. 15 Acyl-hydrazide series of CFTR inhibitors

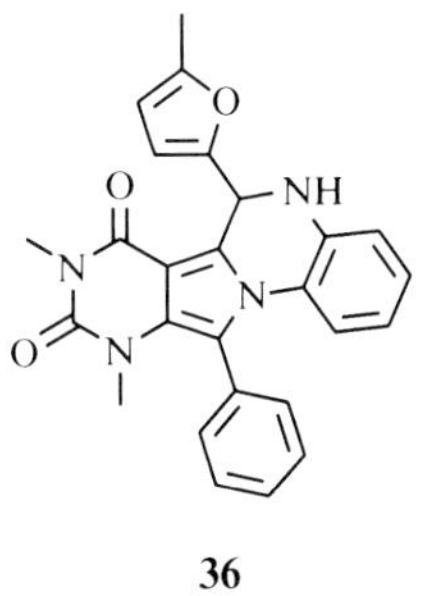

36
PPQ-102

Fig. 16 Pyrazolopyridine series of CFTR inhibitors

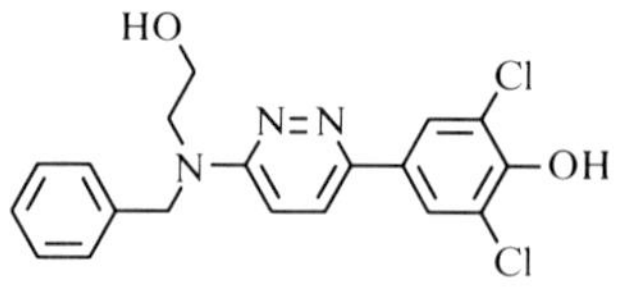

WO 2009/131947
(Compound 98, 0.6μM)
Pyridazine 37

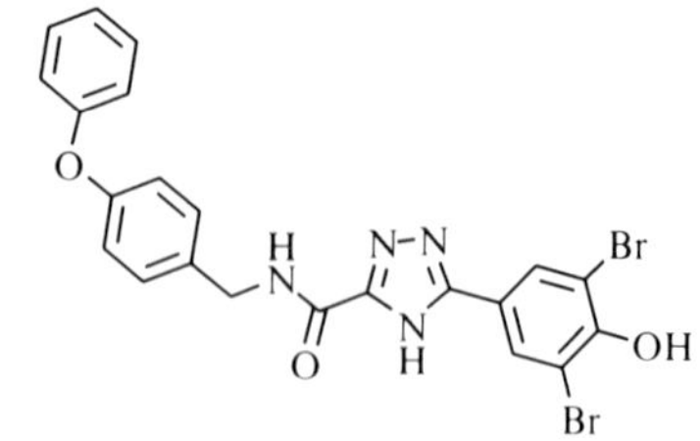

US 2009/0264471 A1
(Compound 81, 1.5μM)
Triazole 38

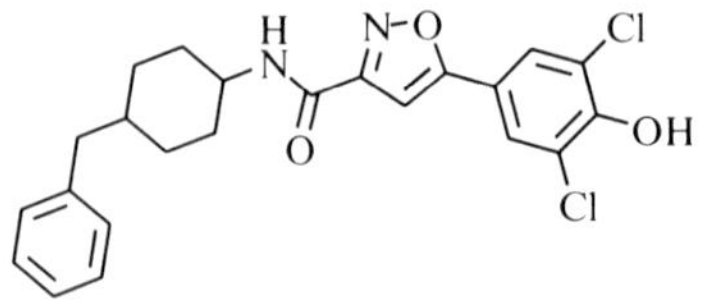

US 2009/0264486 A1
(Compound 12, 6.1μM)
Isoxazole 39

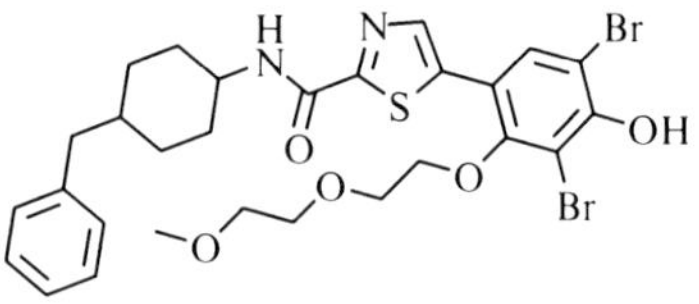

US 2009/131952
(Compound 87, 1.7μM)
Thiazole 40

Fig. 17 Recent patent structures, patent number exemplified with data

previously reported thiazolidinone and acyl hydrazide class. Patch clamp analysis has demonstrated that they bind to the closed state of the channel.

A number of recent patents have further described novel inhibitors of the CFTR chloride channel. These structures exhibit micromolar inhibition of the chloride flux in a fluorescence-based FLIPR assay (Fig. 17). The Institute for One World Health is currently evaluating CFTR inhibitors for the treatment of secretory diarrhea (http://www.oneworldhealth.org/diarrheal_disease).

4.5 Conclusions and Outlook for CTFR Inhibitors for the Treatmentof Secretory Diarrhea

In summary, crofelemer continues to be studied for the treatment of diarrhea as dual activity chloride channel inhibitor, which provides antisecretory effect without the need for oral absorption.

Additionally, significant efforts have identified a number of more potent, more selective CFTR inhibitors from new structural classes. Importantly, some of these compounds have demonstrated efficacy in the mouse closed-loop model of secretion. These novel CFTR inhibitors are from different structural classes and may therefore exhibit different PK–ADME properties and different off-target activities, which may have important implications for efficacy in in vivo models of pharmacology and for their safety and tolerability. One of the very useful observations arising from the discovery of the new thiazolidinone, acyl hydrazide, and the pyrazolopyridine lead series is the different electrophysiological properties (i.e., voltage dependency and use dependency) that these lead series exhibit. This is based on where and how the different lead series interact with alternate binding sites on the CFTR. It will be interesting to see how this translates to in vitro models of secretion and in vivo models of efficacy.

It will be important to understand the site of action of these new CFTR inhibitors to determine whether they can be active in the lumen of the GI tract or whether they require some degree of oral absorption.

5 Role for Opioid Receptor Agonists the Treatment of Secretory Diarrhea

5.1 *Role of Opioid Agonists in Modulating GI Function*

The agonism of opioid receptors in the GI tract has well-documented effects on intestinal motility, secretion, and neuronal transmission [24].

Loperamide (**5**) antagonizes peripheral opioid receptors and does not penetrate the CNS at approved human doses (Immodium, loperamide prescribing information). It is indicated in the United States for the control and symptomatic relief of acute nonspecific diarrhea. It is contraindicated and not recommended in infants below 24 months of age because of reports of paralytic ileus. Electrophysiology studies in the Ussing chamber designed to measure the short circuit current as an indicator of net ion transport across epithelial monolayer have demonstrated that loperamide inhibits chloride secretion induced by bacterial toxins and prostaglandin E2. In vivo jejunal studies indicated that loperamide had no effect on basal transport but inhibited prostaglandin E2 induced secretion. Subsequent studies in humans have demonstrated that loperamide decreases irregular motor activity (phase II) of the migrating motor complex with no demonstrable effects on the transport of water or electrolytes [23].

5.2 *Recent Clinical Safety and Efficacy Studies with Opioid Agonists*

Although loperamide is widely used in adults, the World Health Organization and the American Academy of Pediatrics are concerned about its use in young children because of concerns over its efficacy and safety. In the United States, loperamide is approved by the FDA for use in children older than 2 years. The loperamide prescribing information states use of loperamide for children less than 2 years is not recommended due to rare reports of paralytic ileus associated with abdominal distention.

A recent systemic review and meta-analysis concluded that children who are younger than 3 years, malnourished, moderately or severely dehydrated, systemically ill or have bloody diarrhea, adverse events outweigh benefits even at doses of <0.25 mg/kg/day. In children who are older than 3 years with minimal dehydration, loperamide may be a useful adjunct to oral rehydration and early refeeding [20].

6 Conclusion and Outlook for Antisecretory Medicines for the Treatment of Secretory Diarrhea

The introduction of ORS has significantly reduced mortality in the developing countries from secretory diarrhea. In spite of this advance, there remains significant mortality and morbidity from secretory diarrhea caused by infectious disease. This problem is particularly acute in the developing countries and the pediatric population is most at risk. As a result, there remains a significant medical need for safe antisecretory medicines with improved efficacy for the treatment of infectious diarrhea in the developing countries to supplement the significant improvement that ORS has made on reducing the severity of disease and mortality.

Significant progress is being made in the availability of new antisecretory medicines for the treatment of infectious diarrhea with the approval of the NEP inhibitors Hydrasec and Tiorfan. Crofelemer, the dual pharmacology CFTR and calcium activated chloride channel inhibitor, is currently in advanced clinical studies for a range of different type of diarrhea and is being evaluated in secretory diarrhea. These existing advances in clinical development mean that there is now further clinical understanding on the appropriate clinical use of potent antidiarrheal medicines such as NEP inhibitors e.g., Hydrasec and Tiorfan, opioid agonists e.g., Immodium (loperamide), and luminally restricted antibiotic e.g., Zifaxan (rifaximin).

Additionally, significant understanding has been made in the design of new molecules that target clinically validated mechanisms for reducing secretion in secretory diarrhea. Particularly, noteworthy are the many different classes of NEP inhibitors that create a scientific opportunity to discover a best-in-class antisecretory NEP inhibitor. For the mechanism of CFTR inhibition, there has been an increase in understanding of how to design novel inhibitors with different electrophysiological properties at this chloride channel.

The potential for a luminally active antisecretory medicine is a particularly appealing feature given the potential benefit of improved safety and tolerability. There is the potential for such a medicine in the area of the CFTR inhibitors and NEP inhibitors.

Further studies on new pharmacological mechanisms of secretion and GI function are also providing new mechanisms that target the enteric nervous system e.g., somatostatin agonism, calcium-calmodulin antagonism, 5-HT antagonists, and substance P antagonists. An excellent review outlines the potential of this mechanism to create novel antidiarrheal medicines [51].

Given the medical need in the developing world, the new scientific knowledge that is available around the mechanism of secretion in GI tract, and the novel availability of different chemical classes to modulate secretion mechanisms, priority should be given to provide novel safe and efficacious antisecretory medicines to further reduce the global burden of diarrhea.

References

1. Farthing MJG (2000) Int J Antimicrob Agents 14:65
2. Glass RI, Parashar UM, Estes M (2009) N Engl J Med 361:1776
3. Santosham M (2010) N Engl J Med 362:358
4. Thiagarajah JR, Verkman AS (2005) Trends Pharmacol Sci 26:172
5. Kosek M, Bern C, Guerrant RL (2003) Bull World Health Org 81:197–204
6. World Health Organization, Geneva. (2004) Global burden of disease
7. Parashar UD, Breese JS (2003) Bull World Health Org 81:236
8. World Health Organization, Geneva. (2009) WHO promotes research to avert diarrhea deaths
9. Masoodpoor N, Darakshan R (2008) Pediatrics 121:S111
10. Butta ZA, Black RE, Brown JM (1999) J Pediatr 135:689–697
11. Tormo R, Polanco I, Salazar-Lindo E, Goulet O (2008) Acta Paediatr 97:1008–1015
12. Madhi SA, Cunliffe NA, Steele D, Witte D, Kirsten M, Louw C, Ngwira B, Victor JC, Gillard PH, Cheuvart BB, Han HH, Neuzil KM (2010) N Engl J Med 362:289–298
13. Richardson V, Hernandez-Pichardo J, Quintanar-Solares M, Esparza-Aguilar M, Johnson B, Gomez-Altamirano CM, Parashar U, Patel M (2010) N Engl J Med 362:299–304
14. Sur D, Ochiai RL, Bhattacharaya NK (2009) N Engl J Med 361:335–344
15. Farthing MJG, Casburn-Jones A, Banks MR (2003) Dig Liver Dis 35:378
16. Guttman AJ, Finlay BB (2008) Trends Microbiol 16:535–542
17. Thiagarajah JR, Verkman AS (2003) Curr Opin Pharmacology 3:594–599
18. Crutchley RD, Miller J, Garey KW (2010) Ann Pharmacother 44:878–884
19. Layer P, Andresen V (2010) Aliment Pharmacol Ther 31:1155–1164
20. Li ST, Grossman DC, Cummings P (2007) PLoS Med 4:495–505
21. Turner AJ, Isaac RE, Coates D (2001) Bioessays 23:261–269
22. Roques BP (2000) Trends Pharmacol Sci 21:475
23. Wood JD, Galligan JJ (2004) Neurogastroenterol motil 16:17
24. Holzer P (2009) Reg Pept 155:11
25. Matheson AJ, Noble S (2000) Drugs 59:829
26. Lecomte JM, Costentin J, Vlaiculescu A (1986) J Pharmacol Exp Ther 237:937
27. Oefner C, D'Arcy A, Henning M, Winkler FK, Dale GE (2000) J Mol Biol 296:341
28. Oefner C, Roques BP, Fournie-Zaluski MC, Dale GE (2004) Acta Crystallogr D60:392
29. Baumer P, Dorval ED, Bertrand J, Vetel JM, Schwartz JC, Lecomte JM (1992) Gut 33:753
30. Bergmann JF, Chaussade S, Couturier D, Baumer P, Schwartz JC, Lecomte JM (1992) Aliment Pharmacol Ther 6:305
31. Salazar-Lindo E, Santisteban-Ponce J, Chea-Woo E, Guttierrez M (2000) N Engl J Med 343:463

32. Cezard JP, Duhamel JF, Meyer M, Pharaon I, Bellaiche M, Maurage C, Ginies JL, Vaillant JM, Girardet JP, Lamireau T, Poujol A, Morali A, Sarles J, Olives JP, Whately-Smith C, Audrain S, Lecomte JM (2001) Gastroenterology 120:799
33. Szajewska H, Ruszczynski M, Chmielewska A, Wieczorke J (2007) Aliment Pharmacol Ther 26:807
34. Monteil T, Danvy D, Shiel M, Leroux R, Plaquevent J-C (2002) Mini Rev Med Chem 2:209
35. McDowell G, Nicholls PD (1999) Exp Op Inv Drugs 8:79
36. Gu J, Noe A, Chandra P, Al-Fayoumi S, Ligueros-Saylan M, Sarangapani R, Maahs S, Ksander G, Rigel DF, Jeng AY, Lin T-H, Zheng W, Dole WP (2010) J Clin Pharmacol 50:401
37. Pryde DC, Maw GN, Planken S, Platts MY, Sanderson V, Corless M, Stobie A, Barber CG, Russell R, Poster L, Barker L, Wayman C, Van der Graaf P, Stacey P, Moreen D, Kohl C, Beumont K, Coggon S, Tute M (2006) J Med Chem 49:4409
38. Pryde DC, Cook AS, Burring DJ, Jones LH, Foll S, Patts MY, Sanderson V, Coreless M, Stobie A, Middleton DS, Foster L, Barker L, Van der Graaf P, Stacey P, Khol C, Coggon S, Beaumont K (2006) Bioorg Med Chem Lett 15:142
39. De Lombaert S, Erion MD, Tan J, Blanchard L, El-Chebabi L, Ghai RD, Skane Y, Berry C, Trpanai AJ (1994) J Med Chem 37:489
40. Robl JA, Simpkins LM, Asaad MM (2000) Bioorg Med Chem Lett 10:257
41. Walz AJ, Miller MJ (2002) Org Lett 4:2047–2050
42. Shali S, Stump B, Welti T, Blum-Kaelin D, Aebi JD, Oefner C, Bohm H-J, Diederich F (2004) Chem Bio Chem 5:996
43. Solomon SD, Skali H, Bourgoun M, Fang J, Ghali JK, Martelet M, Wojciechowski D, Ansmite B, Skards J, Laks T, Henry D, Packer M, Pfeffer MA (2005) Am Heart J 150:257
44. Verkman AS, Galietta LJV (2009) Nat Rev Drug Dis 8:153
45. Tradtrantip L, Namkung W, Verkman AS (2009) Mol Pharm 77:69
46. Bardan PK, Sharma ACB (2009) US.-Japan CMPS: 13th International conference on emerging infectious diseases (EID) in the Pacific rim-focused on enteric diseases, 6–9 April 2009, Kolkata, India
47. Sonawane ND, Verkman AS (2008) Bioorg Med Chem Lett 16:8187
48. Verkman AS, Lukacs GL, Galietta LJV (2006) Curr Pharm Des 12:2235
49. Sonawane ND, Zhao D, Moran-Zeggarro O, Galietta LJV, Verkman AS (2008) Chem Biol 15:718
50. Tradtrantip L, Sonawane ND, Namkung W, Verkman AS (2009) J Med Chem 52:6447
51. Farthing MJG (2006) Dig Dis 24:47

Index

D

E

F

Q

R

S

T